From Nano to Space

Michael H. Breitner · Georg Denk · Peter Rentrop

Editors

From Nano to Space

Applied Mathematics
Inspired by Roland Bulirsch

 Springer

Michael H. Breitner
Institut für Wirtschaftsinformatik
Leibniz Universität Hannover
Königsworther Platz 1
30167 Hannover
Germany
breitner@iwi.uni-hannover.de

Peter Rentrop
M2 – Zentrum Mathematik
Technische Universität München
Boltzmannstraße 3
85748 Garching
Germany
toth-pinter@ma.tum.de

Georg Denk
Qimonda AG
Products
81726 München
Germany
Georg.Denk@qimonda.com

ISBN 978-3-540-74237-1   e-ISBN 978-3-540-74238-8

DOI 10.1007/978-3-540-74238-8

Library of Congress Control Number: 2007936374

Mathematics Subject Classification (2000): 65-xx, 68-xx, 97-xx, 94-xx, 34-xx, 35-xx, 49-xx, 90-xx

Cover design: WMX Design GmbH, Heidelberg

Printed on acid-free paper

9 8 7 6 5 4 3 2 1

springer.com

Preface

"Why does a rainbow appear in a certain altitude?" or "How does a calculator compute the sine function?" are typical questions of Mathematics and Engineering students. In general these students want to understand Applied Mathematics to solve everyday problems. In 1986 the professors Rüdiger Seydel and Roland Bulirsch wrote the short book *Vom Regenbogen zum Farbfernsehen* (From a Rainbow to Color Television) published by Springer (cf. Fig. A.3 on page 325). This book collects 14 case studies from our everyday lives where the field of Applied Mathematics is used to explain phenomena and to enable technical progress, balancing motivations, explanations, exercises and their solutions. The numerous exercises are still very interesting and informative, not only for students of Mathematics.

Twenty years later former scholars of Roland Bulirsch working at universities, at research institutions and in industry combine research and review papers in this anthology. Their work is summed up under the title *From Nano to Space – Applied Mathematics Inspired by Roland Bulirsch*. This title reflects the broad field of research which was initiated by the open atmosphere and stimulating interests of Roland Bulirsch and his Applied Mathematics. More than 20 contributions are divided into scales: nano, micro, macro, space, and real world. The contributions provide an overview of current research and present case studies again very interesting and informative for graduate students and postgraduates in Mathematics, Engineering, and the Natural Sciences. The contributions show how modern Applied Mathematics influences our everyday lives.

München, Hannover,
July 2007

Michael H. Breitner
Georg Denk
Peter Rentrop

Contents

Part I Mathematics and Applications in Nanoscale

Part II Mathematics and Applications in Microscale

Part III Mathematics and Applications in Macroscale

Part IV Mathematics and Applications in Real World

Part V Mathematics and Applications in Space

Roland Bulirsch – 75th Birthday

Roland Bulirsch was born on November 10th, 1932 in Reichenberg (Bohemia). From 1947 till 1951 he received his professional training as an engine fitter at the Siemens-Schuckert plant in Nürnberg and worked on the assembly line for industrial transformers. In autumn 1954 he began studying Mathematics and Physics at the Technische Hochschule (later: Universität) München. He finished his Ph.D. in 1961, under his supervisor Prof. Dr. Dr.-Ing. E. h. R. Sauer. Five years later he received the venia legendi for Mathematics at the TU München. At this time the TU München was already one of the leading centers in Numerical Analysis worldwide. The well-known journal *Numerische Mathematik* was established under the influence of Prof.

Fig. 1. Roland Bulirsch

Dr. Dr. h. c. mult. F. L. Bauer. Together with Josef Stoer, Roland Bulirsch contributed the famous extrapolation algorithms. From 1966 to 1967 he served as chair of Mathematics for Engineers at the TU München.

1967 he was offered an Associate Professorship at the University of California at San Diego and worked there through 1969. This period formed his further scientific life. He often returned to the US, especially to San Diego. In 1969 he was offered a chair of Mathematics at the Universität zu Köln, where he worked until 1972. He accepted a full professorship for Mathematics for Engineers and Numerical Analysis at the Department of Mathematics of the TU München in 1972, where he would remain until his retirement in 2001.

The 30 years at the TU München form the core of his scientific and academic career. His lectures focused on the numerical treatment of ordinary and partial differential equations, the calculus of variations and optimal control, and elliptic integrals and special functions in applications. He was responsible for the courses in Mathematics for the engineering sciences. As an excellent and inspiring teacher, and as an enthusiastic researcher, he stimulated and encouraged his numerous Ph.D. students. More than 200 diploma theses, 40 Ph.D. theses and 12 venia legendi theses are signs of his enormous influence.

Approximately 100 papers – among them outstanding works on extrapolation methods, multiple shooting and the theory of special functions – and 11 books prove his extraordinary capability to combine theory with practical applications. Some of his books have been translated into English, Polish, Italian, and even Chinese. He was also engaged in the scientific and organizational administration of the TU München, far from regarding these further duties as an unwanted burden, he was well aware of the opportunities they offered for the development of Mathematics as a discipline. He served as editor for several well-known mathematical journals including *Numerische Mathematik*, the *Journal of Optimization Theory and Applications*, *Optimization* and *Mathematical Modelling*. For more than a decade he was a referee for the Deutsche Forschungsgemeinschaft (German Science Foundation), where he also headed the committee for Mathematics from 1984 to 1988. He was a referee for the Alexander von Humboldt Stiftung and a member of the scientific board of the Mathematical Research Institute of Oberwolfach, where he served several terms Dean of the School of Mathematics. During the last 3 years of active duty he was Senator of the TU München. His outstanding work was recognized with several honorary doctorates: from the Universität of Hamburg in 1992, from the TU Liberec in 2000, from the TU Athens in 2001 and from the University of Hanoi in 2003. He was awarded the Bavarian medal of King Maximilian in 1998 and is a member of the Bavarian Academy of Sciences, where he has chaired the classes in Mathematics and Natural Sciences since 1998. Among numerous further awards, he was decorated with the Alwin-Walther medal by the TU Darmstadt in 2004.

Few scientists have been so able to combine Mathematics with different fields of application. These fields include space technology, astronomy, vehicle dynamics, robotics, circuit simulation and process simulation. In addition to his scientific interests, Roland Bulirsch has always been open to political and cultural questions. He is an expert on orchids, has a deep connection to music, and is highly interested in literature and art: The reader need only recall the speech he held for Daniel Kehlmann's famous book *Die Vermessung der Welt* (Measuring the World) on behalf of the Konrad Adenauer Stiftung in Weimar in 2006. Nor should we forget Roland Bulirsch's political horizons. Having been himself expelled from the Sudetenland, he has always endeavored to bridge the gap between the positions of the Sudetendeutsche community and the Czechoslovakian government. His balanced arguments were recognized with an honorary doctorate from the TU Liberec in 2000 and the Ritter von Gerstner medal in 2004.

Academic Genealogy of Roland Bulirsch

Roland Bulirsch supervised many PhD. theses, so it is no surprise that he has a lot of descendants which themselves have descendants and so on. Currently, the fourth generation is starting to evolve. According to *The Mathematics Genealogy Project* (http://www.genealogy.math.ndsu.nodak.edu/) Bulirsch has supervised 44 students and 199 descendants (as of May 2007). A descendant is defined as a student who has Roland Bulirsch either as first or second advisor of her/his PhD. thesis.

Restricting the list to direct descendants, i. e. where Roland Bulirsch was the first advisor, results in the impressive number of 42 students and 174 descendants, which are given in table below. The indentation of the names indicates the generation level. A picture of his genealogy with portraits of some of his descendants is given in Fig. A.2 on page 324.

Descendant	University	Year
Robert Smith	San Diego	1970
Siegfried Schaible	Köln	1971
Wolfgang Hackbusch	Köln	1973
Götz Hofmann	Kiel	1984
Gabriel Wittum	Kiel	1987
Thomas Dreyer	Heidelberg	1993
Peter Bastian	Heidelberg	1994
Volker Reichenberger	Heidelberg	2004
Christian Wagner	Stuttgart	1995
Nicolas Neuß	Heidelberg	1996
Henrik Rentz-Reichert	Stuttgart	1996
Armin Laxander	Stuttgart	1996
Jürgen Urban	Stuttgart	1996
Reinhard Haag	Stuttgart	1997
Bernd Maar	Stuttgart	1998

Descendant	University	Year
Bernhard Huurdeman	Stuttgart	1998
Klaus Johannsen	Heidelberg	1999
Christian Wrobel	Heidelberg	2001
Sandra Nägele	Heidelberg	2003
Michael Metzner	Heidelberg	2003
Jens Eberhard	Heidelberg	2003
Jürgen Geiser	Heidelberg	2004
Achim Gordner	Heidelberg	2004
Dmitry Logashenko	Heidelberg	2004
Christoph Reisinger	Heidelberg	2004
Achim Gordner	Heidelberg	2005
Michael Lampe	Heidelberg	2006
Walter Kicherer	Heidelberg	2007
Günter Templin	Kiel	1990
Stefan Sauter	Kiel	1992
Rainer Warnke	Zürich	2003
Frank Liebau	Kiel	1993
Thorsten Schmidt	Kiel	1993
Birgit Faermann	Kiel	1993
Jörg Junkherr	Kiel	1994
Christian Lage	Kiel	1996
Ivor Nissen	Kiel	1997
Jörg-Peter Mayer	Kiel	1997
Rainer Paul	Kiel	1997
Jens Wappler	Kiel	1999
Sabine Le Borne	Kiel	1999
Thomas Probst	Kiel	1999
Steffen Börm	Kiel	1999
Lars Grasedyck	Kiel	2001
Maike Löhndorf	Leipzig	2003
Kai Helms	Kiel	2003
Ronald Kriemann	Kiel	2004
Jelena Djokic	Leipzig	2006
Alexander Litvinenko	Leipzig	2006
Helmut Maurer	Köln	1973
Klemens Baumeister	Münster	1986
Laurenz Göllmann	Münster	1996
Christof Büskens	Münster	1998
Wilfried Bollweg	Münster	1998
Thomas Hesselmann	Münster	2001
Dirk Augustin	Münster	2002
Jang-Ho Kim	Münster	2002

Descendant	University	Year
Georg Vossen	Münster	2006
Peter Deuflhard	Köln	1973
Georg Bader	Heidelberg	1983
Eckard Gehrke	Heidelberg	1995
Guntram Berti	Cottbus	2000
Ralf Deiterding	Cottbus	2003
Peter Kunkel	Heidelberg	1986
Ronald Stöver	Bremen	1999
Michael Wulkow	Berlin	1990
Folkmar Bornemann	Berlin	1991
Michael Lintner	München	2002
Susanne Ertel	München	2003
Caroline Lasser	München	2004
Christof Schütte	Berlin	1994
Wilhelm Huisinga	Berlin	2001
Andreas Hohmann	Berlin	1994
Ulrich Nowak	Berlin	1994
Gerhard Zumbusch	Berlin	1995
Achim Schmidt	Augsburg	1998
Tilman Friese	Berlin	1998
Detlev Stalling	Berlin	1998
Peter Nettesheim	Berlin	2000
Martin Weiser	Berlin	2001
Tobias Galliat	Berlin	2002
Evgeny Gladilin	Berlin	2003
Malte Westerhoff	Berlin	2004
Werner Benger	Berlin	2005
Ralf Kähler	Berlin	2005
Marcus Weber	Berlin	2006
Anton Schiela	Berlin	2006
Dietmar Täube	München	1975
Peter Rentrop	München	1977
Utz Wever	München	1989
Bernd Simeon	München	1993
Jörg Büttner	München	2003
Meike Schaub	München	2004
Gabi Engl	München	1994
Michael Günther	München	1995
Andreas Bartel	München	2004
Roland Pulch	München	2004
Gerd Steinebach	Darmstadt	1995
Thomas Neumeyer	Darmstadt	1997

Descendant	University	Year
Christian Schaller	Darmstadt	1998
Yvonne Wagner	Darmstadt	1999
Markus Hoschek	Darmstadt	1999
Robert Gerstberger	Darmstadt	1999
Oliver Scherf	Darmstadt	2000
Martin Schulz	Karlsruhe	2002
Sven Olaf Stoll	Karlsruhe	2003
Kai Arens	München	2006
Michael Lehn	München	2006
Hans Joachim Oberle	München	1977
Andrea Coriand	Hamburg	1992
Helge Baumann	Hamburg	2001
Sigrid Fredenhagen	Hamburg	2001
Rüdiger Seydel	München	1977
Christoph Menke	Ulm	1994
Rolf Neubert	Ulm	1994
Martin Stämpfle	Ulm	1997
Karl Riedel	Köln	2002
Rainer Int-Veen	Köln	2005
Pascal Heider	Köln	2005
Silvia Daun	Köln	2006
Sebastian Quecke	Köln	2007
Wolfgang Busch	München	1977
Hans-Jörg Diekhoff	München	1977
Joachim Steuerwald	München	1977
Hans Josef Pesch	München	1978
Michael Breitner	Clausthal	1995
Henrik Hinsberger	Clausthal	1997
Rainer Lachner	Clausthal	1997
Maximilian Schlemmer	Clausthal	1997
Birgit Naumer	Clausthal	1999
Matthias Gerdts	Bayreuth	2001
Roland Griesse	Bayreuth	2003
Susanne Winderl	Bayreuth	2005
Kati Sternberg	Bayreuth	2007
Peter Lory	München	1978
Thomas Wölfl	Regensburg	2006
Klaus-Dieter Reinsch	München	1981
Maximilian Maier	München	1982

Descendant	University	Year
Michael Gerstl	München	1983
Eckhard Kosin	München	1983
Albert Gilg	München	1984
Ralf Kleinsteuber	München	1987
Bernd Kugelmann	München	1987
Walter Schmidt	München	1987
Claus Führer	München	1988
Cornelia Franke	Ulm	1998
José Díaz López	Lund	2004
Martin Kiehl	München	1989
Stephan Franz	Darmstadt	2003
Thomas Speer	München	1989
Ulrich Leiner	München	1989
Peter Hiltmann	München	1990
Rainer Callies	München	1990
Georg Denk	München	1992
Werner Grimm	München	1992
Kurt Chudej	München	1994
Oskar von Stryk	München	1994
Thomas Kronseder	Darmstadt	2002
Uwe Rettig	Darmstadt	2003
Angela Baiz	Darmstadt	2003
Torsten Butz	Darmstadt	2004
Robert Höpler	Darmstadt	2004
Markus Glocker	Darmstadt	2005
Jutta Kiener	Darmstadt	2006
Stefan Miesbach	München	1995
Francesco Montrone	München	1995
Markus Alefeld	München	1998
Alexander Heim	München	1998
Oliver Mihatsch	München	1998
Petra Selting	München	1998
Jörg Haber	München	1999
Martin Kahlert	München	2000

Descendant	University	Year
Christian Penski	München	2000
Dietmar Tscharnuter	München	2002

Mathematics and Applications in Nanoscale

Circuit Simulation for Nanoelectronics

Georg Denk and Uwe Feldmann

Qimonda AG, Am Campeon 1–12, 85579 Neubiberg
georg.denk@qimonda.com, uwe.feldmann@qimonda.com

Summary. Though electronics is a quite young field, it is found almost everywhere in nowadays life. It became an important industrial sector within a short time frame. Behind this success story, very advanced research is necessary to push further the possibilities of the technology. This led to decreasing dimensions, from millimeters in the 1950s to nanometers in current products. To emphasize the new challenges due to the small sizes, the term "nanoelectronics" was coined. One important field of nanoelectronics is circuit simulation which is strongly connected to numerical mathematics. This paper highlights with some examples the interaction between actual and future problems of nanoelectronics and their relation to mathematical research. It is shown that without significant progress of mathematics the simulation problems showing up cannot be solved any more.

1 Introduction

Since more than 100 years semiconductor devices are used in industrial applications, starting with the discovery of the rectifier effect of crystalline sulfides by Ferdinand Braun in 1874. But it took time until 1940, that the first semiconductor device was developed at Bell Labs. In 1951 Shockley presented the first junction transistor, and ten years later the first integrated circuit (IC) was presented by Fairchild. It contained only a few elements, but with this IC an evolution was initiated which to our knowledge is unique in the technical world. Due to technological progress and larger chip sizes, it was possible to integrate more and more devices on one chip, up to more than one billion transistors on advanced memory chips or microprocessors of today.

The increasing number of elements was already predicted in 1965 and is now well-known as "Moore's law" (cf. Fig. 1). It states that the number of transistors per chip doubles every 2 years, and although being nothing else than an experimental observation in the beginning, it turned out to hold for more than 30 years now. The most essential outcomes of higher integration of functionality on chip are continuously decreasing cost per function by $\approx 30\%$

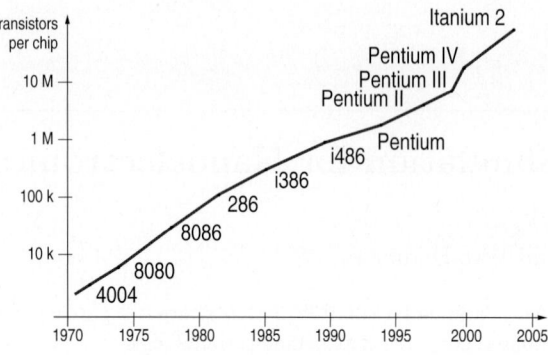

Fig. 1. Moore's law

per year, and continuously increasing processing bandwith by a factor of 2 every 2 years.

The development of more complex and faster chips would not have been possible without computer aided design tools, pioneered by circuit simulation. In 1967 one of the first simulation programs called BIAS was developed by Howard. So circuit simulation is nearly as old as the design of integrated circuits itself. In the sequel a simulation program CANCER was developed by Nagel in the research group of Rohrer. Later on the development of the programs was brought together under supervision of Pederson, and in 1972 the first version of SPICE was released at Berkeley University.

The breakthrough of circuit simulation tools came with higher integration of more functions on one chip, since it was no longer possible to get reliable prediction of circuit behavior just by breadboarding. One key factor for their success – and in particular of SPICE – was the development of simple but accurate compact models for the spatially distributed integrated semiconductor devices, thus mapping semiconductor equations onto a set of Kirchhoff's equations for a specific configuration of circuit primitives like resistors, capacitors, and controlled sources. The models of Ebers-Moll [10] and Gummel-Poon [16] for bipolar transistors, and of Shichman-Hodges [25] and Meyer [20] for MOS transistors have to be mentioned here.

Numerical mathematics played another key enabling role for success of circuit simulation. Through the progress in handling stiff implicit ODE systems it became easily possible to set up mathematical models directly from inspecting the circuit netlist, and to solve them with flexible and robust integration methods. First of all, Gear's pioneering work [13] has to be mentioned here, which by the way was mainly devoted to circuit simulation problems and consequently was published in an electronics journal. Later on the introduction of sparse linear solvers was another major step for the success of circuit simulation with its ever increasing complexity of problems. And finally, transition to charge/flux oriented formulation of circuit equations brought physical consistency, as well as numerical benefits.

The device models, a rather robust numerical kernel, dynamic memory management providing flexibility, and the free availability made SPICE to become widely used and some kind of an industry standard (see [23] for further information about the history of SPICE). Meanwhile SPICE-like simulators are known to yield realistic and reliable simulation results. They are universally applicable and often used for "golden" simulations. Over the years the methods and codes have received a high level of robustness, which allows to use them in mass applications, without requiring much knowledge about the mathematical methods employed.

It is one purpose of this contribution to demonstrate with some examples that, beyond Gear's work, numerical mathematics can significantly contribute to progress in circuit simulation. This has been surely true in the past, but it is even more necessary for the future. So the second purpose of this contribution is to highlight some disciplines of mathematical research, which will be of interest to solve the simulation problems of nanoelectronics in the future. In Sect. 2, the impact of progress in micro- and nanoelectronics onto mathematical problems is sketched, as far as circuit simulation topics are concerned. In Sect. 3, 4, and 5 some of the topics are discussed in more detail. Section 6 gives a glance on further important topics, and Sect. 7 presents some conclusions.

2 Mathematics for Nanoelectronics

The permanently high rate of progress in semiconductor technology is based on intense research activities, and would not have been possible without support of academic research institutes. This is in particular true for circuit simulation. Since the models and modeling techniques are mostly standardized worldwide in this area, competition is mostly with respect to user interfaces – which we do not consider here –, simulation capabilities, and numerics. This gives mathematical disciplines a key role for the success of such tools. Table 1 lists some of the topics, where our university partners contributed to solve upcoming challenges for circuit simulation in the past. Similar efforts have been undertaken on topics from layout and lithography, test, and verification.

The first block in Table 1 deals with the impact of *downscaling feature size*. The second one considers the growing importance of *Radio Frequency (RF), analog, and mixed signal* circuitry. The third block mirrors attempts to maximize circuit *robustness and yield*, and the last one is related to attempts to improve *designers efficiency*.

Looking at current trends in semiconductor industry, we can clearly identify as the main driving forces [19]:

- Shrinking of feature size brings reduced production cost, higher performance or less power consumption, and saving of space in the final product
- Integration of more functionality, like data sensing, storing and processing on the same chip

Table 1. Mathematical topics, derived from past trends in microelectronics

Trend	Circuit simulation problem	Mathematical topic
downscaling feature size		
increasing circuit complexity	efficiency improvements	hierarchical solvers
	parallel simulation	partitioners
	latency exploitation	multirate integration
impact of parasitic elements	large and dense linear subnets	model order reduction
RF, analog and mixed signal		
radio frequency (RF) circuits	nonlinear frequency analysis	shooting methods, Harmonic Balance
	multiscale time domain analysis	MPDAE solvers
slowly settling oscillators	envelope analysis	mixed time/frequency methods
high gain / high resolution	transient noise analysis	stochastic DAEs
robustness and yield		
increase of design robustness	stability analysis	Lyapunov theory
stabilizing feedback structures	multiple operating points	continuation methods
yield improvement	yield estimation, optimization	sensitivity analysis
designers efficiency		
complex design flows	robustness	higher index DAEs, semi-implicit integration
top-down/bottom-up design	functional circuit description	solvability and diagnosis

- Enabling of innovative products like GSM, radar sensors for automotive applications, or a one chip radio

So there is strong pressure to integrate complete systems on one chip, or in one common package. As a consequence, effects which were isolated before now tend to interact, and need to be considered simultaneously. Other consequences are that the spatial extension of devices becomes more and more important, and that new materials with previously unknown properties have to be modeled and included into simulation. All this imposes new challenges onto the simulators, which only can be solved with advanced numerical methods.

To get a more detailed view we take the International Technology Roadmap for Semiconductors (ITRS) [17] from the year 2005 as a reference, and try to identify the most important topics of mathematical research from the 10 years trends presented there.

Downscaling feature size: Moore's law will continue, occasionally at a slightly lower rate:
  - 30% feature size reduction per year
  - 2 times more functions per chip every 3 years
  - 25% less cost per function every year
  - 2 times more instructions per second every 2 years

   As a consequence, channel length of transistors is expected to be ≈ 15nm in 2015, and may come down to its limits of 9nm some 5 years later.
RF, analog and mixed signal: Moore's law is restricted to data storing and processing, but not applicable to power, high frequency applications, sensing, and actuating.
Robustness and yield: Thermal budgets require further decrease of supply voltage, with high impact on noise margins.
Designers efficiency: Technology cycle time will stay at ≈ 4 years. This requires extremely fast design manufacturing ramp up, in order to get compensation for high technology invest in short time.

From Table 2 we can see how these trends map onto new circuit simulation problems, and how the latter might be addressed by advanced numerical methods.

So there is a clear trend in circuit simulation towards coupling with various domains, described by partial differential equations in 1, 2 or even 3 dimensions, and with probably very different time scales. This will require flexible and efficient coupling schemes in general, which exploit the multiscale property of the problems by multirate integration or some different way of latency concepts. Furthermore, to cope with the strongly increasing complexity, application of linear and nonlinear Model Order Reduction (MOR) techniques will be inevitable. Third, the paradigm of harmonic basis functions in the high frequency domain will become questionable, and finally stochastic methods and multiobjective optimization will get high priority to increase robustness of designs and to improve designers efficiency.

3 Device Modeling

Following Moore's law is that the devices on a chip will get smaller and smaller while the number of devices increases. This shrinking allows a reduction of the production costs per transistor, and due to the smaller devices a faster switching. The effect of shrinking on die size can be seen in Fig. 2, where the same memory chip is shown for different shrinking sizes. But shrinking has drawbacks, too: The compact modeling used for devices like MOSFETs is no longer accurate enough, and the power density on the chip increases. Both issues have their impact on circuit simulation.

Table 2. Mathematical topics, derived from future trends in nanoelectronics

Trend	Circuit simulation problem	Mathematical topic
downscaling feature size		
non-planar transistors	limits of compact modeling	coupled circuit/device simulation
quantum and atomistic effects	hierarchy of models	hierarchical solver strategies
new materials and device concepts	more flexible modeling	(not yet clear)
impact of interconnect	efficiency, memory	linear MOR hybrid/split linear solver
multi-layer dielectrics	coupled multiconductor lines	coupled T-line solvers
impact of selfheating	electro-thermal simulation	coupling circuit/thermal
huge complexity	improve adaptivity	multirate integration, hierarchical solver, latency
	parallel simulation	scalable parallelism
ultra-high RF	non-quasistationary models	coupling circuit/device
RF, analog and mixed signal		
RF: analog goes for digital	RF with pulse shaped signals envelope analysis	wavelet based Galerkin envelope methods
robustness and yield		
reliability	electro migration analysis	statistical failure analysis
reduced signal/noise ratio	statistical variation of devices	yield estimation incremental analysis
design for yield	critical parameter identification	statistical methods
	yield improvement	multiobjective optimization
designers efficiency		
design productivity	design reuse for analog	hierarchical MOR, parameterized MOR
top-down/bottom-up design	analog verification	system identification

With decreasing device geometry more and more effects have to be considered by the compact models used for MOSFETs and other devices. The modeling of these effects are packed onto the existing model which gets more and more complicated. A sign of this complexity is the number of model parameters – the BSIM4 model from Berkeley University [3], which is used as a quasi standard, has more than 800 parameters! For simulation the parameters have to be chosen in such a way that the model reflects the current technology. But due to the high fitting dimension this task is quite difficult and may lead to reasonable matching characteristics with unphysical param-

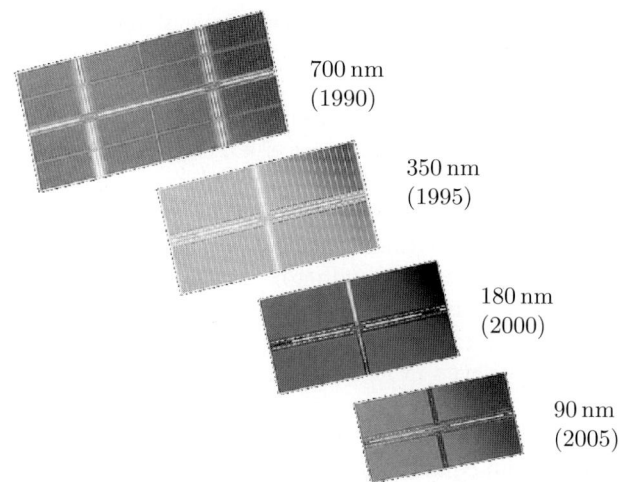

700 nm
(1990)

350 nm
(1995)

180 nm
(2000)

90 nm
(2005)

Fig. 2. Shrinking of a DRAM chip for different technologies

eter settings, which cause numerical problems during simulation. In addition, the device characteristics are heavily affected by transistor geometry and even by neighborhood relations which can not be reflected by compact models.

A remedy for this situation is to switch back to the full set of semiconductor equations for critical MOSFETs (like high frequency settings, MOSFETs used for electrostatic discharge (ESD) protection). This approach allows full accuracy at the cost of a large computational effort for the device evaluation. Even when compared to complex compact models, there are some orders of magnitude between compact modeling and solving the semiconductor equations. According to the ITRS roadmap [17], device simulation will become more and more important for future devices which may show quantum effects or may have no planar structures anymore.

Device simulation itself is almost as old as circuit simulation, and – similar to circuit simulation – there is still need for improvement. Three-dimensional device simulation was made productive in the 1980s, e. g. MINIMOS [29]. A main problem in 3D-device simulation is the grid generation, this has also been investigated at Bulirsch's chair by Montrone [21]. Additional research was done within the Bavarian Consortium for High Performance Scientific Computing (FORTWIHR) under the direction of R. Bulirsch. Besides improving the efficiency of a device simulation, modeling aspects continue to be very important. Due to the smaller dimension of devices, additional effects have to be considered which are reflected in the models. As the planar MOSFET device will no longer be sufficient, new geometries have to be developed. For these new devices, there are no compact models available increasing the need for a 3D-device simulation [6, 31].

In the context of circuit simulation, device simulation is just some kind of model evaluation. Due to the large computational effort needed for a device simulation, it is impossible to treat all MOSFETs of a large design with the full accuracy provided by device simulation. Considering the layout of a complex chip, for instance a GSM transceiver (cf. Fig. A.4 on page 326) with its different parts like memory cells, transceiver, digital signal processing, and so on, there are different parts of the chip with different requirements regarding accuracy of the MOSFET models. Therefore, some criteria are needed for the decision which MOSFETs can be simulated using the compact model and which devices require the semiconductor equations for appropriate modeling. Currently the decision is made by the designer; a more automatic selection can be achieved on the basis of sensitivity analysis, which can easily identify the "critical" devices even in large circuits.

Assuming that it is known which model fits to which device, there is still an open issue: how should the coupling be done between the circuit simulation (differential-algebraic equations) and the device simulation (partial differential equations)? Using both simulators as black box may work, but there is a need to analyze the interaction of them. This will allow also a more efficient coupling, especially when transient simulations are performed. The analysis requires the extension of the index definition of the DAE and of the computation of consistent initial values. A theoretical foundation for this are abstract differential-algebraic systems (ADASs) [28]. A main research area of the European project COMSON [5] is to bring this theoretical foundation into a demonstrator which enables a development environment to easily implement different approaches [7].

4 Interconnect Modeling

During the design phase of a chip the connections between the elements are treated as ideal, i. e. it is assumed that the elements do not influence each other and there are no delays between devices. But this is not true on a real chip, due to the length and the neighborhood of the wires there is interference like crosstalk, and the elements themselves suffer from the actual placement on silicon. In Fig. 3 the layout of a small portion of a waver is given consisting of four MOSFETs. In order to simulate the real behavior of the chip, the circuit is extracted from of the layout, and this post-layout circuit containing all the parasitic elements has to be simulated and cross-checked against the original design. An example of a parasitic subnet representing the interconnect is shown in Fig. 4. Currently it is still sufficient in many cases that only resistances and capacitances are used to model the parasitic effects. However, to get all relevant effects due to the small structures and currents present on a chip, the accuracy of the extraction has to be improved by extracting more elements and using additional element types like inductances.

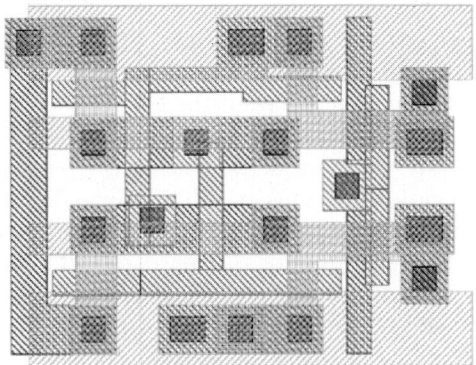

Fig. 3. Layout extraction

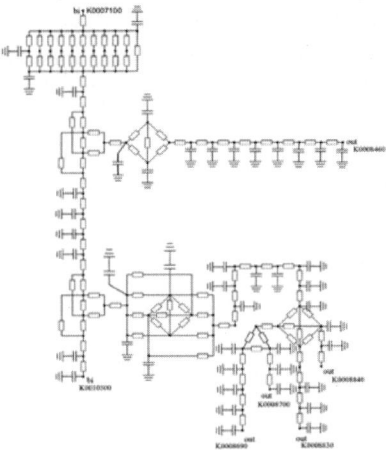

Fig. 4. Extracted parasitics

Of course, the quantity of parasitic elements has an impact on circuit simulation: Due to the parasitics, the number of nodes present in a circuit increases significantly, which increases the dimension of the system to be solved. This effect is amplified by the fact that due to fill-in the sparsity of the system decreases, therefore the effort necessary for solving the underlying linear equation system becomes dominant and the simulation's run times are no longer acceptable. To speedup the simulation it is necessary to perform a parasitics reduction. Several approaches are possible: One possibility is to rearrange the parasitic elements in such a way that the influence on the circuit is unchanged but the number of nodes is reduced, possibly at the cost of additional elements. A typical example for this is the so-called star-delta conversion. Another way is to remove and rearrange parasitic elements which do not significantly contribute to the circuit's response, for example long RC trees are replaced by shorter ones. As some errors are introduced by this reduction, a reliable error control is necessary for this approach.

The first step of a reduction process is the separation of the linear elements from the circuit, leaving the non-linear elements like MOSFETs untouched. The linear elements are then assembled into matrices which transfer them into the mathematical domain. One possibility for reduction is the spectral decomposition, which was investigated at Bulirsch's chair by Kahlert [18]. This method computes the eigenvalues of the matrices and removes insignificant eigenvalues. A very important property of this approach is that the passivity and stability of the reduced system is guaranteed.

Model order reduction (MOR) is still an active field of research. One of the problems is the number of inputs and outputs of the linear subnets. Standard methods for MOR work best if the number of inputs and outputs is quite small, which is not true in general for applications in circuit simulation. Especially the simulation of power grids is investigated, where the huge number of inputs and outputs renders established MOR methods unusable. Though the system's dimension is reduced, the sparsity of the matrices is lost yielding an even larger effort necessary for solving. First results in maintaining the sparsity can be found in [2, 12].

Model order reduction may not applied only to linear subnets but also to non-linear subcircuits including MOSFETs and other semiconductor devices. This requires non-linear MOR methods which are another field of research, e. g. within COMSON [5].

5 Noise Simulation

One mean to accomplish smaller devices and higher frequency is the reduction of the power-supply voltage. While supplies of 5 V have been used in the past, the supply voltage has been reduced down to 1 V or even below. The advantages of this approach is the avoidance of breakthrough and punch in the small devices, and – as the voltage swing is reduced – a higher switching frequency. The lower supply voltages also help in the field of mobile applications. But reducing the power supply has also a drawback: The signal-to-noise ratio decreases which means that parasitic effects and noise become more and more significant and can no longer be omitted from circuit simulation.

Reduced signal-to-noise ratio means that the difference between the wanted signal and noise is getting smaller. A consequence of this is that the circuit simulation has to take noise into account. Usually noise simulation is performed in the frequency domain, either as small-signal noise analysis in conjunction with an AC analysis or as large-signal noise analysis as part of an harmonic balance or shooting method. These noise analyses are well-established in the meantime. But noise analysis is also possible in the context of transient noise analysis for non-oscillatory circuits. For an implementation of an efficient transient noise analysis in an analog simulator, both an appropriate modeling and integration scheme is necessary.

Modeling of transient noise

A noisy element in transient simulation is usually modeled as an ideal, non-noisy element and a stochastic current source which is shunt in parallel to the ideal element. As a consequence of this approach the underlying circuit equations

$$A \cdot \dot{q}(x(t)) + f(x(t), t) + s(t) = 0 \tag{1}$$

are extended by an additional stochastic part, which extends the DAE to a stochastic differential-algebraic equation (SDAE)

$$A \cdot \dot{q}(x(t)) + f(x(t), t) + s(t) + G(x(t), t) \cdot \xi = 0 \tag{2}$$

(for details refer to [30]). Here, $x(t)$ is the vector of unknowns (node voltages, some branch currents), $q(x(t))$ are the terminal charges and branch fluxes, A is an incidence matrix describing the topology of the circuit, $f(x(t), t)$ describes the static part of the element equations, and $s(t)$ holds the independent sources. The matrix $G(x(t), t)$ contains the noise densities, and ξ is a vector of noise sources.

Depending on the cause of noise there are mainly three different noise models in use: thermal noise, shot noise and flicker noise. While the stochastic current source for thermal and shot noise can be simulated using Gaussian white noise, this is not possible for flicker noise. The memory of this process – corresponding to the $1/f^{\beta}$ dependency for low frequencies f and $\beta \approx 1$ – does not allow a white noise modeling where the increments are stochastically independent. One possibility to represent flicker noise for transient analysis is to use fractional Brownian motion (fBm) for $0 < \beta < 1$. Fractional Brownian motion is a Gaussian stochastic process, and the increments of the fBm required for a transient simulation can be realized with normal-distributed random numbers, for details see [8].

Integration of stochastic differential-algebraic equations

The modeling of transient noise is only one part of a transient noise simulation, the other one is the integration of the SDAEs. Though there are some numerical schemes available for stochastic differential equations (SDEs), they do not fit properly to the context of circuit simulation. Besides the fact that the standard schemes are for SDEs and not for SDAEs, they mostly require high-order Itô integrals and/or higher order derivatives of the noise densities. Both is not possible in circuit simulation, this is either too expensive or even not available. For efficiency reasons we need specialized integration schemes which exploit the special structure of the equations.

Restricting the noise sources to thermal noise and shot noise, it is possible to solve the stochastic part using the Itô calculus, while some adaptions are necessary to incorporate the charge-oriented structure of the system [22]. A main drawback of integration methods for SDAEs is the low convergence order – due to the stochastic part and the multiplicative character of the noise, the theoretical order is 1/2. Fortunately, even in current chip designs, the noise level is still smaller in magnitude compared to the wanted signal, which can be used for the construction of efficient numerical schemes.

Numerical experiments [9] have shown that the effective integration order of the integration scheme depends on the error tolerances and the magnitude of the noise sources compared to the deterministic part. The effective order is a combination of the orders of the integration schemes used for the deterministic part $A \cdot \dot{q}(x(t)) + f(x(t), t) + s(t)$ of (2) and the stochastic part $G(x(t), t) \cdot \xi$, respectively. For relaxed tolerances and small noise sources, the integration order of the deterministic part is dominant. This order can be made high quite easily. For strict error tolerances and large noise sources, however, the integration order of the stochastic part is dominant, and it is almost impossible to construct schemes of high order for the stochastic part in the context of circuit simulation. For the drift-implicit BDF-2 scheme [4, 26] with constant stepsize h

$$A \cdot (q_k - \frac{4}{3}q_{k-1} + \frac{1}{3}q_{k-2}) + \frac{2}{3}h(f(x_k, t_k) + s(t_k))$$
$$+ G(x_{k-1}, t_{k-1}) \cdot \Delta W_k - \frac{1}{3}G(x_{k-2}, t_{k-2}) \cdot \Delta W_{k-1} = 0,$$

the integration order is $\mathcal{O}(h^2 + \epsilon h + \epsilon^2 h^{1/2})$ assuming small noise densities $\tilde{G}(x(t), t) = \epsilon G(x(t), t)$. Here, q_k denotes the numerical approximation of $q(t_k)$ at time $t_k = k \cdot h$, and ΔW_k is a vector of $N(0, h)$-distributed random numbers describing the increment of the Brownian motion at time t_k. So the integration order is 1/2 from a theoretical point of view, while it is 2 in practical applications. Of course, the stepsize control has to be adapted to benefit from this property.

Normally it is not sufficient to compute a single path of the transient noise but several paths are necessary to get reliable stochastic conclusions. It helps to improve the efficiency of a transient noise analysis, when these paths are computed simultaneously. This requires a new stepsize control which incorporates the stochastic nature of the paths. This has the advantage of averaging out the outliers and results in a smoother stepsize sequence [30].

6 Further Topics

Besides the challenges described above, there are many other issues which need to be solved for a successful and efficient circuit simulation. Most of the results have in common that they incorporate methods which are especially adapted to the settings of circuit simulation.

A very time-consuming part is the time-domain simulation of circuits. It is possible to use multirate integration [14, 27] where different parts of the circuit with different level of activity are computed with different stepsizes. Depending on the application this may lead to a significant speedup. An even more advantageous approach is using warped multirate partial-differential equations [24] where the different time scales of the circuit equations are represented by different independent variables. This transforms the differential-algebraic system into a partial-differential equation.

Not only the coupling with device simulation as described above is of importance but also the coupling to other domains. Thermal effects are getting more and more important, so it becomes necessary to incorporate thermal effects into circuit simulation [1]. Also interconnect issues may be resolved by coupling the telegrapher's equation to the circuit equations [15]. Both lead to the coupling of differential-algebraic equations with partial-differential equations and require specialized coupling approaches and corresponding numerical methods.

The complexity of current designs requires an hierarchical design approach. However, it is not only the design but also the solver and the modeling part require hierarchy. It will no longer be possible to treat the overall system as one single system, it has to be split into several subsystems, possibly from different domains. This shows the need for hierarchical solvers especially adapted to the situation of circuit simulation. For the models of a circuit a hierarchy of models is needed which are adapted to the current need of the simulation. They may range from device models with full accuracy to simplified digital models for large subcircuits. As these models should describe the same effects, an automated verification method is needed. Though verification is a standard approach in the digital field, it is just at its beginning for analog circuits.

During the design process there will be some errors contained within the circuit. These errors may result e. g. in convergence problems. As it becomes more and more difficult to find the root of the problems, there is a need for tools which map the numerical issues to the circuitry [11] and thus help the designer to find a solution. This requires a connection of numerical mathematics to other fields like graph theory emphasizing again the need for coupling.

7 Conclusion

Though circuit simulation is a rather old field of research, it will remain active for the future. A major trend in circuit simulation is that it is no longer possible to neglect the distributed structures of circuits. While it was sufficient in the beginning to treat the devices as 0-dimensional, more and more 1-, 2-, and even 3-dimensional effects have to be regarded in circuit simulation. It was possible in the past to treat these effects separately but now the interaction must be considered. This leads to the need for co-simulation where different aspects have to be coupled in a proper manner.

The coupling of more and more different effects in an increasing accuracy requires more adaptivity. As it is not possible to incorporate all effects at all levels, the numerical methods must be adapted to the designers' need. The schemes have to consider as much effects as necessary while they have to be as fast as possible. This requires specialized approaches like multirate schemes or locally adapted solvers. In addition, parallel simulation helps to increase the efficiency.

As the time-to-market is decreasing, it is necessary to speedup the design process. While compact models have been hand-crafted in the past, automated procedures are needed for the future. This might be done by model-order reduction of discretized models thus generating a hierarchy of models with controlled errors. While reuse is already a valuable tool for design, it might become also important for simulation. In addition, yield and reliability of circuits have to be considered in circuit simulation.

A consequence of the increasing need for co-simulation and coupling is an increasing need for cooperation between the mathematical research groups. Each group brings in its own field of knowledge which has to be combined to enable successful circuit simulation for future applications. It is very important to encourage the joint research in formerly separated areas, otherwise the coupled simulation will not be as effective as needed for the coming needs. It is also crucial to keep the application in mind while doing research. This enables good opportunities for especially tailored methods which feature high efficiency. This was always a very important concern for Roland Bulirsch which has been handed on to his students. We highly appreciate that this attitude has led to successful results which have found their way into industrial applications. We have indicated in this paper that there is still a lot to do, so we hope that this approach will be continued by Bulirsch's students.

References

[1] Bartel, A.: Partial differential-algebraic models in chip design – thermal and semiconductor problems. PhD. thesis, University Karlsruhe. Fortschritt-Berichte VDI, Reihe 20, Nr. 391, VDI-Verlag Düsseldorf, 2004.

[2] Berry, M., Pulatova, S., Stewart, G.: Algorithm 844: Computing sparse reduced-rank approximations to sparse matrices. ACM Trans. Math. Software **31**, 252–269 (2005)

[3] BSIM4 manual. Department of Electrical Engineering and Computer Science, University of California, Berkeley (2000). http://www-device.EECS.Berkeley.EDU/~bsim/

[4] Buckwar, E., Winkler, R.: Multi-step methods for SDEs and their application to problems with small noise. SIAM J. Num. Anal., (44), 779-803 (2006)

[5] COMSON – COupled Multiscale Simulation and Optimization in Nanoelectronics. Homepage at http://www.comson.org/

[6]   de Falco, C.: Quantum corrected drift-diffusion models and numerical simu-
        lation of nanoscale semiconductor devices. PhD. thesis, University of Milano,
        Italy, 2005.
[7]   de Falco, C., Denk, G., Schultz, R.: A demonstrator platform for coupled mul-
        tiscale simulation. In: Ciuprina, G., Ioan, D. (eds) Scientific Computing in
        Electrical Engineering, 63–71, Springer, 2007.
[8]   Denk, G., Meintrup, D., Schäffler, S.: Transient noise simulation: Modeling
        and simulation of $1/f$-noise. In: Antreich, K., Bulirsch, R., Gilg, A., Rentrop,
        P. (eds) Modeling, Simulation and Optimization of Integrated Circuits. ISNM
        Vol. 146, 251–267, Birkhäuser, Basel, (2003)
[9]   Denk, G., Schäffler, S.: Adams methods for the efficient solution of stochastic
        differential equations with additive noise. Computing, **59**, 153–161 (1997)
[10]   Ebers, J.J., Moll, J.L.: Large signal behaviour of junction transistors. Proc.
        IRE, **42**, 1761–1772 (1954)
[11]   Estévez Schwarz, D., Feldmann, U.: Actual problems in circuit simulation. In:
        Antreich, K., Bulirsch, R., Gilg, A., Rentrop, P. (eds) Modeling, Simulation
        and Optimization of Integrated Circuits. ISNM Vol. 146, 83–99, Birkhäuser,
        Basel, (2003)
[12]   Feldmann, P., Liu, F.: Sparse and efficient reduced order modeling of linear
        subcircuits with large number of terminals. In: Proc. Intl. Conf. on CAD, San
        Jose, November 2004, 88–92 (2004)
[13]   Gear, W.: Simultaneous numerical solution of differential-algebraic equations.
        IEEE Trans. Circuit Theory, **CT-18**, 89–95 (1971)
[14]   Günther, M., Rentrop, P.: Multirate ROW methods and latency of electric
        circuits. Appl. Numer. Math., **13**, 83–102 (1993)
[15]   Günther, M.: Partielle differential-algebraische Systeme in der numerischen
        Zeitbereichsanalyse elektrischer Schaltungen. network equations in chip design.
        Fortschritt-Berichte VDI, Reihe 20, Nr. 343, VDI-Verlag Düsseldorf, 2001.
[16]   Gummel, H.K., Poon, H.C.: An Integral Charge Control Model of Bipolar
        Transistors. Bell Syst. Techn. J., **49**, 827–852 (1970)
[17]   International Technology Roadmap for Semiconductors (ITRS), Edition 2005.
        Available at http://www.itrs.net/
[18]   Kahlert, M.: Reduktion parasitärer Schaltungselemente unter Verwendung der
        Spektralzerlegung. PhD. thesis, Technische Universität München, Germany,
        2002.
[19]   Marmiroli, A., Carnevale, G., Ghetti, A.: Technology and device modeling in
        micro and nano-electronics: Current and future challenges. In: Ciuprina, G.,
        Ioan, D. (eds) Scientific Computing in Electrical Engineering, 41–54, Springer,
        2007.
[20]   Meyer, J.E.: MOS models and circuit simulation. RCA Rev. **32**, 42–63 (1971)
[21]   Montrone, F.: Ein robustes adaptives Verfahren zur numerischen Lösung
        der partiellen Differentialgleichungen bei elektronischen Bauelementen in drei
        Raumdimensionen. PhD. thesis, Technische Universität München, Germany
        1995.
[22]   Penski, Chr.: Numerische Integration stochastischer differential-algebraischer
        Gleichungen in elektrischen Schaltungen. PhD. thesis, Technische Universität
        München, Germany, 2002.
[23]   Perry, T.: Donald O. Pederson. IEEE Spectrum, June 1998, 22–27
[24]   Pulch, R.: Transformation qualities of warped multirate partial differential
        algebraic equations. This issue, 27–42.

[25] Shichman, H., Hodges, D.A.: Insulated-gate field-effect transistor switching circuits. IEEE J. Solid State Circuits, **SC-3**, 285–289 (1968)

[26] Sickenberger, T., Winkler, R.: Efficient transient noise analysis in circuit simulation. PAMM, **6**, 55–58 (2006)

[27] Striebel, M.: Hierarchical mixed multirating for distributed integration of DAE network equations in chip design. PhD. thesis, University of Wuppertal. Fortschritt-Berichte VDI, Reihe 20, Nr. 404, VDI-Verlag Düsseldorf, 2006.

[28] Tischendorf, C.: Coupled systems of differential algebraic and partial differential equations in circuit and device simulation. Modeling and numerical analysis. Habil. thesis, Humboldt-Universität, Berlin, 2003.

[29] Thurner, M., Selberherr, S.: The extension of MINIMOS to a three dimensional simulation program. Proc. NASECODE V Conf, 327–332 (1987)

[30] Winkler, R.: Stochastic differential algebraic equations of index 1 and applications in circuit simulation. J. Comp. Appl. Math., **157**, 477–505 (2003)

[31] Wong, H.-S.: Beyond the conventional transistor. IBM Journal of Research and Development, **46**, 133–167 (2002)

Transformation Qualities of Warped Multirate Partial Differential Algebraic Equations

Roland Pulch

Lehrstuhl für Angewandte Mathematik und Numerische Mathematik, Bergische
Universität Wuppertal, Gaußstr. 20, D-42119 Wuppertal, Germany.
Email: pulch@math.uni-wuppertal.de

Summary. Radio frequency (RF) applications exhibit oscillating signals, where
the amplitude and the frequency change slowly in time. Numerical simulations can
be performed by a multidimensional model involving systems of warped multirate
partial differential algebraic equations (MPDAEs). Consequently, a frequency modulated signal demands a representation via a function in two variables as well as a
univariate frequency function. The efficiency of this approach depends essentially on
the determination of appropriate frequencies. However, the multidimensional representation is not specified uniquely by the corresponding RF signal. We prove that
choices from a continuum of functions are feasible, which are interconnected by
a specific transformation. Existence theorems for solutions of the MPDAE system
demonstrate this degree of freedom. Furthermore, we perform numerical simulations
to verify the transformation properties, where a voltage controlled oscillator is used.

1 Introduction

The modified nodal analysis represents a well established strategy for modelling electric circuits, see [4]. This network approach yields systems of differential algebraic equations (DAEs), which describe the transient behaviour of
all node voltages and some branch currents. In particular, the determination
of quasiperiodic solutions is a well known problem in radio frequency (RF) applications. These signals are characterised by a specific oscillating behaviour
at several time scales, where the respective magnitudes differ significantly.
Consequently, a transient analysis of the DAE system becomes inefficient,
since the size of time steps is limited by the fastest oscillation, whereas the
slowest time scale determines the time interval of the simulation. Time and
frequency domain methods have been constructed for the direct computation
of quasiperiodic solutions, see [2] and [12], for example. However, drawbacks
may occur in these techniques in view of widely separated time scales or strong
nonlinearities.

A multivariate model enables an alternative approach by decoupling the
time scales of such signals and thus generates an efficient representation of

amplitude modulated signals. Based on this strategy, Brachtendorf et al. [1] remodelled the DAEs into multirate partial differential algebraic equations (MPDAEs). Narayan and Roychowdhury [7] generalised the approach in view of frequency modulation and introduced corresponding warped MPDAEs. Multiperiodic solutions of this system reproduce quasiperiodic signals satisfying the DAEs. The determination of a suitable local frequency function arising in the model is crucial for the efficiency of the technique.

In this article, we analyse transformations of multivariate representations for frequency modulated signals. Thereby, the emphasis is on quasiperiodic functions. We perform investigations in the most frequent case of two time scales. Nevertheless, generalisations to several time rates are straightforward. Firstly, a transformation of multivariate functions is examined, where all associated representations reproduce the same signal. Secondly, we retrieve the same degree of freedom for solutions of the warped MPDAE system, too. Moreover, a transformation to a representation with constant local frequency is feasible. Thus solutions can be transferred to the ordinary MPDAE system. Consequently, we can apply theorems referring to the ordinary system, which are given by Roychowdhury [10], to obtain analogue results for the warped system.

The article is organised as follows. In Sect. 2, we outline the multidimensional model for signals. Consequently, we define quasiperiodic signals via corresponding multivariate functions. A specific transformation of multivariate representations is constructed, which illustrates an intrinsic degree of freedom in the model. We introduce the warped MPDAE model briefly in Sect. 3. An underlying information transport along characteristic curves is described, which reflects the transformation properties. Accordingly, we formulate theorems concerning transformations of solutions, which represent the analogon of the results on the level of the signals. Based on these implications, the specification of additional boundary conditions is discussed. Finally, Sect. 4 includes numerical simulations using the warped MPDAE system, which demonstrate the degrees of freedom in the representations.

2 Signal Model

2.1 Multivariate Signal Representation

To illustrate the multidimensional signal model, we consider the purely amplitude modulated signal

$$y(t) := \left[1 + \alpha \sin\left(\tfrac{2\pi}{T_1}t\right)\right] \sin\left(\tfrac{2\pi}{T_2}t\right) . \qquad (1)$$

The parameter $\alpha \in (0,1)$ determines the amount of amplitude modulation. Figure 1a depicts this function qualitatively. Hence many time steps are required to resolve all oscillations within the interval $[0, T_1]$ if $T_1 \gg T_2$ holds. Therefore we describe each separate time scale by an own variable and obtain

$$\hat{y}(t_1, t_2) := \left[1 + \alpha \sin\left(\frac{2\pi}{T_1}t_1\right)\right] \sin\left(\frac{2\pi}{T_2}t_2\right) . \tag{2}$$

The new representation is called the *multivariate function (MVF)* of the signal (1). In the example, the MVF is biperiodic and thus given uniquely by its values in the rectangle $[0, T_1[\times[0, T_2[$. Figure 1b shows this function. Since the time scales are decoupled, the MVF exhibits a simple behaviour. Accordingly, we need a relatively low number of grid points to represent the MVF sufficiently accurate. Yet the original signal (1) can be completely reconstructed by its MVF (2), because it is included on the diagonal, i.e., it holds $y(t) = \hat{y}(t, t)$. In general, straightforward constructions of MVFs for purely amplitude modulated signals yield efficient representations.

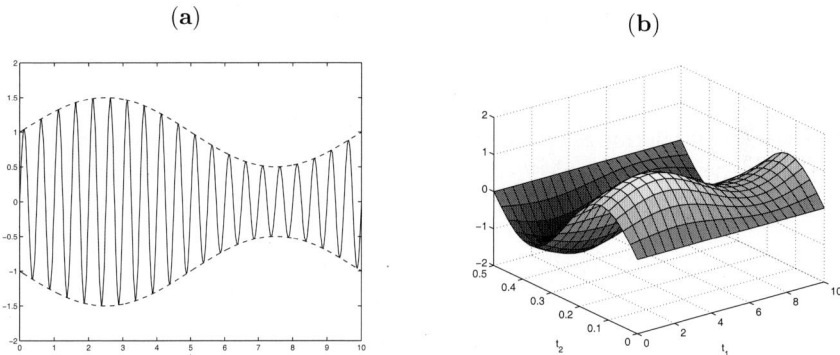

(a)   **(b)**

Fig. 1. Amplitude modulated signal y **(a)** and corresponding MVF $\hat{y}$ **(b)**.

The situation becomes more difficult in case of frequency modulation. For example, we examine the modified signal

$$x(t) := \left[1 + \alpha \sin\left(\frac{2\pi}{T_1}t\right)\right] \sin\left(\frac{2\pi}{T_2}t + \beta \cos\left(\frac{2\pi}{T_1}t\right)\right) , \tag{3}$$

where the parameter $\beta > 0$ specifies the amount of frequency modulation. This signal is illustrated in Fig. 2a. A direct transition to a biperiodic MVF is also feasible in this situation and we obtain

$$\hat{x}_1(t_1, t_2) := \left[1 + \alpha \sin\left(\frac{2\pi}{T_1}t_1\right)\right] \sin\left(\frac{2\pi}{T_2}t_2 + \beta \cos\left(\frac{2\pi}{T_1}t_1\right)\right) . \tag{4}$$

However, the MVF exhibits many oscillations in the rectangle $[0, T_1[\times[0, T_2[$, too, see Fig. 2b. The number of oscillations depends on the amount of frequency modulation β. Hence this multidimensional description is inappropriate.

Narayan and Roychowdhury [7] proposed a readjusted model to achieve an efficient representation. Thereby, the MVF incorporates only the amplitude modulation part

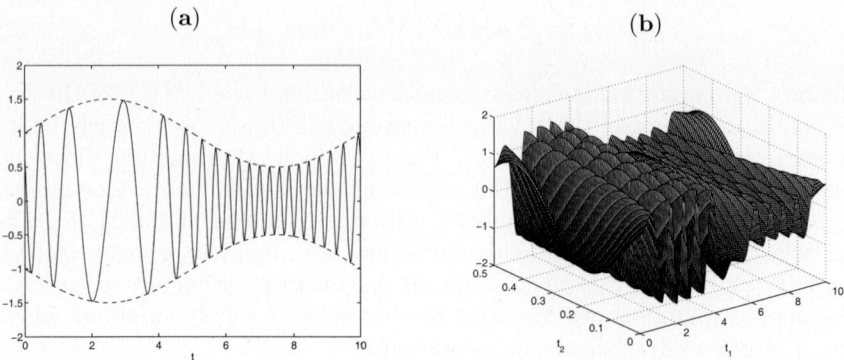

Fig. 2. Frequency modulated signal x (a) and unsophisticated MVF $\hat{x}_1$ (b).

$$\hat{x}_2(t_1, t_2) := \left[1 + \alpha \sin\left(\tfrac{2\pi}{T_1} t_1\right)\right] \sin\left(2\pi t_2\right) , \tag{5}$$

where the second period is transformed to 1. In the rectangle $[0, T_1[\times [0, 1[$, this function has the same simple form as the MVF (2). The frequency modulation part is described by an additional time-dependent *warping function*

$$\Psi(t) := \frac{t}{T_2} + \frac{\beta}{2\pi} \cos\left(\tfrac{2\pi}{T_1} t\right) . \tag{6}$$

We take the derivative of the warping function as a corresponding *local frequency function* $\nu(t) := \Psi'(t)$, which represents an elementary T_1-periodic function in this example. Since we assume $T_1 \gg T_2$, it holds $\nu(t) > 0$ for a broad range of parameters β. The reconstruction of the original signal (3) reads $x(t) = \hat{x}_2(t, \Psi(t))$, where the warping function stretches the second time scale. Hence we obtain a powerful model for signals, which feature amplitude as well as frequency modulation with largely differing rates.

2.2 Definition of Quasiperiodic Signals

Commonly, a univariate function $x : \mathbb{R} \to \mathbb{C}$ is said to be two-tone quasiperiodic, if it can be represented by a two-dimensional Fourier series of the form

$$x(t) = \sum_{j_1, j_2 = -\infty}^{+\infty} X_{j_1, j_2} \exp\left(\mathrm{i}\left(\tfrac{2\pi}{T_1} j_1 + \tfrac{2\pi}{T_2} j_2\right) t\right) \tag{7}$$

with rates $T_1, T_2 > 0$ and coefficients $X_{j_1, j_2} \in \mathbb{C}$, where $\mathrm{i} := \sqrt{-1}$ denotes the imaginary unit. However, we have to specify the kind of convergence in the arising series. The continuity of signals is guaranteed via absolute convergence, i.e.,

$$\sum_{j_1, j_2 = -\infty}^{+\infty} |X_{j_1, j_2}| < \infty , \tag{8}$$

which implies also a uniform convergence. Moreover, the series becomes well defined with respect to permutations of the terms. In particular, an interchange of j_1 and j_2 is allowed. Assuming (8), the biperiodic MVF $\hat{x} : \mathbb{R}^2 \to \mathbb{C}$ of (7)

$$\hat{x}(t_1, t_2) := \sum_{j_1, j_2 = -\infty}^{+\infty} X_{j_1, j_2} \exp \left(\mathrm{i} \left(\tfrac{2\pi}{T_1} j_1 t_1 + \tfrac{2\pi}{T_2} j_2 t_2 \right) \right) . \tag{9}$$

is continuous, too. For locally integrable functions, weaker concepts of convergence are feasible. However, our aim is to compute solutions of differential equations and thus smooth functions are required. Using the representation (7), just sufficient conditions can be formulated. To omit the discussion of convergence properties, an alternative definition of quasiperiodic functions becomes reasonable.

Definition 1. *A function* $x : \mathbb{R} \to \mathbb{C}$ *is* two-tone quasiperiodic *with rates* T_1 *and* T_2 *if a* (T_1, T_2)-*periodic function* $\hat{x} : \mathbb{R}^2 \to \mathbb{C}$ *exists satisfying the relation* $x(t) = \hat{x}(t, t)$.

If $\hat{x}$ exhibits some smoothness, then the function x inherits the same smoothness. For signals of the form (7), the multivariate representation is given by (9), which implies quasiperiodicity in view of this definition, too.

Our aim is to analyse frequency modulated signals. Following the modelling in the previous subsection, we consider functions $x : \mathbb{R} \to \mathbb{C}$ of the form

$$x(t) = \sum_{j_1, j_2 = -\infty}^{+\infty} X_{j_1, j_2} \exp \left(\mathrm{i} \left(\tfrac{2\pi}{T_1} j_1 t + 2\pi j_2 \Psi(t) \right) \right) , \tag{10}$$

where $\Psi : \mathbb{R} \to \mathbb{R}$ represents a warping function. The above discussion of convergence applies also to this case. Alternatively, we formulate a characterisation according to Definition 1.

Definition 2. *A function* $x : \mathbb{R} \to \mathbb{C}$ *is* frequency modulated two-tone quasiperiodic *with rate* T_1 *if there exists a* $(T_1, 1)$-*periodic function* $\hat{x} : \mathbb{R}^2 \to \mathbb{C}$ *and a function* $\Psi : \mathbb{R} \to \mathbb{R}$ *with* T_1-*periodic derivative* Ψ' *such that it holds* $x(t) = \hat{x}(t, \Psi(t))$.

In this definition, it is not necessary that $\nu(t) := \Psi'(t) > 0$ holds for all t. However, frequency modulated signals, which arise in electric circuits, feature fast oscillations, whose frequencies vary slowly in time. Thus the positivity of the local frequency function ν and the demand $T_1 > \nu(t)^{-1}$ for all t is often required to obtain an efficient representation.

We recognise that all specifications of quasiperiodic functions imply the existence of corresponding MVFs. Hence quasiperiodic signals exhibit an inherent multidimensional structure and the transition to the multivariate model becomes natural.

Furthermore, the slow time scale may be aperiodic. In this case, we obtain a Fourier expansion of the type

$$x(t) = \sum_{j_2=-\infty}^{+\infty} X_{j_2}(t) \exp\left(\mathrm{i}2\pi j_2 \Psi(t)\right) \tag{11}$$

with time-dependent coefficients $X_{j_2} : \mathbb{R} \to \mathbb{C}$, which are called the envelopes. The envelopes introduce an amplitude modulation, whereas the warping function again describes a frequency modulation. A formal definition is given below.

Definition 3. *A function $x : \mathbb{R} \to \mathbb{C}$ is called* envelope modulated *if a function $\hat{x} : \mathbb{R}^2 \to \mathbb{C}$, which is periodic in the second variable with rate 1, and a function $\Psi : \mathbb{R} \to \mathbb{R}$ exist, where $x(t) = \hat{x}(t, \Psi(t))$ holds.*

We take $\nu := \Psi'$ as a local frequency of the signal again. The generalisation of the above definitions to vector-valued functions $\mathbf{x} : \mathbb{R} \to \mathbb{C}^k$ consists in demanding the conditions in each component separately.

2.3 Transformation of Signal Representations

For modelling a frequency modulated signal, the employed MVF and corresponding warping function is not unique. We obtain a fundamental result already for general signals, which do not necessarily feature periodicities in the time scales.

Theorem 1. *If the signal $x : \mathbb{R} \to \mathbb{C}$ is represented by the MVF $\hat{x} : \mathbb{R}^2 \to \mathbb{C}$ and the warping function $\Psi : \mathbb{R} \to \mathbb{R}$, i.e., $x(t) = \hat{x}(t, \Psi(t))$, then the MVF*

$$\hat{y} : \mathbb{R}^2 \to \mathbb{C}, \quad \hat{y}(t_1, t_2) := \hat{x}\left(t_1, t_2 + \Psi(t_1) - \Phi(t_1)\right) \tag{12}$$

satisfies $x(t) = \hat{y}(t, \Phi(t))$ for an arbitrary function $\Phi : \mathbb{R} \to \mathbb{R}$.

Thus if a representation of a signal exists using some MVF and warping function, then we can prescribe a new warping function and transform to another MVF, which yields the same information as before. Hence we apply the warping function or the associated local frequency function as free parameters to obtain an efficient multivariate representation.

In the following, we discuss the model on the level of local frequency functions. Considering $\nu := \Psi'$, $\mu := \Phi'$ and $\Psi(0) = \Phi(0) = 0$, the transformation (12) reads

$$\hat{y} : \mathbb{R}^2 \to \mathbb{C}, \quad \hat{y}(t_1, t_2) := \hat{x}\left(t_1, t_2 + \int_0^{t_1} \nu(s) - \mu(s)\,\mathrm{d}s\right). \tag{13}$$

In case of envelope modulated signals, see Definition 3, the fast time scale is periodic, whereas the slow time scale may be aperiodic. The transformation (13) preserves the periodicity in the second variable. Specifying an arbitrary local frequency function yields a corresponding MVF in view of Theorem 1. For quasiperiodic signals, the slow time scale is periodic, too. Due to Definition 2, the corresponding MVFs have to be biperiodic. Thus transformations are feasible only if they preserve the periodicities. To analyse the feasibility, we define an average frequency.

Definition 4. *If $\nu : \mathbb{R} \to \mathbb{R}$ represents a T-periodic locally integrable frequency function, then the average frequency is given by the integral mean*

$$\overline{\nu} := \frac{1}{T} \int_0^T \nu(s) \, \mathrm{d}s \; . \tag{14}$$

Theorem 1 implies the following result for transformations in the quasiperiodic case.

Theorem 2. *Let $\nu, \mu : \mathbb{R} \to \mathbb{R}$ be T_1-periodic locally integrable functions with the property $\overline{\nu} = \overline{\mu}$. If $\hat{x} : \mathbb{R}^2 \to \mathbb{C}$ is a $(T_1, 1)$-periodic function, then the function $\hat{y} : \mathbb{R}^2 \to \mathbb{C}$ defined by (13) is $(T_1, 1)$-periodic, too. Furthermore, if $x(t) = \hat{x}(t, \int_0^t \nu(s) \, \mathrm{d}s)$ holds for all $t \in \mathbb{R}$, then it follows $x(t) = \hat{y}(t, \int_0^t \mu(s) \, \mathrm{d}s)$ for all $t \in \mathbb{R}$.*

Hence a frequency modulated quasiperiodic signal implies a continuum of representations via MVFs and respective local frequency functions, which all exhibit the same average frequency. We remark that the transformation (13) can be discussed on the level of the representation (10), too, using the Fourier coefficients.

An important consequence of Theorem 2 is that transformations to representations with constant local frequency can be performed. Given a biperiodic MVF and periodic local frequency function ν, the constant frequency $\mu \equiv \overline{\nu}$ enables a corresponding transformation, which is feasible in view of $\overline{\mu} = \overline{\nu}$. Thus the following corollary unifies the definition of quasiperiodic signals with respect to different local frequencies.

Corollary 1. *A frequency modulated quasiperiodic function characterised by Definition 2 features a representation as quasiperiodic function with constant rates according to Definition 1 and vice versa.*

Hence no qualitative difference between quasiperiodic signals involving constant and stretched time scales exists. Therefore we will just speak of quasiperiodic functions in the following. Given a frequency modulated signal, it does not make sense to assign a constant fast rate. However, an according interpretation as an average rate is reasonable.

The choice of a local frequency function and according multivariate representation is important for the efficiency of the multidimensional signal model. Regarding quasiperiodic functions, we consider a MVF corresponding to a constant frequency $\overline{\nu}$. Now we may transform the model to any local frequency function of the type

$$\mu(s) := \overline{\nu} + \xi(s) \quad \text{with} \quad \overline{\xi} = 0 \ . \tag{15}$$

The function ξ is the degree of freedom in our design of multidimensional models for quasiperiodic signals.

3 Warped MPDAE Model

3.1 Derivation of the Model

The numerical simulation of electric circuits employs a network approach, which yields systems of *differential algebraic equations* (*DAEs*), see [3]. Thereby, the system describes the transient behaviour of all node voltages and some branch currents. We write such a system in the general form

$$\frac{d\mathbf{q}(\mathbf{x})}{dt} = \mathbf{f}(\mathbf{x}(t)) + \mathbf{b}(t) \ , \tag{16}$$

where $\mathbf{x} : \mathbb{R} \to \mathbb{R}^k$ represents the unknown voltages and currents. The functions $\mathbf{q}, \mathbf{f} : \mathbb{R}^k \to \mathbb{R}^k$ correspond to a charge and a resistive term, respectively. Predetermined input signals are included in the time-dependent function $\mathbf{b} : \mathbb{R} \to \mathbb{R}^k$. We demand $\mathbf{f}, \mathbf{b} \in C^0$ and $\mathbf{q} \in C^1$, since smooth solutions of (16) are desired. DAEs cause theoretical and numerical particularities like the index concept or the need of consistent initial values, see [6, 11].

Given some two-tone quasiperiodic input $\mathbf{b}$ with rates T_1 and T_2, a forced oscillation arises, which often leads to amplitude modulated signals. We assume that the solution inherits the time scales, i.e., $\mathbf{x}$ is also quasiperiodic with same rates. Consequently, we obtain the (T_1, T_2)-periodic MVFs $\hat{\mathbf{b}}$ and $\hat{\mathbf{x}}$, which represent the corresponding quasiperiodic signals. Brachtendorf et al. [1] introduced the according system of *multirate partial differential algebraic equations* (*MPDAEs*)

$$\frac{\partial \mathbf{q}(\hat{\mathbf{x}})}{\partial t_1} + \frac{\partial \mathbf{q}(\hat{\mathbf{x}})}{\partial t_2} = \mathbf{f}(\hat{\mathbf{x}}(t_1, t_2)) + \hat{\mathbf{b}}(t_1, t_2) \ , \tag{17}$$

which results from a transformation of the DAEs (16) with respect to multivariate functions. An arbitrary solution of the system (17) yields a solution of the system (16) via the reconstruction

$$\mathbf{x}(t) = \hat{\mathbf{x}}(t, t) \ . \tag{18}$$

The proof is straightforward and can be found in [10], for example.

If the MVF is (T_1, T_2)-periodic, then the reconstructed signal is two-tone quasiperiodic with rates T_1, T_2. Thus the determination of quasiperiodic signals leads to the boundary value problem

$$\hat{\mathbf{x}}(t_1, t_2) = \hat{\mathbf{x}}(t_1 + T_1, t_2) = \hat{\mathbf{x}}(t_1, t_2 + T_2) \quad \text{for all} \ t_1 \in \mathbb{R}, \ t_2 \in \mathbb{R} \ . \tag{19}$$

If the slow time scale is aperiodic, then we obtain envelope modulated signals by solutions of the MPDAE, too. Hence a mixture of initial and boundary conditions is considered, namely

$$\hat{\mathbf{x}}(0, t_2) = \mathbf{h}(t_2), \quad \hat{\mathbf{x}}(t_1, t_2 + T_2) = \hat{\mathbf{x}}(t_1, t_2) \quad \text{for all } t_1 \geq 0, \ t_2 \in \mathbb{R}, \quad (20)$$

where $\mathbf{h} : \mathbb{R} \to \mathbb{R}^k$ represents a prescribed T_2-periodic function, whose values have to be consistent with respect to the DAEs (16). The choice of appropriate initial values influences the efficiency of this approach. Note that the reconstructed signal (18) depends on the value $\mathbf{h}(0)$ only. For further details, we refer to [10].

Now we assume that the input signals exhibit a slow time scale only. Nevertheless, the DAE system (16) shall feature an inherent fast time scale. Consequently, multitone signals arise, which may be amplitude modulated as well as frequency modulated. Narayan and Roychowdhury [7] generalised the MPDAE model to this problem. The transition to MVFs implies a system of *warped multirate partial differential algebraic equations*, namely

$$\frac{\partial \mathbf{q}(\hat{\mathbf{x}})}{\partial t_1} + \nu(t_1) \frac{\partial \mathbf{q}(\hat{\mathbf{x}})}{\partial t_2} = \mathbf{f}(\hat{\mathbf{x}}(t_1, t_2)) + \mathbf{b}(t_1) \quad (21)$$

with the unknown solution $\hat{\mathbf{x}} : \mathbb{R}^2 \to \mathbb{R}^k$. Since we assume that the input $\mathbf{b}$ just acts on the slow time scale, a multivariate description is not necessary here. A local frequency function $\nu : \mathbb{R} \to \mathbb{R}$ arises, which depends on the same variable as $\mathbf{b}$ if the input causes the frequency modulation. In the following, we assume $\nu \in C^0$, since smooth solutions are discussed. An appropriate choice for the local frequencies is unknown a priori. Furthermore, the system (21) is autonomous in the second variable t_2, since the fast time scale is not forced by the input but inherent.

Solving the warped MPDAEs (21) for some given local frequency function, we obtain a solution of the DAEs (16) via

$$\mathbf{x}(t) = \hat{\mathbf{x}}(t, \Psi(t)) \quad \text{with} \quad \Psi(t) := \int_0^t \nu(s) \, ds \, . \quad (22)$$

The proof operates similar to the case of constant time scales. Again periodicities are necessary to solve the system (21) in a bounded domain. We always consider a periodic fast time scale, where the period is standardised to $T_2 = 1$. The magnitude of the fast rate is included in the local frequency function. Initial-boundary value problems (20) of the system (21) determine envelope modulated signals, which exhibit frequency modulation.

If the input signals are T_1-periodic, then a $(T_1, 1)$-periodic MVF in addition to a T_1-periodic local frequency function yield a two-tone quasiperiodic signal in the reconstruction (22). Thus the boundary value problem (19) is considered. However, given a biperiodic solution $\hat{\mathbf{x}}$, the shifted function

$$\hat{\mathbf{y}}(t_1, t_2) := \hat{\mathbf{x}}(t_1, t_2 + c) \quad \text{for } c \in \mathbb{R} \quad (23)$$

also satisfies the system including the same local frequency function. Hence a solution of the MPDAE reproduces a continuum of signals solving the underlying DAE. Therefore the multidimensional approach reveals a degree of freedom by shifting, which is not directly transparent in the according DAE model (16). In a corresponding numerical method for the problem (19),(21), a specific solution has to be isolated from the continuum (23).

The system (21) is underdetermined, since the local frequency function represents a degree of freedom. A priori, we do not have sufficient knowledge to prescribe a local frequency function appropriately. Thus alternative conditions are added to the system (21) in order to fix a suitable solution. Several strategies are feasible like criteria based on specific minimisations, cf. [5]. The use of multidimensional phase conditions for this purpose will be discussed in Sect. 3.4.

3.2 Characteristic Curves

The warped MPDAE system (21) exhibits a specific transport of information, see [9]. The corresponding *characteristic system* reads

$$\frac{\mathrm{d}}{\mathrm{d}\tau}t_1(\tau) = 1 \ , \quad \frac{\mathrm{d}}{\mathrm{d}\tau}t_2(\tau) = \nu(t_1(\tau)) \ , \quad \frac{\mathrm{d}}{\mathrm{d}\tau}\mathbf{q}(\tilde{\mathbf{x}}(\tau)) = \mathbf{f}(\tilde{\mathbf{x}}(\tau)) + \mathbf{b}(t_1(\tau)) \ \ (24)$$

with t_1, t_2 as well as $\tilde{\mathbf{x}}$ depending on a parameter τ. Solutions of the system (24) are called *characteristic curves*. For fixed local frequency function ν, we solve the part with respect to the variables t_1, t_2 explicitly and obtain the *characteristic projections*

$$t_2 = \Psi(t_1) + c \quad \text{with} \quad \Psi(t_1) := \int_0^{t_1} \nu(s) \, \mathrm{d}s \ , \tag{25}$$

where $c \in \mathbb{R}$ represents an arbitrary constant. Hence the characteristic projections form a continuum of parallel curves in the domain of dependence. Figure 3 illustrates this property.

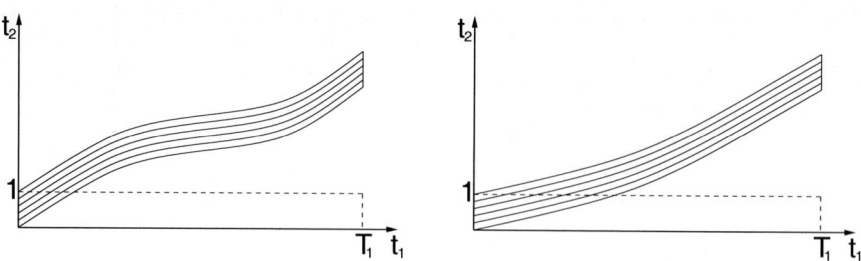

Fig. 3. Characteristic projections for two different choices of local frequency functions with identical average.

In the initial-boundary value problem (20) belonging to (21), we consider the initial manifold $\mathcal{F}_0 := \{0\} \times [0,1[$. The solution of this problem can be obtained via solving a collection of initial value problems corresponding to the characteristic system (24), see Fig. 3. However, this approach is not efficient, since each initial value problem demands the same amount of work as solving the original DAE system (16).

Likewise, the determination of biperiodic solutions can be discussed with respect to characteristic curves. Considering an arbitrary $(T_1, 1)$-periodic solution of (21), its initial values in $\mathcal{F}_0$ reproduce the complete solution via (24). Consequently, a biperiodic solution of (21) is already fixed by its initial values in the manifold $\mathcal{F}_0$ and the corresponding local frequency function ν. Solving the systems (24) yields final values in $\mathcal{F}_1 := \{T_1\} \times [\Psi(T_1), \Psi(T_1)+1[$. The last equation in system (24) does not involve the local frequency. Using another T_1-periodic local frequency μ and identical initial state in $\mathcal{F}_0$ produces the same final state in $\mathcal{F}_2 := \{T_1\} \times [\Phi(T_1), \Phi(T_1) + 1[$ with $\Phi(t_1) = \int_0^{t_1} \mu(s)\, ds$. Hence the MVF corresponding to μ is biperiodic, too, if $\Phi(T_1) = \Psi(T_1)$ holds, i.e., $\overline{\nu} = \overline{\mu}$. Note that the reconstructed signal (22) is the same for all choices of the local frequency. This behaviour of the characteristic system motivates a transformation of solutions, which is investigated in the next subsection.

In case of widely separated time rates $T_1 \gg \overline{\nu}^{-1}$, solving initial value problems of the characteristic system (24) in the large time interval $[0, T_1]$ demands the computation of a huge number of oscillations. Consequently, methods of characteristics for solving the initial-boundary value problem (20) are inefficient. The analysis of MPDAE solutions via the characteristic system with initial values in $\mathcal{F}_0$ represents just a theoretical tool. On the other hand, initial values in $\mathcal{G}_0 := [0, T_1[\times \{0\}$ determine a biperiodic solution completely, too. Characteristic projections starting in $\mathcal{G}_0$ can be used to construct an efficient numerical method to solve the boundary value problem (19), see [8, 9]. Thereby, a boundary value problem for a finite number of DAEs (24) arises. If we tackle this boundary value problem by a shooting method, cf. [11], each corresponding initial value problem of DAEs has to be solved only in a relatively short time interval.

3.3 Transformation of Solutions

In Sect. 2.3, Theorem 1 demonstrates that the local frequency of an according signal is optional. Just the efficiency of a multivariate representation depends on an appropriate choice. We recover this arbitrariness of local frequency functions in the MPDAE model, too.

Theorem 3. *Let $\hat{\mathbf{x}} : \mathbb{R}^2 \to \mathbb{R}^k$ and $\nu : \mathbb{R} \to \mathbb{R}$ satisfy the warped MPDAE system (21). Given a function $\mu : \mathbb{R} \to \mathbb{R}$, the MVF $\hat{\mathbf{y}} : \mathbb{R}^2 \to \mathbb{R}^k$ specified by (13) represents a solution of the warped MPDAE system (21) corresponding to the local frequency function μ.*

Hence we can transform a solution of the MPDAE corresponding to a specific local frequency to another solution with arbitrary local frequency. Note that the deformation (13) does not change the initial manifold at $t_1 = 0$. In particular, we can transform a multivariate model to a corresponding constant local frequency $\mu \equiv 1$. Consequently, the left-hand sides of (17) and (21) coincide. Therefore solutions of both PDAE models can be transformed in each other provided that the input signals involve just one time scale.

Corollary 2. *If $\hat{\mathbf{x}} : \mathbb{R}^2 \to \mathbb{R}^k$ and $\nu : \mathbb{R} \to \mathbb{R}$ satisfy the warped MPDAE system (21), then the MVF given by (13) with $\mu \equiv 1$ represents a solution of the MPDAE system (17) including the input $\hat{\mathbf{b}}(t_1, t_2) := \mathbf{b}(t_1)$. Vice versa, if the system (17) with $\hat{\mathbf{b}}(t_1, t_2) := \mathbf{b}(t_1)$ exhibits a solution $\hat{\mathbf{x}} : \mathbb{R}^2 \to \mathbb{R}^k$, then using $\nu \equiv 1$ in (13) results in a solution of the warped system (21) for arbitrary local frequency function $\mu : \mathbb{R} \to \mathbb{R}$.*

The transformation (13) changes neither the initial values at $t_1 = 0$ nor the periodicity in t_2. Thus transforming a solution of the initial boundary value problem (21),(20) yields another solution of the problem. In case of quasiperiodic signals, some restrictions with regard to the used transformation arise again, which are necessary to preserve the periodicities. Theorem 2 and Theorem 3 imply an according corollary.

Corollary 3. *The following assumptions shall hold:*

(i) $\nu : \mathbb{R} \to \mathbb{R}$ is T_1-periodic,   (iii) $\overline{\nu} = \overline{\mu}$,

(ii) $\mu : \mathbb{R} \to \mathbb{R}$ is T_1-periodic,   (iv) $\hat{\mathbf{x}} : \mathbb{R}^2 \to \mathbb{R}^k$ is $(T_1, 1)$-periodic.

If $\hat{\mathbf{x}}$ and ν fulfil the warped MPDAE system (21), then the MVF $\hat{\mathbf{y}} : \mathbb{R}^2 \to \mathbb{R}^k$ obtained by (13) represents a $(T_1, 1)$-periodic solution of the warped MPDAE system (21) including the local frequency function μ.

We note that a $(T_1, 1)$-periodic solution of (21) for constant frequency $\overline{\nu}$ can be transformed in a (T_1, T_2)-periodic solution of (17) with $T_2 = \overline{\nu}^{-1}$. This connection is caused by the standardisation of the second period in the warped system.

Moreover, these corollaries allow to apply results from the MPDAE model with constant rates in the context of the warped MPDAE model. Especially, it is proved that the existence of a quasiperiodic solution of the DAE system (16) implies the existence of a corresponding biperiodic solution of the MPDAE system (17), see [10]. Together with Corollary 2 and Corollary 3, we obtain the following important property.

Corollary 4. *Let the system of DAEs (16) have a two-tone quasiperiodic solution $\mathbf{x} : \mathbb{R} \to \mathbb{R}^k$ with rates T_1, T_2. Given an arbitrary T_1-periodic local frequency function $\nu : \mathbb{R} \to \mathbb{R}$ with $\overline{\nu}^{-1} = T_2$ in (21), the system of warped MPDAEs features a $(T_1, 1)$-periodic solution $\hat{\mathbf{x}} : \mathbb{R}^2 \to \mathbb{R}^k$, where $\mathbf{x}(t) = \hat{\mathbf{x}}(t, \int_0^t \nu(s)\,\mathrm{d}s)$ holds for all t.*

Based on Corollary 3, we can transform any multidimensional representation of a quasiperiodic signal to a reference solution including constant frequency $\bar{\nu}$. However, the corresponding MVF yields an inefficient model for frequency modulated signals as we have seen in Sect. 2.1.

The degree of freedom in the transformation of biperiodic solutions is given by the function ξ in (15) again. Now a numerical technique for solving the biperiodic boundary value problem of the MPDAE system (21) is imaginable, where we prescribe a periodic function ξ with $\bar{\xi} = 0$ and determine the scalar unknown $\bar{\nu}$. Unfortunately, the problem becomes extremely sensitive in case of widely separated time scales, i.e., $T_1\bar{\nu} \gg 1$. Tiny changes in the local frequency function result in significant deformations of the corresponding MVF. Thus we do not expect an a priori specification of the local frequency to generate a suitable solution.

3.4 Continuous Phase Conditions

If the local frequency is an unidentified function in the MPDAE (21), then we need an additional postulation to determine the complete solution. Houben [5] specifies an optimisation condition, which minimises oscillatory behaviour in MVFs and thus yields simple representations. Another possibility consists in demanding a continuous phase condition, cf. [7]. Thereby, the phase in each cross section with $t_1 = $ const. is controlled, which often produces elementary MVFs, too. Without loss of generality, we choose the first component of $\hat{\mathbf{x}} = (\hat{x}_1, \ldots, \hat{x}_k)^\top$. Examples for continuous phase conditions are

$$\hat{x}_1(t_1, 0) = \eta(t_1) \quad \text{for all } t_1 \in \mathbb{R} \tag{26}$$

or

$$\left.\frac{\partial \hat{x}_1}{\partial t_2}\right|_{t_2=0} = \eta(t_1) \quad \text{for all } t_1 \in \mathbb{R} \tag{27}$$

including a predetermined slowly varying function $\eta : \mathbb{R} \to \mathbb{R}$. We add such a phase condition as additional boundary condition in time domain. In general, we do not have a priori knowledge about the specification of the function η. Nevertheless, applying constant choices, i.e., $\eta \equiv$ const., is often successful. Note that (26),(27) represent conditions for scalar functions depending on t_1, which agrees to the structure of the unknown parameters $\nu(t_1)$.

The degree of freedom in transformations of MPDAE solutions can be used to justify the existence of solutions satisfying some phase condition. For example, we discuss the constraint (27) setting $\eta \equiv 0$. We assume the existence of a biperiodic solution $\hat{\mathbf{x}} \in C^1$ of (21). The smoothness and periodicity implies that, for each $t_1 \in \mathbb{R}$, a corresponding $\Theta(t_1) \in \mathbb{R}$ exists such that $\frac{\partial \hat{x}_1}{\partial t_2}(t_1, \Theta(t_1)) = 0$. For $t_1 = 0$, we select some $\Theta(0) \in [0, 1[$. Motivated by the implicit function theorem, we postulate that a certain choice $t_2 = \Theta(t_1)$ with $\Theta \in C^1$ exists, which forms an isolated curve in the domain of dependence. The periodicity of $\hat{\mathbf{x}}$ produces a T_1-periodic function Θ. Now we transform the curve $t_2 = \Theta(t_1)$ onto the line $t_2 = 0$, which yields the new MVF

$$\hat{\mathbf{y}}(t_1, t_2) := \hat{\mathbf{x}}(t_1, t_2 + \Theta(t_1)) = \hat{\mathbf{x}}\left(t_1, t_2 + \int_0^{t_1} \Theta'(s)\, \mathrm{d}s + \Theta(0)\right). \qquad (28)$$

The involved translation by $\Theta(0)$ is of the type (23) and thus results in a solution again. Furthermore, the periodicity yields

$$\overline{\Theta'} = \frac{1}{T_1} \int_0^{T_1} \Theta'(s)\, \mathrm{d}s = \frac{\Theta(T_1) - \Theta(0)}{T_1} = 0. \qquad (29)$$

Hence Theorem 3 implies that the function (28) represents a biperiodic solution of the MPDAE satisfying the phase condition (27) with $\eta \equiv 0$.

4 Illustrative Example

To demonstrate the application of the multidimensional strategy, we consider the tanh-based LC oscillator in Fig. 4. This circuit is similar to an example investigated in [7]. The node voltage u and the branch current $\imath := I_L$ are unknown time-dependent functions. The current-voltage relation of the nonlinear resistor reads

$$I_R = g(u) := (G_0 - G_\infty)U_0 \tanh(u/U_0) + G_\infty u \qquad (30)$$

including parameters are $G_0 = -0.1\,\mathrm{A/V}$, $G_\infty = 0.25\,\mathrm{A/V}$, $U_0 = 1\,\mathrm{V}$. An independent input signal b specifies the capacitance C, which produces a voltage controlled oscillator. We write the arising mathematical model in the form

$$\dot{u} = (-\imath - g(u))/(C_0 w), \qquad i = u/L, \qquad 0 = w - b(t) \qquad (31)$$

with the unknown functions $u, \imath, w$ and parameters $C_0 = 100\,\mathrm{nF}$, $L = 1\,\mu\mathrm{H}$. Hence system (31) represents a semi-explicit DAE of index 1. For constant input, the system exhibits a periodic oscillation of a high frequency. We choose the input signal $b(t) = 1 + 0.9\sin(2\pi t/T_1)$ with slow rate $T_1 = 30\,\mu\mathrm{s}$, which produces frequency modulated signals. Consequently, we apply the respective warped MPDAE model (21) and determine a biperiodic solution satisfying the phase condition (27). A finite difference method employing centred differences on a uniform grid yields a numerical approximation.

Firstly, we select the function $\eta \equiv 0$ in phase condition (27). Figure A.5 on page 327 illustrates the results of the multidimensional model. According to the input signal, the local frequency is high in areas, where the capacitance is low. Since this behaviour is typical for LC oscillators, the used phase condition identifies a physically reasonable frequency function. The MVFs $\hat{u}$ and $\hat{\imath}$ exhibit a weak and a strong amplitude modulation, respectively. The MVF $\hat{w}$ reproduces the input signal.

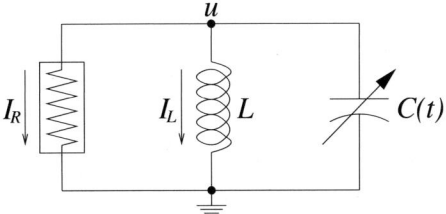

Fig. 4. Circuit of voltage controlled LC oscillator.

Now we employ the solution of the MPDAE to reconstruct a corresponding quasiperiodic response of the DAE (31) via (22). For comparison, an initial value problem of the DAE is solved by trapezoidal rule. Figure A.6 on page 327 shows the resulting signals. We observe a good agreement of both functions, which confirms the relations between univariate and multivariate model.

Secondly, we perform the simulation using the same techniques but considering $\eta(t_1) = 0.8 \sin(2\pi t_1/T_1)$ in phase condition (27). We recognise that the difference $\Delta\nu := \mu - \nu$ with respect to the previous frequency, see Fig. A.7 on page 328, satisfies $\overline{\Delta\nu} = 0$, i.e., $\overline{\nu} = \overline{\mu}$. However, the function $\Delta\nu$ is nontrivial in comparison to the difference between the two choices of the function η. Figure A.7 illustrates the phase condition in the MVF $\hat{u}$, too. Considering the previous MVFs and applying the transformation (13) with the updated frequency μ yields the same solution (except for numerical errors). These results clarify that the solutions belonging to the two different phase conditions are located in the same continuum of transformed solutions. Moreover, the simple choice $\eta \equiv 0$ in the phase condition (27) already yields an efficient representation.

Finally, we simulate a case of widely separated rates with $T_1 = 100\,\mathrm{ms}$ in the input signal. The used phase condition is (27) with $\eta \equiv 0$. To validate the applicability of the inherent MPDAE structure outlined in Sect. 3.2, a method of characteristics produces the numerical solution now, see [9]. Involved DAE systems are discretised by trapezoidal rule. Nevertheless, the finite difference method, which has been applied in the previous simulations, yields the same solution (except for numerical errors). Figure A.8 on page 328 depicts the results of the MPDAE model. Local frequency as well as MVFs exhibit the same behaviour as in the previous case. Just the amplitude modulation in $\hat{u}$ disappears. The corresponding DAE solution features about 66000 oscillations during the period T_1 here, which can be reconstructed by these results.

5 Conclusions

Multidimensional models of a quasiperiodic signal are not unique but interconnected by a transformation formula. We obtain quasiperiodic solutions of DAEs via determining multiperiodic solutions of MPDAEs. Accordingly, the warped MPDAE system exhibits a continuum of solutions, which reproduce the same quasiperiodic response of the underlying DAE system. However, the average frequency of all representations coincides. Thus solutions of the warped system can be transformed to solutions of the ordinary system, which enables the use of respective theoretical results. In particular, the existence of quasiperiodic solutions of DAEs implies the existence of corresponding multiperiodic solutions of warped MPDAEs for a broad class of local frequency functions. Moreover, the arising transformation formula allows for the analysis of additional conditions for the determination of a priori unknown local frequency functions. Numerical simulations based on the warped MPDAE system confirm the transformation qualities of the multidimensional model.

References

[1]   Brachtendorf, H. G.; Welsch, G.; Laur, R.; Bunse-Gerstner, A.: Numerical steady state analysis of electronic circuits driven by multi-tone signals. Electrical Engineering, **79**, 103–112 (1996)

[2]   Chua, L. O.; Ushida, A.: Algorithms for computing almost periodic steady-state response of nonlinear systems to multiple input frequencies. IEEE CAS, **28(10)**, 953–971 (1981)

[3]   Günther, M.; Feldmann, U.: CAD based electric circuit modeling in industry I: mathematical structure and index of network equations. Surv. Math. Ind., **8**, 97–129 (1999)

[4]   Ho, C.W.; Ruehli, A.E.; Brennan, P.A.: The modified nodal approach to network analysis. IEEE Trans. CAS, **22**, 505–509 (1975)

[5]   Houben, S.H.M.J.: Simulating multi-tone free-running oscillators with optimal sweep following. In: Schilders, W.H.A., ter Maten, E.J.W., Houben, S.H.M.J. (eds.): Scientific Computing in Electrical Engineering, Mathematics in Industry, **4**, 240–247, Springer, Berlin (2004)

[6]   Kampowsky, W.; Rentrop, P.; Schmitt, W.: Classification and numerical simulation of electric circuits. Surv. Math. Ind., **2**, 23–65 (1992)

[7]   Narayan, O.; Roychowdhury, J.: Analyzing oscillators using multitime PDEs. IEEE Trans. CAS I, **50(7)**, 894–903 (2003)

[8]   Pulch, R.: Finite difference methods for multi time scale differential algebraic equations. Z. Angew. Math. Mech., **83(9)**, 571–583 (2003)

[9]   Pulch, R.: Multi time scale differential equations for simulating frequency modulated signals. Appl. Numer. Math., **53(2-4)**, 421–436 (2005)

[10]   Roychowdhury, J.: Analysing circuits with widely-separated time scales using numerical PDE methods. IEEE Trans. CAS I, **48(5)**, 578–594 (2001)

[11]   Stoer, J.; Bulirsch, R.: Introduction to Numerical Analysis. (3rd ed.) Springer, New York (2002)

[12]   Ushida, A.; Chua, L. O.: Frequency-domain analysis of nonlinear circuits driven by multi-tone signals. IEEE CAS, **31(9)**, 766–779 (1984)

An Improved Method to Detect Riblets on Surfaces in Nanometer Scaling Using SEM

E. Reithmeier and T. Vynnyk

Institute of Measurement and Automatic Control, Nienburgerstr. 17, 30167, Hannover, Germany, `eduard.reithmeier@imr.uni-hannover.de`

Summary. An improved photometric method for recording a 3D-microtopogrpahy of technical surfaces will be presented. The suggested procedure employs a scanning electron microscope (SEM) as multi-detector system. The improvement in measurement is based on an extended model of the electron detection in order to evaluate the detectors signals in a different way compared to known approaches. The method will be applied on a calibration sphere in order to demonstrate the accuracy of the current approach.

1 Introduction

The micro-topography of technical surfaces (Fig. 1) plays an important role for the functional behaviour of many mechanical parts in machinery. For example a large part of the energy-loss in consequence of turbulent air-flow in aviation occurs due to friction between the hull of the plane and the air flowing around it.

The losses can be reduced considerably by ribbed structures aligned to the flow; so-called riblets. A structure of a trapezoidal shape has been chosen as base geometry for this research project. The height of these riblet structures is about 20 μm, the approximate period length is about 40 μm, the tip radius

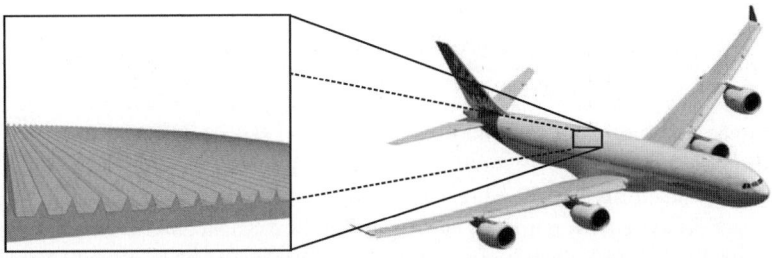

Fig. 1. Riblet structures on airplane

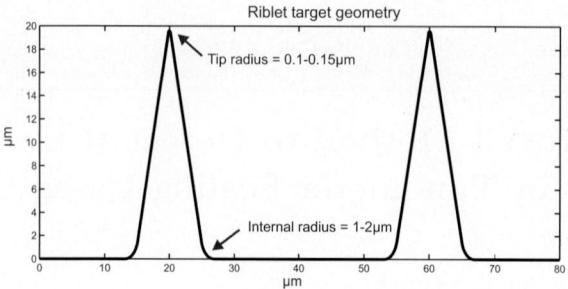

Fig. 2. Trapezoidal base geometry for riblet structures

lies between 0.1 μm and 0.15 μm and the internal radius between 1–2 μm (see Fig. 2).

To ensure the quality of the production process, it is required to measure the manufactured surfaces with a horizontal and vertical resolution up to 20 nm. Due to the large areas of measurement these kind of surfaces needs to be measured fast, sufficiently accurate and, if possible, in non contact mode. Due to the fact that the requested lateral/vertical resolution lies within a range from 10 to 20 nm and that the shape of the surface consists of convex and concave regions, the measurement problem can not be solved using standard measuring tactile or optical methods. For instance, the problem of using a tactile measuring system lies in the insufficient sharpness of the stylus tip. Typically, the tip radius is bigger then 2 μm and the cone angle greater than 60°, see Fig. 3.

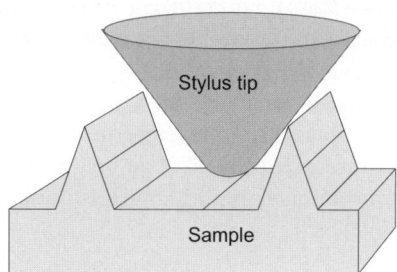

Fig. 3. Problem with the measurement of the concave areas

Due to the high aspect ratio of the riblet structures it turned out that the morphologic filter effects distorts the riblet structures substantially. This distortion is caused by the shape of the stylus tip. A morphologic deconvolution in case of known tip geometry does not lead to an accurate reconstruction of the profile. Thus tactile measurement technique is only suitable for riblet structures in a very restricted way.

The stylus tip of the most accurate tactile measurement device, namely atomic force microscope, is much smaller, but the maximal measurement range differs between 10 μm and 15 μm. Also, scanning time is too long for the measurement of large areas.

The usage of optical measuring systems is restricted due to the fact that these systems are based on the analysis of the light reflected by the surface of the measured object (see Fig. 4).

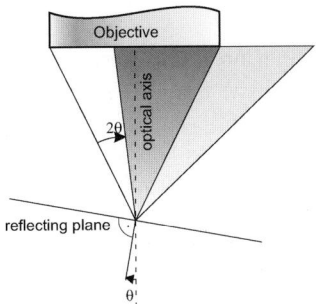

Fig. 4. Problem of direct light reflection at the lateral faces of riblet structures

With increasing tilt angle of the even but rough surface the part of the reflected light cone captured by the objective decreases. Reaching the critical angle the signal to noise ratio finally becomes so bad that the surface cannot be detected anymore. In addition, the horizontal resolution of the best 100x objectives with a numerical aperture NA=0.95 is $d = \frac{\lambda}{2 \cdot NA} \approx \frac{0,55\mu m}{2 \cdot 0,95} = 0,29\mu m$ leads to a value which does not meet the requirements for measurement of the riblet structures.

Employing a scanning electron microscope(SEM) uses an electron beam to take images of the object surface. The electron beam scans the object surface in vacuum. A detector catches the electrons emitted by the surface. In this case the resulting image has much higher optical depth of the field as well as higher lateral resolution. The advantage of the SEM is clearly the higher lateral resolution (approx. 5 nm) in relation to other measuring techniques. The 3D-Surface detection can be realized via stereometric evaluation of the image sequence received at different tilt angles of the specimen. For evaluation all methods require at least 2 images taken at different inclination angles of the specimen. The vertical resolution depends on the lateral resolution and on the tilt angle according to:

$$R_v = \frac{R_h}{\sin(\lambda)} \tag{1}$$

Here R_v is the vertical resolution, R_h is the horizontal resolution, and λ states the tilt angle.

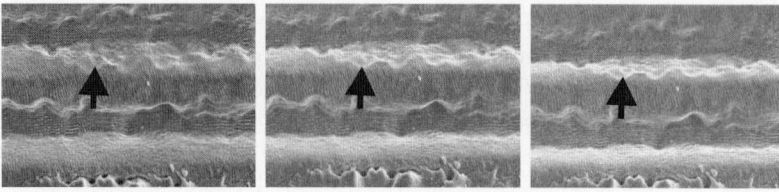

Fig. 5. SEM-images with different tilt angles. Arrows mark identical points in the topography whose similarity is no more recognizable, particularly in the right image

Unfortunately, the inclination angle of the lateral riblet faces is restricted to approximately 10°, due to the limited contrast ration of the SEM images. (Fig. 5: The arrows mark identical points on the topography, which may not be detected in the right image anymore). The typical tilt angle in which images are still similar, lies within the range of 5–7°. By these angles the vertical resolution is approximately 10 times larger than the lateral resolution. To reach a vertical resolution of 0.05 μm, the lateral resolution must lie within the range of 5 nm. Such a high lateral resolution is close to the limit of the SEM-recording technology. Additionally, the final 3d images would have a maximum size of 10 μm x 10 μm, which is insufficient for the measurement of riblet structures due to their period of 40 μm.

2 Solution: Photometric Method

The reconstruction of surfaces from scanning electron microscope (SEM) images is object of extensive research [1], [2], [4], but fundamental image processing problems are still unsolved. The approach, described in this paper, uses an alternative model for the image formation. The method is based on the analysis of the secondary electrons registered by the Everhart-Thornley SE detector (Fig. 6).

Before the algorithm will be introduced, let us discuss the physical processes occurring during irradiation of the specimen's surface with primary electrons. The energy of the primary electrons lies in the range of 20–30kV. Electrons, penetrated in the surface, create an electron diffusion region. De-

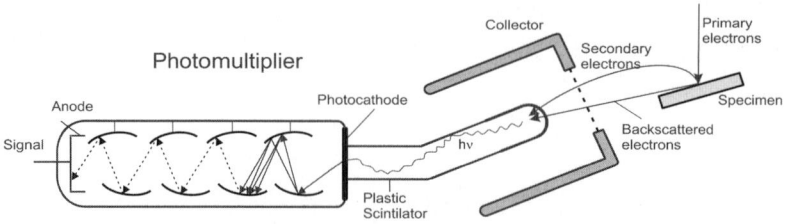

Fig. 6. Everhart-Thornley SE-detector [1]

pending on the activation type, free electrons of different energy levels are being produced. Inelastic scattering in the area close to the injection hole (0.1–1 nm) yields creation of the low-energy (0–50eV), so called, pure secondary electrons (SE). Deeper areas also produce SE, but their energy is too small to be emitted from the surface, therefore only high-energy backscattered electrons(BSE) leave the specimen. The typical energy distribution of electrons emitted from the surface of the specimen is shown on Fig. 7. Due to the fact that most part of emitted electrons are SE, only these electrons are essential for 3D analysis of the SEM images. Furthermore, electrons with high energy can be distinguished using SEM images with a negative bias voltage as shown below.

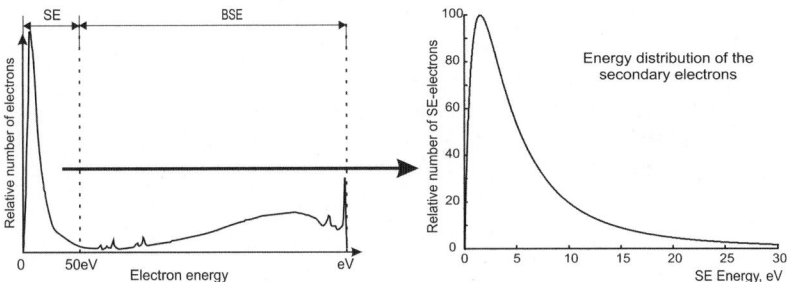

Fig. 7. Energy distribution of the secondary electron's yield(left), low energy distribution(right) [1]

The typical energy distribution of SE's is shown on Fig. 7. According to [1] the probability density of secondary electrons $f(U_e)$ as function of the electron's energy can be calculated as:

$$f(U_e) = K(U_p)\frac{U_e}{(U_e + W_e)^4}, \quad K(U_p) = \frac{6W_e^2 (U_p + W_e)^3}{U_p^3 + 3U_p^2 W_e} \tag{2}$$

Here W_e is so called electron's work function, U_p is the energy of the primary electrons and U_e is the energy of the emitted electrons.

According to Lambert's cosine law, the angle distribution of the probability density $g(\varphi)$ may be expressed by:

$$g(\varphi) = \frac{1}{2}\cos(\varphi) \tag{3}$$

φ states the angle between the normal vector of the surface and the emitting direction of the electron.

Using distributions (2) and (3), the distribution of secondary electrons registered by the SE-detector may be expressed by a function of the specimen's tilt angle φ. For this purpose the SE-detector is modelled by an electric charge, homogeneously distributed on a sphere with a radius r_0. The electron

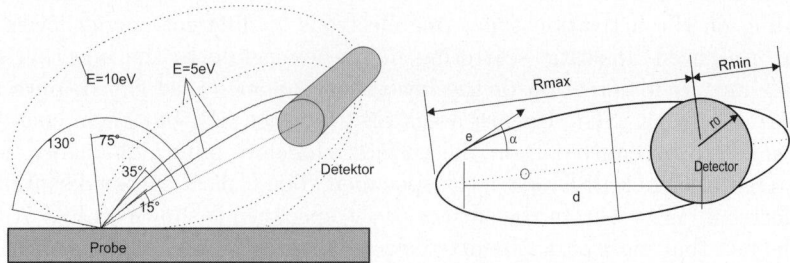

Fig. 8. Possible orbits of electrons

is considered to be registered by SE-detector, if the minimal distance from the centre of the spherical charge is less then r_0. In fact the interaction with the probe also needs to be considered; for simplification however, the influence of the specimen's conducting material is neglected. Due to the nature of electric fields all closed orbits of the electrons have an elliptic shape. According to the laws of energy and momentum conservation, this ellipse is defined by the following parameters:

$$R_{\min} = \frac{-U_d r_0 e + \sqrt{U_d^2 r_0^2 e^2 + 4\left(U_e - U_d r_0 e/d\right) U_e d^2 \sin^2(\alpha)}}{2\left(U_e - U_d r_0 e/d\right)}$$

$$R_{\max} = \frac{-U_d r_0 e - \sqrt{U_d^2 r_0^2 e^2 + 4\left(U_e - U_d r_0 e/d\right) U_e d^2 \sin^2(\alpha)}}{2\left(U_e - U_d r_0 e/d\right)}$$

(4)

Obviously, $R_{\min}$ and $R_{\max}$ depend on the angle α, which is the angle between the vector of the initial velocity $v_0 = \sqrt{\frac{2U_e}{m_e}}$ of the electron and the centre of the SE-detector, m_e is the mass of the electron, U_d is the detector bias, U_e is the energy of the emitted electron, r_0 is the radius of the detector, d is the distance to the detector and e is the charge of the electron

If $R_{\max} < r_0$, the electron is considered to have been registered by the SE-detector. Due to the properties of the sine function, for each initial velocity vector v_0 there exists an angle $\alpha_{\max}(v_0, U_d, d)$, for which all electrons with a start angle of $[-\alpha_{\max}(v_0, U_d, d)...\alpha_{\max}(v0, U_d, d)], [\pi - \alpha_{\max}(v_0, U_d, d)...\pi + \alpha_{\max}(v_0, U_d, d)]$ will be collected by the SE detector. The amount of registered electrons as function of the angle between surface and detector can be represented as follows:

$$f(\beta) = \int\limits_0^{U\max(=50V)} \frac{U_e}{(U_e + W_e)^4} \sigma(U_e, \beta) dU_e$$

(5)

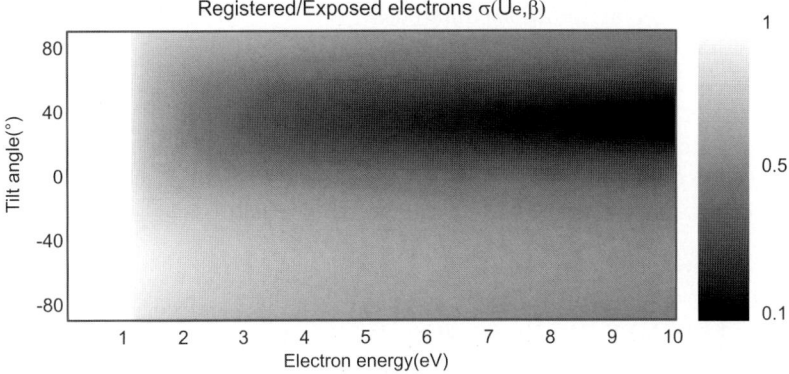

Fig. 9. Ratio of the registered to exposed electrons

Here $\sigma(U_e, \beta) := (4\pi)^{-1} \int\limits_{\beta-\alpha_{max}}^{\beta-\alpha_{max}} |\cos(\varphi)| \sqrt{\alpha_{max}^2 - (\varphi - \beta)^2} d\varphi$

Figure 9 shows the ration $\sigma(U_e, \beta)$ of the registered to exposed electrons. Low energy electrons are collected by the detector at each tilt angle (left part of the image). With the increasing velocity v_0 of the electrons, $\sigma(U_e, \beta)$ becomes sinusoidal (right part).

As a result $f(\beta)$ and its corresponding Fourier transformation is shown in Fig. 10:

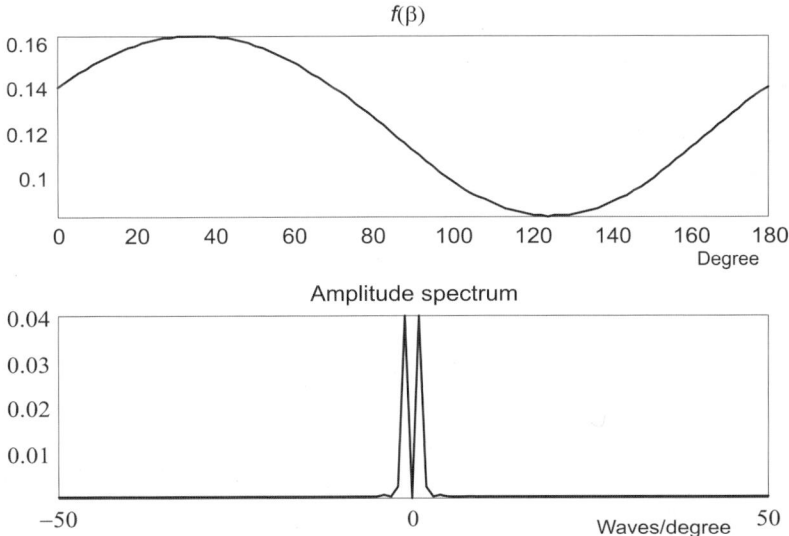

Fig. 10. Angle distribution f_β of the registered electrons and its Fourier transformation

That is $f(\beta)$ may be approximated by

$$f(\beta) = A - B\cos^2(\beta) \tag{6}$$

without loss of the necessary accuracy. This approximation is used for further calculations. Now let us describe the angle β via differential characteristics of the surface with respect to the position of the detector:

$$\mathbf{n}_d = \begin{bmatrix} \cos(\varphi_0) \\ 0 \\ \sin(\varphi_0) \end{bmatrix}, \mathbf{n} = \begin{bmatrix} -\dfrac{\partial Z(x,y)}{\partial x} \\ \dfrac{\partial Z(x,y)}{\partial y} \\ 1 \end{bmatrix}, \mathbf{n}_z = \begin{bmatrix} 0 \\ 0 \\ 1 \end{bmatrix}$$

Here, n_d is the vector to the SE-detector, n the normal to the surface and n_z is the vector to SEM electron gun (see Fig. 11).

The angle between n_d and n vectors can be expressed by:

$$\cos(\beta(x,y)) = \frac{\cos(\varphi_0)\dfrac{\partial Z(x,y)}{\partial x} + \sin(\varphi_0)}{\sqrt{1 + \dfrac{\partial Z(x,y)}{\partial x}^2 + \dfrac{\partial Z(x,y)}{\partial y}^2}} \tag{7}$$

At this stage the registered portion by the SE-detector will be used for further calculations.

The last step is to define the intensity I_e of emitted SE's as a function of the tilt angle $\alpha(x,y)$:

$$I_e(\alpha(x,y)) = \frac{I_0}{\cos(\alpha(x,y))} = I_0\sqrt{1 + \frac{\partial Z(x,y)}{\partial x}^2 + \frac{\partial Z(x,y)}{\partial y}^2} \tag{8}$$

Here α is the angle between the tangent plane on the surface mounted in point (x, y) and the direction to the electron gun. Finally, the intensity I_e of electrons, registered by the SE-detector results in

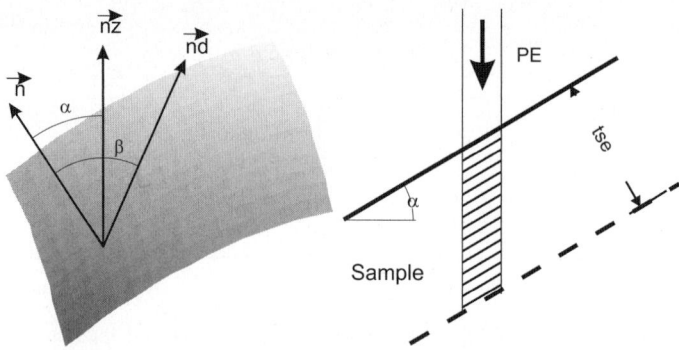

Fig. 11. Normal vectors (left), tilt of the specimen (right)

$$I = \sqrt{1+\frac{\partial Z(x,y)}{\partial x}^2+\frac{\partial Z(x,y)}{\partial y}^2}\left(C_1 - C_2\frac{\left(-\cos(\varphi_0)\frac{\partial Z(x,y)}{\partial x}+\sin(\varphi_0)\right)^2}{1+\frac{\partial Z(x,y)}{\partial x}^2+\frac{\partial Z(x,y)}{\partial y}^2}\right)$$

(9)

This expression will be used for the gradient algorithm, which is presented in the following section. For the surface reconstruction four Everhart-Thornley SE-detectors are used. They are symmetrically positioned in the X-Z and Y-Z planes. The tilt angle of each detector is φ_0 (see Fig. 12).

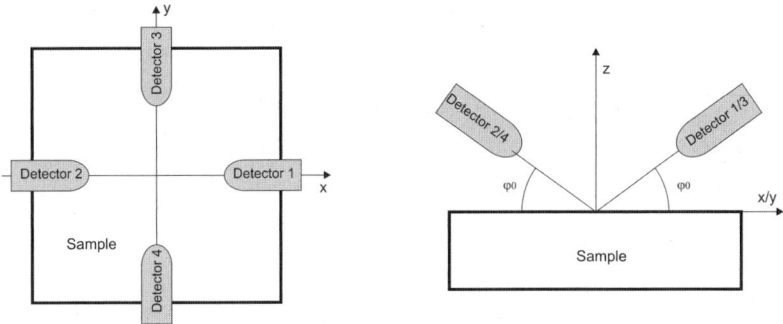

Fig. 12. Measurement setup

2.1 Algorithm

Step 1: Elimination of the BSE electrons. For this purpose each detector will be used twice. Firstly SEM image I^* should be created using a detector bias voltage of -20V. Then the bias voltage must be set to approx. 50V and image I^{**} will be produced. The difference $I = I^{**} - I^*$ serves as basis for further calculations and should be produced sequentially for each of the four detectors.

Step 2: Evaluation. According to equation (9) intensities of the collected electrons are estimated as follows:

$$I_{1,2} = \sqrt{1+\frac{\partial Z(x,y)}{\partial x}^2+\frac{\partial Z(x,y)}{\partial y}^2}\left(C_1 - C_2\frac{\left(-\cos(\varphi_0)\frac{\partial Z(x,y)}{\partial x}\pm\sin(\varphi_0)\right)^2}{1+\frac{\partial Z(x,y)}{\partial x}^2+\frac{\partial Z(x,y)}{\partial y}^2}\right)$$

(10)

$$I_{3,4} = \sqrt{1 + \frac{\partial Z(x,y)}{\partial x}^2 + \frac{\partial Z(x,y)}{\partial y}^2} \left(C_1 - C_2 \frac{\left(-\cos(\varphi_0)\frac{\partial Z(x,y)}{\partial y} \pm \sin(\varphi_0) \right)^2}{1 + \frac{\partial Z(x,y)}{\partial x}^2 + \frac{\partial Z(x,y)}{\partial y}^2} \right)$$

(11)

Hence, for each point (x, y) there exist four equations with four unknowns. The Jacobian of the equation system (10)–(11) takes the following view:

$$J = \frac{16 \cdot \cos^4(\varphi_0) \cdot \left(\left(\frac{\partial Z}{\partial x}\right)^2 - \left(\frac{\partial Z}{\partial y}\right)^2 \right) \cdot C_2^2 \cdot \sin^2(\varphi_0)}{1 + \left(\frac{\partial Z}{\partial x}\right)^2 + \left(\frac{\partial Z}{\partial y}\right)^2}$$

(12)

According to the inverse function theorem $\frac{\partial Z}{\partial x}$ and $\frac{\partial Z}{\partial y}$ can be uniquely determined, if $J \neq 0$, i.e. $\left|\frac{\partial Z}{\partial x}\right| \neq \left|\frac{\partial Z}{\partial y}\right|$. In that case, combining of I_1, I_2 and I_3, I_4, respectively, yields to:

$$I_1 - I_2 = \frac{-2C_2 \frac{\partial Z(x,y)}{\partial x} \sin(2\varphi_0)}{\sqrt{1 + \frac{\partial Z(x,y)}{\partial x}^2 + \frac{\partial Z(x,y)}{\partial y}^2}}$$

(13)

$$I_3 - I_4 = \frac{-2C_2 \frac{\partial Z(x,y)}{\partial y} \sin(2\varphi_0)}{\sqrt{1 + \frac{\partial Z(x,y)}{\partial x}^2 + \frac{\partial Z(x,y)}{\partial y}^2}}$$

(14)

Also:

$$I_1 + I_2 - I_3 - I_4 = \frac{2C_2 \left(\frac{\partial Z(x,y)}{\partial y}^2 - \frac{\partial Z(x,y)}{\partial x}^2 \right) \cos^2(\varphi_0)}{\sqrt{1 + \frac{\partial Z(x,y)}{\partial x}^2 + \frac{\partial Z(x,y)}{\partial y}^2}}$$

(15)

Combining of (13)–(15) leads to the following linear system of equations:

$$\begin{cases} \dfrac{I_2 + I_1 - (I_3 + I_4)}{I_1 - I_2 + I_3 - I_4} = \dfrac{\frac{\partial Z(x,y)}{\partial y} - \frac{\partial Z(x,y)}{\partial x}}{2\tan(\varphi_0)} \\[4ex] \dfrac{I_2 + I_1 - (I_3 + I_4)}{I_3 - I_4 - I_1 + I_2} = \dfrac{\frac{\partial Z(x,y)}{\partial y} + \frac{\partial Z(x,y)}{\partial x}}{2\tan(\varphi_0)} \end{cases}$$

(16)

Finally the partial derivatives of the surface are determined by:

$$
\begin{cases}
\dfrac{\partial Z(x,y)}{\partial x} = 2\tan(\varphi_0)\dfrac{I_2 + I_1 - I_3 - I_4}{(I_3 - I_4)^2 - (I_1 - I_2)^2}(I_1 - I_2) \\[3mm]
\dfrac{\partial Z(x,y)}{\partial y} = 2\tan(\varphi_0)\dfrac{I_2 + I_1 - I_3 - I_4}{(I_3 - I_4)^2 - (I_1 - I_2)^2}(I_3 - I_4)
\end{cases}
\tag{17}
$$

For the case that $J = 0$, the constant C_2 needs to be predefined. $\frac{\partial Z}{\partial x}$ and $\frac{\partial Z}{\partial y}$ can be determined in this case from (13)–(14)

$$
\frac{\partial z}{\partial y}^2 = \frac{(I_3 - I_4)^2}{-(2C_2\sin(2\varphi_0))^2 + (I_3 - I_4)^2 + (I_1 - I_2)^2}
\tag{18}
$$

$$
\frac{\partial z}{\partial x}^2 = \frac{(I_1 - I_2)^2}{-(2C_2\sin(2\varphi_0))^2 + (I_3 - I_4)^2 + (I_1 - I_2)^2}
\tag{19}
$$

A more realistic description of the problem may be obtained by involving a random function in I_{1-4}. In that sense the signal to noise ratio as well as discretization errors in the measurement of I_{1-4} could be taken into consideration. For $(I_1 - I_2)^2 - (I_3 - I_4)^2 \to 0$, it is recommended to determine constant C_2 via calibration methods and use (18)–(19) for the definition of the partial derivatives .

The main objective, namely to determine the surface $Z(x,y)$ of the measured specimen, can be carried out by numerical integration [3] of the partial derivatives $Z_x(x,y)$ and $Z_y(x,y)$. This holds for a certain range of derivatives.

3 Comparing Simulation Results with a Real Data

A calibration sphere of a radius r=775 μm was used for the verification of the described approach. The results of the simulation were compared with real data. As measurement device SEM EVO 60 XVP (Fa. Carl Zeiss) was used. This SEM has 1 SE-detector, but includes the feature to rotate around z-axis. This capability was used for taking four images as if there exist four SE-detectors (Fig. 13).

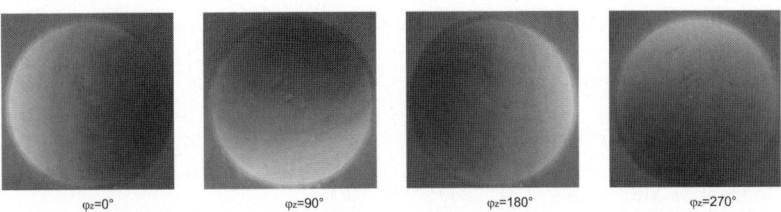

$\varphi_z=0°$ $\varphi_z=90°$ $\varphi_z=180°$ $\varphi_z=270°$

Fig. 13. SEM images from 4 SE-detectors

The measurements detected the spherical structure, with a radius of r=780 μm. The average deviation from the spherical surface lies in the range of $\pm 10 \mu$m. The maximal deviation, however, is about 30 μm. Such a big error occurs due to neglecting of the higher frequencies in (6) as well as insufficient precision of I_{1-4} functions (max 256 levels).

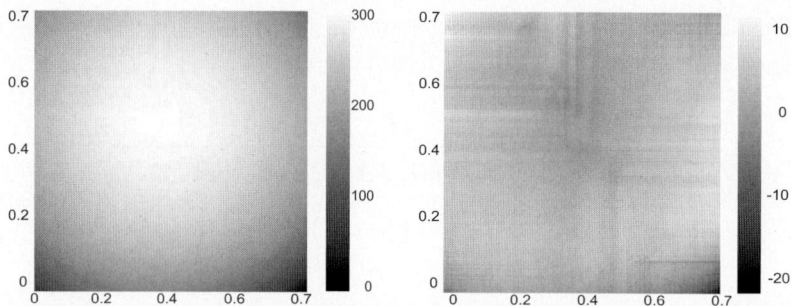

Fig. 14. SEM images from 4 SE-detectors

4 Conclusion

For simplification it was assumed that the angle distribution of the registered electrons takes the simple form of equation (6). However, this approximation is not exact and should be expanded. Furthermore, in reality the detector is not spherical, as well as the influence of the specimen's material is not being negligible. As a result the SE intensity distribution should have a more sophisticated form. Nevertheless, these results are very promising and improving the approximations, mentioned above, will certainly lead to even better results.

References

[1] Reimer, L.: Scanning Electron Microscopy. Physics of Image Formation and Microanalysis. Springer, Springer, Berlin Heidelberg New York (1998)
[2] Beil, W., Carlsen, I.C.: Surface Reconstruction from Stereoscopy and "Shape from Shading" in SEM Images, Macine Vision and Applications. Vol. 4, 271–285 (1991)
[3] Bulirsch, R., Stoer,J.: Introduction to Numerical Analysis, Springer-Verlag, New York (2002)
[4] Paluszynkski, J., Slowko, W.: Compensation of the shadowing error in three-dimensional imaging with a mutltiple detector scanning electron microscope, Journal of Microscopy. Vol. 224, 93–96 (2006)
[5] Paluszynkski, J., Slowko, W.: Surface reconstruction with the photometric method in SEM, Vacuum. Vol. 78, 533–537 (2005)

Mathematics and Applications in Microscale

Part II

Mathematics and Applications in Hergoula

Numerical Simulation of a Molten Carbonate Fuel Cell by Partial Differential Algebraic Equations

K. Chudej[1], M. Bauer[1], H.J. Pesch[1], and K. Schittkowski[2]

[1] Lehrstuhl für Ingenieurmathematik, Universität Bayreuth, 95440 Bayreuth
[2] Fachgruppe Informatik, Universität Bayreuth, 95440 Bayreuth

Summary. The dynamical behavior of a molten carbonate fuel cell (MCFC) can be modeled by systems of partial differential algebraic equations (PDEAs) based on physical and chemical laws. Mathematical models for identification and control are considered as valuable tools to increase the life time of the expensive MCFC power plants, especially to derive control strategies for avoiding high temperature gradients and hot spots. We present numerical simulation results for a load change of a new one-dimensional counterflow MCFC model consisting of 34 nonlinear partial and ordinary differential algebraic-equations (PDEAs) based on physical and chemical laws. The PDAE system is discretized by the method of lines (MOL) based on forward, backward, and central difference formulae, and the resulting large system of semi-explicit differential-algebraic equations is subsequently integrated by an implicit DAE solver.

1 Introduction

Molten carbonate fuel cells (MCFCs) are a challenging new technology for stationary power plants, see e.g. Bischoff and Huppmann [2], Rolf [11], or Winkler [19]. They allow internal reforming of a fuel gas, for example methane, inside the cell with an operating temperature of about 650° C, and have the advantage of producing clean exhaust gases. The dynamical behavior of MCFCs can be modeled by one- and two dimensional systems of partial differential algebraic equations, see Heidebrecht and Sundmacher [8, 9, 10]. One of these particular models was recently validated for a real fuel cell operated at the power plant of the Magdeburg university hospital, see Gundermann, Heidebrecht, and Sundmacher [5].

The following main assumptions are made to derive the MCFC model equations, see also Heidebrecht [7]. First, plug flow conditions for the gas phase in anode and cathode are assumed, where different phases may have different temperatures and may exchange heat. All solid parts of the cell are lumped to one phase with respect to enthalpy balance. The temperatures of the two gas phases are calculated separately, and the MCFC is operated at nearly ambient pressure. We do not consider pressure drops across the gas channels, i.e., isobaric conditions are assumed. All cells in the stack behave alike, so that the simulation of a single cell is sufficient taking insulation conditions at boundaries of neighboring cells into account. All chemical substances have the same heat capacity, which is independent of temperature. Reaction enthalpies are independent of the temperature. The temperature dependent chemical equilibrium constants and standard open circuit voltages are approximated by affine-linear functions. Ideal gas law is applied. Reforming reactions in the anode gas channel are modeled as quasi-homogeneous gas phase reactions using volume-related reaction rates. Methane steam reforming and water gas shift reactions are considered. Their heat of reaction is fully transferred to the gas phase. Diffusion in flow direction is negligible compared to convective transport. Heat exchange between electrode and gas phase is described by a linear function. The corresponding heat exchange coefficient also includes the effect of thermal radiation in a linearized form.

In this paper, we present a new one-dimensional counterflow model, see Bauer [1], which is derived from the two-dimensional crossflow model of Heidebrecht and Sundmacher [7, 10]. In contrast to a previous model presented in Chudej et al. [3], this new one has differential time index $\nu_t = 1$ and MOL index $\nu_{MOL} = 1$, which are easily derived from the two-dimensional analysis of Chudej, Sternberg, and Pesch [4]. The resulting dynamical system consists of twenty partial differential equations, four partial algebraic or steady-state equations, respectively, nine coupled ordinary differential and one coupled ordinary algebraic equation.

The spatial derivatives are approximated at grid points by forward, backward, and central difference formulae depending on the mathematical structure of the differential equation, especially the transport direction. An equidistant grid is used. Neumann boundary conditions are set for the heat equation and Dirichlet boundary conditions for the transport equations. Together with the additional coupled system of DAEs, the method of lines leads to a large set of differential algebraic equations, where the initial values are derived from the given initial values of the PDAE system. The procedure is called numerical method of lines, see Schiesser [12] or Schittkowski [15].

The nonlinear model equations are implemented in the model language PCOMP, see Schittkowski [17], under the interactive software system EASY-FIT, see Schittkowski [15, 16]. For a further set of practical PDAE models see Schittkowski [18].

Section 2 contains a brief summary of the technological background and the detailed mathematical equations. In Sect. 3, we describe the numerical procedures in more detail and present numerical simulation results for a load change of the fuel cell.

2 The Molten Carbonate Fuel Cell Model

2.1 Technological Description and Mathematical Variables

The model under investigation describes the dynamical behavior of a MCFC in a counterflow configuration with respect to anode and cathode gas streams, see Fig. 1.

In the anode channel, the feed gas, usually methane CH_4, and water H_2O are converted to hydrogen H_2, carbon monoxide CO, and carbon dioxide CO_2, see Fig. 1 (ref1, ref2). This reforming process consumes heat, and only high temperature fuel cells offer the opportunity for internal reforming.

$$\begin{aligned} \text{(ref1)} && CH_4 + H_2O &\rightleftharpoons CO + 3\,H_2\,, \\ \text{(ref2)} && CO + H_2O &\rightleftharpoons CO_2 + H_2\,. \end{aligned}$$

Simultaneously, two electro-chemical reactions consume hydrogen and carbon monoxide to produce electrons in the porous anode electrode. Both consume carbonate ions, CO_3^{2-}, from the electrolyte layer which is located between the anode and cathode electrodes,

$$\begin{aligned} \text{(ox1)} && H_2 + CO_3^{2-} &\rightleftharpoons H_2O + CO_2 + 2e^-\,, \\ \text{(ox2)} && CO + CO_3^{2-} &\rightleftharpoons 2\,CO_2 + 2e^-\,. \end{aligned}$$

At the porous cathode electrode the reduction reaction produces new carbonate ions from oxygen, O_2 and carbon dioxide, CO_2,

$$\text{(red)} \qquad \frac{1}{2}O_2 + CO_2 + 2e^- \rightleftharpoons CO_3^{2-}\,.$$

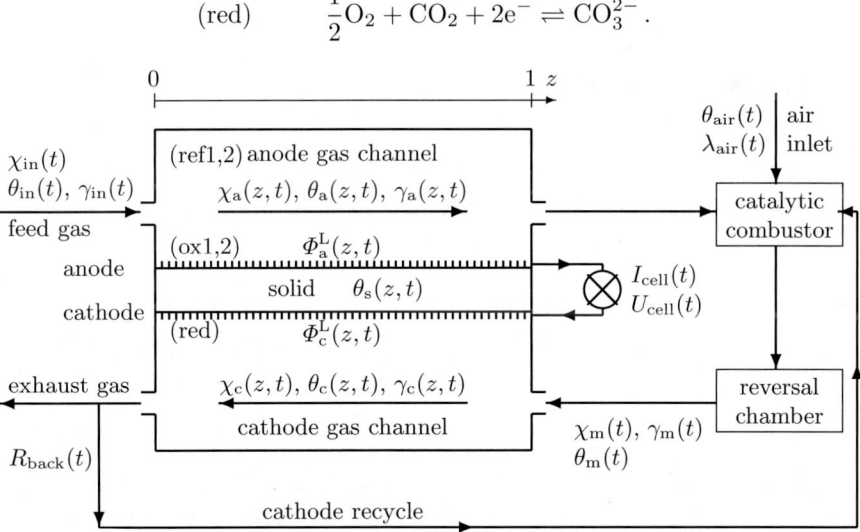

Fig. 1. 1D counterflow model of a molten carbonate fuel cell with mathematical variables and considered chemical reactions

Hereby, the carbonate ions produced at the cathode electrode are transferred through the electrolyte to the anode electrode. Electric current I_{cell} and cell voltage U_{cell} can be collected at the electrodes, see Fig. 1.

The mathematical model uses the following mathematical variables, see again Fig. 1, where $j \in \{a, c, m, in\}$:

$\chi_{i,j}$ - molar fractions, $i \in I \overset{\text{def}}{=} \{CH_4, H_2O, H_2, CO, CO_2, O_2, N_2\}$,

χ_j $= (\chi_{i,j})_{i \in I}$,

γ_j - molar flow densities,

θ_s - solid temperature,

θ_j - gas temperatures,

Φ_a^L, Φ_c^L - electrical potentials at anode and cathode,

U_{cell} - cell voltage,

I_{cell} - cell current,

θ_{air} - temperature of air,

λ_{air} - air number,

R_{back} - cathode recycle switch.

The definition of molar fractions intrinsically leads to the balance equation $\sum_{i \in I} \chi_{i,j} \equiv 1$, $j \in \{a, c, m, in\}$, since the model equations fulfill $\sum_{i \in I} \frac{\partial \chi_{i,j}}{\partial t} \equiv 0$, $j \in \{a, c\}$. Therefore global minimal coordinates, i.e. a suitable subset of the molar fractions depending on the boundary conditions, can be applied.

The main variable is the solid temperature θ_s which is needed to define a heat equation. The solution of the solid temperature should avoid high temperature gradients and especially hot spots, which drastically shorten service life and eventually damage the fuel cell irreparably.

2.2 The Dynamical Model

All variables are dimensionless. One unit of the dimensionless time corresponds to 12.5 seconds. Input variable of the system is the cell current $I_{cell}(t)$, which is given as a step function for modeling a load change of the fuel cell.

For performing numerical simulations, the boundary conditions for the gas temperature θ_{in}, the molar fractions χ_{in}, and the molar flow density γ_{in} are given in form of time dependent functions at the anode inlet ($z = 0$). The quantities λ_{air} and θ_{air} are also prescribed as time dependent functions, see Fig. 1.

Heat equation in the solid:

$$\frac{\partial \theta_s}{\partial t} = \lambda \frac{\partial^2 \theta_s}{\partial z^2} + f_1(\theta_s, \theta_a, \theta_c, \chi_a, \chi_c, \Phi_a^L, \Phi_c^L, U_{cell}), \quad \frac{\partial \theta_s}{\partial z}\Big|_{z \in \{0,1\}} = 0.$$

Advection equations in the gas streams:

$$\frac{\partial \chi_a}{\partial t} = -\gamma_a \theta_a \frac{\partial \chi_a}{\partial z} + f_2(\theta_s, \theta_a, \chi_a, \Phi_a^L), \quad \chi_a|_{z=0} = \chi_{in}(t),$$

$$\frac{\partial \chi_c}{\partial t} = +\gamma_c \theta_c \frac{\partial \chi_c}{\partial z} + f_3(\theta_s, \theta_c, \chi_c, \Phi_c^L, U_{cell}), \quad \chi_c|_{z=1} = \chi_m(t),$$

$$\frac{\partial \theta_a}{\partial t} = -\gamma_a \theta_a \frac{\partial \theta_a}{\partial z} + f_4(\theta_s, \theta_a, \chi_a, \Phi_a^L), \quad \theta_a|_{z=0} = \theta_{in}(t)$$

$$\frac{\partial \theta_c}{\partial t} = +\gamma_c \theta_c \frac{\partial \theta_c}{\partial z} + f_5(\theta_s, \theta_c, \chi_c, \Phi_c^L, U_{cell}), \quad \theta_c|_{z=1} = \theta_m(t).$$

Steady-state equations for molar flow densities:

$$0 = \frac{\partial(\gamma_a \theta_a)}{\partial z} - f_6(\theta_s, \theta_a, \chi_a, \Phi_a^L), \quad \gamma_a|_{z=0} = \gamma_{in}(t),$$

$$0 = \frac{\partial(\gamma_c \theta_c)}{\partial z} - f_7(\theta_s, \theta_c, \chi_c, \Phi_c^L, U_{cell}), \quad \gamma_c|_{z=1} = \gamma_m(t).$$

Equations for potentials and current density:

$$\frac{\partial \Phi_a^L}{\partial t} = [i_a(\theta_s, \chi_a, \Phi_a^L) - i]/c_a,$$

$$\frac{\partial \Phi_c^L}{\partial t} = [i_a(\theta_s, \chi_a, \Phi_a^L) - i]/c_a + [i_e(\Phi_a^L, \Phi_c^L) - i]/c_c,$$

$$i = (c_a^{-1} + c_e^{-1} + c_c^{-1})^{-1} \left(\frac{i_a - I_a}{c_a} + \frac{i_e - I_e}{c_e} + \frac{i_c - I_c}{c_c} \right) + I_{cell}$$

Integro differential algebraic equations:

$$\frac{dU_{cell}}{dt} = \frac{I_a - I_{cell}}{c_a} + \frac{I_e - I_{cell}}{c_e} + \frac{I_c - I_{cell}}{c_c}$$

$$I_a(t) = \int_0^1 i_a(\theta_s, \theta_a, \chi_a, \Phi_a^L) \, dz,$$

$$I_c(t) = \int_0^1 i_c(\theta_s, \theta_c, \chi_c, \Phi_c^L, U_{cell}) \, dz,$$

$$I_e(t) = \int_0^1 i_e(\Phi_a^L, \Phi_c^L) \, dz,$$

$$\frac{d\chi_m}{dt} = f_8(\theta_m, \chi_m, \theta_a|_{z=1}, \chi_a|_{z=1}, \gamma_a|_{z=1}, \theta_c|_{z=0}, \chi_c|_{z=0}, \gamma_c|_{z=0},$$
$$\lambda_{air}, \theta_{air}, R_{back}),$$

$$\frac{d\theta_m}{dt} = f_9(\theta_m, \theta_a|_{z=1}, \chi_a|_{z=1}, \gamma_a|_{z=1}, \theta_c|_{z=0}, \chi_c|_{z=0}, \gamma_c|_{z=0},$$
$$\lambda_{air}, \theta_{air}, R_{back}),$$

$$\gamma_m = f_{10}(\theta_m, \theta_a|_{z=1}, \chi_a|_{z=1}, \gamma_a|_{z=1}, \theta_c|_{z=0}, \chi_c|_{z=0}, \gamma_c|_{z=0},$$
$$\lambda_{air}, \theta_{air}, R_{back}).$$

Initial values:

$$\theta_s|_{t=0} = \theta_{s,0}(z), \quad \theta_a|_{t=0} = \theta_{a,0}(z), \quad \theta_c|_{t=0} = \theta_{c,0}(z), \quad \theta_m|_{t=0} = \theta_{m,0}(z),$$
$$\chi_a|_{t=0} = \chi_{a,0}(z), \quad \chi_c|_{t=0} = \chi_{c,0}(z), \quad \chi_m|_{t=0} = \chi_{m,0}(z),$$
$$\Phi_a^L|_{t=0} = \Phi_{a,0}^L(z), \quad \Phi_c^L|_{t=0} = \Phi_{c,0}^L(z), \quad U_{\text{cell}}|_{t=0} = U_{\text{cell},0}$$

Constraints:

$$\gamma_a > 0, \quad \gamma_c > 0, \quad \theta_a > 0, \quad \theta_c > 0,$$
$$0 \le \chi_{i,j} \le 1, \quad i \in I, \quad j \in \{a, c, m, in\},$$
$$\sum_{i \in I} \chi_{i,j} \equiv 1, \quad j \in \{a, c, m, in\}.$$

The dynamical behavior at the outlet of the reversal chamber is described by a DAE, which couples the outlet of the anode and the inlet of the cathode and, if the cathode recycle is switched on, with the outlet of the cathode. To summarize, we obtain a coupled PDAE/DAE system consisting of 34 nonlinear equations.

2.3 The Details

The missing functions are to be specified subsequently for the matter of completeness.

Right hand sides of the PDAEs:

$$f_1 = -\sum_{\substack{i=H_2,CO \\ j=ox1,2 \\ \nu_{i,j}<0}} (\theta_a - \theta_s)\nu_{i,j} Da_j r_j - \sum_{\substack{i=CO_2,O_2 \\ \nu_{i,red}<0}} (\theta_c - \theta_s)\nu_{i,red} Da_{red} r_{red}$$
$$+ q_{\text{solid}} - (\theta_s - \theta_a)St_{as} - (\theta_s - \theta_c)St_{cs},$$

$$\frac{f_{2,i}}{\theta_a} = \sum_{j=ox1,2,ref1,2} \left(\nu_{i,j} - \chi_{i,a} \sum_{k \in I} \nu_{k,j} \right) Da_j r_j,$$

$$\frac{f_4}{\theta_a} = \sum_{\substack{i=H_2O,CO_2 \\ j=ox1,2 \\ \nu_{i,j}>0}} (\theta_s - \theta_a)\nu_{i,j} Da_j r_j + \sum_{j=ref1,ref2} -\Delta_R h_j^0 \frac{Da_j r_j}{c_p}$$
$$+ \frac{(\theta_s - \theta_a)St_{as}}{c_p},$$

$$f_6 = \frac{f_4}{\theta_a} + \theta_a \sum_{\substack{j=ox1,ox2,ref1,ref2 \\ i \in I}} \nu_{i,j} Da_j r_j,$$

$$\frac{f_{3,i}}{\theta_c} = \left(\nu_{i,red} - \chi_{i,c} \sum_{j \in I} \nu_{j,red} \right) Da_{red} r_{red},$$

Table 1. Constants

c_p	4.5	F	$\frac{3.5}{8}$	λ	0.666/2.5
$c_{p,s}$	10000	κ_e	1	θ_u	1

St_{as} 80.0	c_a 0.00001	λ_{air}	2.2
St_{cs} 120.0	c_e 0.00001	$\chi_{O_2,air}$	0.21
St_m 1.0	c_c 0.00001	θ_{air}	1.5

j	Da_j	Arr_j	θ_j^0	$\Delta_R h_j^0$	α_j^+	α_j^-	n_j
ref1	25.0	84.4	2.93	90.5			
ref2	100.0	6.2	2.93	-14.5			
ox1	5.0	21.6	2.93	56.0	0.5		2.0
ox2	5.0	21.6	2.93	42.0	0.5		2.0
red	0.3	31.6	2.93	156.0	2.5	0.5	2.0

i	$\Delta_c h_i^0$	$\nu_{i,ref1}$	$\nu_{i,ref2}$	$\nu_{i,ox1}$	$\nu_{i,ox2}$	$\nu_{i,red}$
CH_4	-323.85	-1.0	0.0	0.0	0.0	0.0
H_2O	0.0	-1.0	-1.0	1.0	0.0	0.0
H_2	-97.62	3.0	1.0	-1.0	0.0	0.0
CO	-114.22	1.0	-1.0	0.0	-1.0	0.0
CO_2	0.0	0.0	1.0	1.0	2.0	-1.0
O_2	0.0	0.0	0.0	0.0	0.0	-0.5
N_2	0.0	0.0	0.0	0.0	0.0	0.0

$$\frac{f_5}{\theta_c} = \frac{St_{cs}}{c_p}(\theta_s - \theta_c), \quad f_7 = \frac{f_5}{\theta_c} + \theta_c \sum_{i \in I} \nu_{i,red} Da_{red} r_{red},$$

$$i_a = F \sum_{j=ox1,ox2} n_j Da_j r_j, \quad i_e = (\Phi_a^L - \Phi_c^L)\kappa_e,$$

$$i_c = -F\, n_{red} Da_{red} r_{red},$$

$$q_{solid} = \sum_{j=ox1,ox2} [-\Delta_R h_j^0 + n_j(\Phi_a^S - \Phi_a^L)] Da_j r_j$$

$$+ [-\Delta_R h_j^0 + n_{red}(\Phi_c^S - \Phi_c^L)] Da_{red} r_{red} + (\Phi_a^L - \Phi_c^L) i_e / F.$$

Reaction kinetics:

$$r_{ref1} = e^{Arr_{ref1}\left(\frac{1}{\theta_{ref1}^0} - \frac{1}{\theta_a}\right)} \cdot \left(\chi_{CH_4,a}\chi_{H_2O,a} - \frac{\chi_{H_2,a}^3 \chi_{CO,a}}{K_{ref1}}\right),$$

$$r_{ref2} = e^{Arr_{ref2}\left(\frac{1}{\theta_{ref2}^0} - \frac{1}{\theta_a}\right)} \cdot \left(\chi_{CO,a}\chi_{H_2O,a} - \frac{\chi_{H_2,a}\chi_{CO_2,a}}{K_{ref2}}\right).$$

$$r_{ox1} = e^{Arr_{ox1}\left(\frac{1}{\theta_{ox1}^0} - \frac{1}{\theta_s}\right)} \left[\chi_{H_2,a} e^{\frac{\alpha_{ox1}^+ n_{ox1}(-\Phi_a^L - \Delta\Phi_{ox1}^0)}{\theta_s}} \right.$$

$$\left. - \chi_{H_2O,a} \chi_{CO_2,a} e^{\frac{-(1-\alpha_{ox1}^+)n_{ox1}(-\Phi_a^L - \Delta\Phi_{ox1}^0)}{\theta_s}} \right],$$

$$r_{ox2} = e^{Arr_{ox2}\left(\frac{1}{\theta_{ox2}^0} - \frac{1}{\theta_s}\right)} \left[\chi_{CO,a} e^{\frac{\alpha_{ox2}^+ n_{ox2}(-\Phi_a^L - \Delta\Phi_{ox2}^0)}{\theta_s}} \right.$$

$$\left. - \chi_{CO_2,a}^2 e^{\frac{-(1-\alpha_{ox2}^+)n_{ox2}(-\Phi_a^L - \Delta\Phi_{ox2}^0)}{\theta_s}} \right].$$

$$r_{red} = e^{Arr_{red}\left(\frac{1}{\theta_{red}^0} - \frac{1}{\theta_c}\right)} \left[\chi_{CO_2,c}^{-2} e^{\frac{\alpha_{red}^+(U_{cell} - \Phi_c^L - \Delta\Phi_{red}^0)}{\theta_c}} \right.$$

$$\left. - \chi_{O_2,c}^{0.75} \chi_{CO_2,c}^{-0.5} e^{\frac{-\alpha_{red}^-(U_{cell} - \Phi_c^L - \Delta\Phi_{red}^0)}{\theta_c}} \right].$$

$$K_{ref1}(\theta_a) = \exp(30.19 - 90.41/\theta_a), \quad K_{ref2}(\theta_a) = \exp(-3.97 + 14.57/\theta_a),$$
$$\Delta\Phi_{ox1}^0(\theta_s) = 28.26 - 19.84\theta_s, \quad \Delta\Phi_{ox2}^0(\theta_s) = 20.98 - 17.86\theta_s,$$
$$\Delta\Phi_{red}^0(\theta_s) = 78.00 - 23.06\theta_s.$$

Catalytic combustor and reversal chamber:

$$\gamma_{air} = \gamma_a|_{z=0} \frac{\lambda_{air}}{\chi_{O_2,a}|_{z=0}} \left(2\chi_{CH_4,a}|_{z=0} + 0.5\chi_{CO,a}|_{z=0} + 0.5\chi_{H_2,a}|_{z=0}\right)$$
$$\gamma_{back} = R_{back}(t)\gamma_c|_{z=0}$$

In the following equations θ_a, χ_a, γ_a are abbreviations for $\theta_a|_{z=1}, \chi_a|_{z=1}, \gamma_a|_{z=1}$.

$$\gamma_b = \gamma_a \left(1 - \frac{1}{2}\chi_{H_2,a} - \frac{1}{2}\chi_{CO,a}\right) + \gamma_{air} + \gamma_{back},$$

$$\chi_{CH_4,b} = \chi_{H_2,b} = \chi_{CO,b} = 0,$$

$$\chi_{H_2O,b} = \frac{\gamma_a}{\gamma_b}(2\chi_{CH_4,a} + \chi_{H_2O,a} + \chi_{H_2,a}) + \frac{\gamma_{back}}{\gamma_b}\chi_{H_2O,c}|_{z=0},$$

$$\chi_{CO_2,b} = \frac{\gamma_a}{\gamma_b}(\chi_{CH_4,a} + \chi_{CO,a} + \chi_{CO_2,a}) + \frac{\gamma_{back}}{\gamma_b}\chi_{CO_2,c}|_{z=0},$$

$$\chi_{O_2,b} = \frac{\gamma_{air}}{\gamma_b}\chi_{O_2,air} - \frac{\gamma_a}{\gamma_b}\left(2\chi_{CH_4,a} + \frac{1}{2}\chi_{H_2,a} + \frac{1}{2}\chi_{CO,a}\right) + \frac{\gamma_{back}}{\gamma_b}\chi_{O_2,c}|_{z=0},$$

$$\chi_{N_2,b} = \frac{\gamma_{air}}{\gamma_b}(1 - \chi_{O_2,air}) + \frac{\gamma_{back}}{\gamma_b}\chi_{N_2,c}|_{z=0},$$

$$\theta_b = \theta_u + \frac{\gamma_a}{\gamma_b}\left(\sum_{i\in I}\frac{-\Delta_c h_i^0}{c_p}\chi_{i,a} + \theta_a - \theta_u\right) + \frac{\gamma_{air}}{\gamma_b}(\theta_{air} - \theta_u)$$

$$+ \frac{\gamma_{back}}{\gamma_b}(\theta_c|_{z=0} - \theta_u),$$

$$f_{8,i} = \gamma_b(\chi_{i,b} - \theta_{i,m})\theta_m,$$

$$f_9 = \gamma_b(\theta_b - \theta_m)\theta_m - \frac{St_m}{c_p}(\theta_m - \theta_u)\theta_m,$$

$$f_{10} = \gamma_b + \gamma_b\frac{\theta_b - \theta_m}{\theta_m} - \frac{St_m}{c_p}\frac{\theta_m - \theta_u}{\theta_m}.$$

Non-zero boundary conditions:

$$\chi_{CH_4,a} = 1/3.5, \ \chi_{H_2O,a} = 2.5/3.5, \ \theta_a = 3, \ \gamma_a = 1, \ R_{back} = 0.$$

Non-zero initial conditions at $t = 0$:

$$\chi_{CH_4,a,0} = 1/3.5, \ \chi_{H_2O,a,0} = 2.5/3.5, \ \chi_{H_2O,c,0} = 0.2, \ \chi_{CO_2,c,0} = 0.1,$$
$$\chi_{O_2,c,0} = 0.1, \quad \chi_{N_2,c,0} = 0.6, \quad \theta_{a,0} = 3, \quad \theta_{c,0} = 3,$$
$$\theta_{m,0} = 3, \quad \theta_{s,0} = 3.1, \quad \gamma_{a,0} = 1, \quad \gamma_{c,0} = 6,$$
$$\Phi_{a,0}^S = 32.3, \quad \Phi_{c,0}^S = 32.3, \quad U_{cell,0} = 32.3, \ \chi_{H_2O,m,0} = 0.2,$$
$$\chi_{CO_2,m,0} = 0.1, \quad \chi_{O_2,m,0} = 0.1, \quad \chi_{N_2,m,0} = 0.6.$$

3 Numerical Methods and Results

A widely used idea is to transform partial differential equations into a system of ordinary differential algebraic equations by discretizing the model functions subject to the spatial variable z. This approach is known as the numerical method of lines (MOL), see for example Schiesser [12]. We define a uniform grid of size n_g and get a discretization of the whole space interval from $z = 0$ to $z = 1$. To approximate the first and second partial derivatives of the state

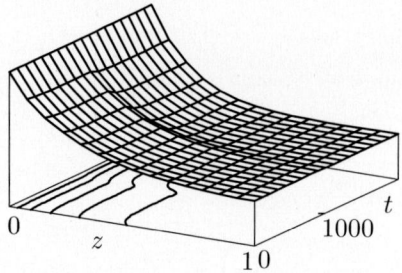

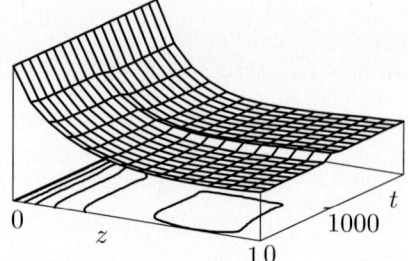

Fig. 2. Molar fraction $\chi_{CH_4,a}$ in anode gas channel

Fig. 3. Molar fraction $\chi_{H_2O,a}$ in anode gas channel

variables subject to the spatial variable at a given point z_k, $k = 1, \ldots, n_g$, several different alternatives have been implemented in the code PDEFIT, see Schittkowski [14] for more details, which is applied for the numerical tests of this paper.

Spatial derivatives of the heat variable $\theta_s(z, t)$ are approximated by a three-point-difference formulae for first derivatives, which is recursively applied to get second derivative approximations. The difference formulae are adapted at the boundary to accept given function and gradient values. First derivatives of the remaining transport variables are approximated by simple forward and backward differences, so-called upwind formulae, where the wind direction is known in advance. Ordinary differential and algebraic equations are added to the discretized system without any further modification.

In case of algebraic partial or ordinary differential equations, boundary conditions have to satisfy the algebraic equations. Consistent initial values are computed internally proceeding from given starting values for the nonlinear programming algorithm NLPQL of Schittkowski [13]. It is known that the system of PDAEs possesses the index one.

The method of lines leads to a large system of differential algebraic equations, where the size depends on the number of grid points, i.e., on n_g. The system equations are integrated by the implicit code RADAU5, see Hairer and Wanner [6]. Because of the non-continuous input variable

$$I_{cell}(t) := \begin{cases} 3 \, , & \text{if } t \leq 1000 \\ 3.5 \, , & \text{if } t > 1000 \end{cases} ,$$

the cell current, the right-hand side of some equations become non-continuous subject to integration time. Thus, it is necessary to restart the integration of the DAE at $t = 1000$.

Because of a complex input structure, the code PDEFIT is called from a GUI called EASY-FIT, see Schittkowski [15, 16], to facilitate modeling, execution, and interpretation of results. Model functions are interpreted based on a language called PCOMP similar to Fortran, see Schittkowksi [17]. To give

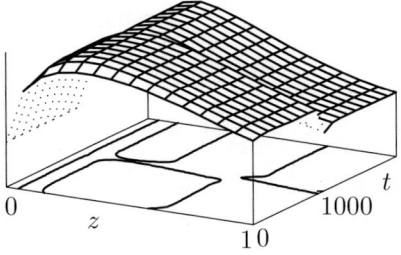

Fig. 4. Molar fraction $\chi_{H_2,a}$ in anode gas channel

Fig. 5. Molar fraction $\chi_{CO,a}$ in anode gas channel

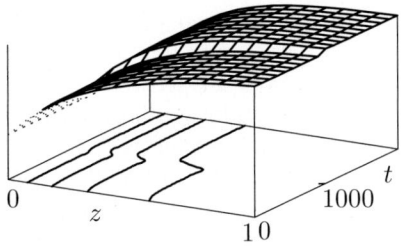
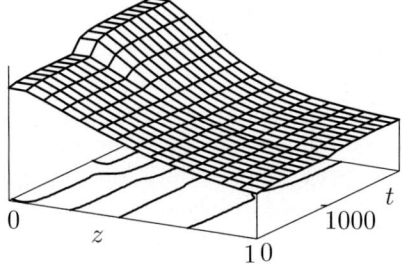

Fig. 6. Molar fraction $\chi_{CO_2,a}$ in anode gas channel

Fig. 7. Molar fraction $\chi_{H_2O,c}$ in cathode gas channel

an example, consider the parabolic heat equation which is implemented in the form

```
*        FUNCTION Ts_t
         sigmaTred = (-hrred + nred *(phicS - phicL))*Dared*rred
         sigmaTox  = (-hrox1 + nox1*(phiaS - phiaL))*Daox1*rox1
/                  + (-hrox2 + nox2*(phiaS - phiaL))*Daox2*rox2
         qsolid    = sigmaTox + sigmaTred - (phicL - phiaL)/F
         hsc       = cpm*(-ncsCO2 - ncsO2)*(Tc - Ts)
         hsa       = cpm*(-nasH2 - nasCO2)*(Ta - Ts)
         qcond     = Ts_zz*12/Pes
         Ts_t      = (qcond + hsa + hsc - qas - qcs + qsolid)/cps
```

Here, `Ts_zz` denotes the second partial derivative of $\theta_s(z,t)$ subject to the spatial variable z.

The MOL is applied with $n_g = 15$ lines. For our numerical tests, the absolute stopping tolerance of RADAU5 is $\epsilon = 10^{-5}$. The DAE is integrated from $t = 0$ to $t = 2000$ with a break at $t = 1000$.

The numerical solution of some selected components is shown in Figs. 2–13 to illustrate the qualitative behavior of the state variables.

Note that at $t \approx 800$ the stationary solution for $I_{cell} = 0.3$ is reached. At $t = 1000$ the load change happens. At $t \approx 1800$ the new stationary solution

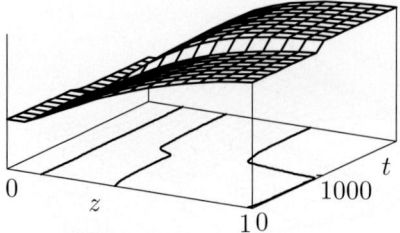

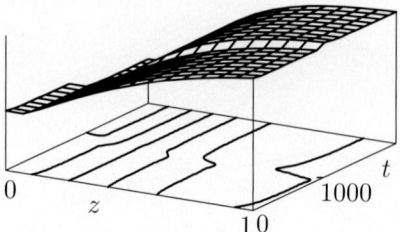

Fig. 8. Molar fraction $\chi_{CO_2,c}$ in cathode gas channel

Fig. 9. Molar fraction $\chi_{O_2,c}$ in cathode gas channel

for $I_{cell} = 0.35$ is reached. The molar fraction of methane and water depicted in Figs. 2 and 3 decrease over the entire spatial width due to the methane and water consuming endothermic reforming reaction. In accordance with the methane behavior, hydrogen is produced in the endothermic reforming reaction immediately after the anode inlet, to become subsequently consumed in the exothermic oxidation reaction, see Fig. 4. At $t = 1000$ the load change takes place. More electrones are needed, therefore due to (ox1) the molar fraction of hydrogen decreases (Fig. 4). In Figs. 5 and 6, the behavior of the molar fractions of carbone monoxide and dioxide are plotted. The molar fraction of water increases in flow direction of the cathode gas channel, Fig. 7, whereas the molar fractions of carbon dioxide and oxygen decrease, see Figs. 8 and 9.

The most important results concern the temperatures shown in Figs. 10 to 11. They directly influence the reaction rates. Moreover, the solid temperature of Fig. 12 must be particularly observed to avoid so-called hot spots leading to material corrosion and consequently to a reduced service life. The temperature distribution in the anode channel coincides with the heat demand and the heat release of the reactions therein, see Fig. 10. Initially we have the endothermic reforming reaction and thus, the temperature declines. Subsequently, the anode gas temperature is increased again by heat exchange with the solid, which is heated by the electro chemical reactions. Since the cathode gas is heated up by the solid phase along the channel, the temperature continuously increases, see Fig. 11. Finally, the current density i_a is plotted in Fig. 13. i_a is almost similar to i_e and i_c.

4 Conclusions

A complex mathematical model describing realistically the dynamical behavior of a molten carbonate fuel cell, has been presented. The semi-discretisation in space of the large scale partial differential-algebraic equation system together with its nonstandard boundary conditions including an Integro-DAE system yields a large system of differential-algebraic equations by the method

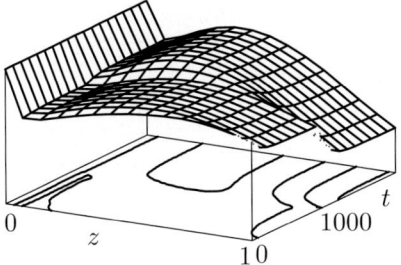

Fig. 10. Temperature θ_a in anode gas channel

Fig. 11. Temperature θ_c in cathode gas channel

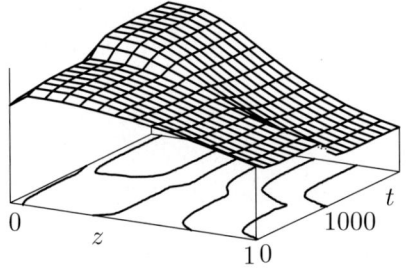

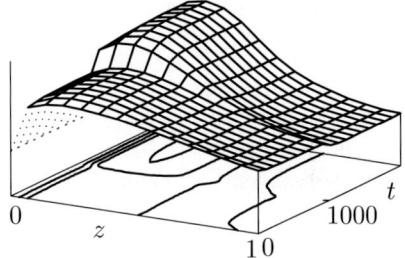

Fig. 12. Temperature θ_s in solid

Fig. 13. Current density i_a

of lines. The obtained numerical results correspond to the practical experiences of engineers with real fuel cells of the type investigated. The model will be used in future for optimal boundary control purposes.

Acknowledgement. The research project was partly funded by the German Federal Ministry of Education and Research within the WING-project "Optimierte Prozessführung von Brennstoffzellensystemen mit Methoden der Nichtlinearen Dynamik". We are indebted to Dr.-Ing. P. Heidebrecht and Prof. Dr.-Ing. K. Sundmacher from the Max-Planck-Institut für Dynamik Komplexer Technischer Systeme, Magdeburg, for providing us with the fuel cell model, to Dr. K. Sternberg from the University of Bayreuth for her advice and to Dipl.-Ing. J. Berndt and Dipl.-Ing. M. Koch from the management of IPF Heizkraftwerksbetriebsges. mbH, Magdeburg, for their support.

References

[1]   M. Bauer, *1D-Modellierung und Simulation des dynamischen Verhaltens einer Schmelzcarbonatbrennstoffzelle*, Diplomarbeit, Universität Bayreuth, 2006.

[2]   M. Bischoff, G. Huppmann, *Operating Experience with a 250 kW_{el} Molten Carbonate Fuel Cell (MCFC) Power Plant*, Journal of Power Sources, **105**, 2 (2002), 216–221.

[3]   K. Chudej, P. Heidebrecht, V. Petzet, S. Scherdel, K. Schittkowski, H.J. Pesch, K. Sundmacher, *Index Analysis and Numerical Solution of a Large Scale Nonlinear PDAE System Describing the Dynamical Behaviour of Molten Carbonate Fuel Cells,* Z. angew. Math. Mech., **85**, 2 (2005), 132–140.

[4]   K. Chudej, K. Sternberg, H.J. Pesch, *Simulation and Optimal Control of Molten Carbonate Fuel Cells,* In: I. Troch, F. Breitenecker (Eds.), Proceedings 5th Mathmod Vienna, Argesim Report No. 30, Argesim Verlag, Wien, 2006.

[5]   M. Gundermann, P. Heidebrecht, K. Sundmacher, *Validation of a Mathematical Model Using an Industrial MCFC Plant,* Journal of Fuel Cell Science and Technology **3** (2006), 303-307.

[6]   E. Hairer, G. Wanner, *Solving Ordinary Differential Equations II, Stiff and Differential-Algebraic Problems,* Springer, Berlin, 1996, 2nd. rev. ed.

[7]   P. Heidebrecht, *Modelling, Analysis and Optimisation of a Molten Carbonate Fuel Cell with Direct Internal Reforming (DIR-MCFC),* VDI Fortschritt Berichte, Reihe 3, Nr. 826, VDI Verlag, Düsseldorf, 2005.

[8]   P. Heidebrecht, K. Sundmacher, *Dynamic Modeling and Simulation of a Countercurrent Molten Carbonate Fuel Cell (MCFC) with Internal Reforming,* Fuel Cells **3–4** (2002), 166–180.

[9]   P. Heidebrecht, K. Sundmacher, *Molten carbonate fuel cell (MCFC) with internal reforming: model-based analysis of cell dynamics,* Chemical Engineering Science **58** (2003), 1029–1036.

[10]   P. Heidebrecht, K. Sundmacher, *Dynamic Model of a Cross-Flow Molten Carbonate Fuel Cell with Direct Internal Reforming.* Journal of the Electrochemical Society, **152** (2005), A2217–A2228.

[11]   S. Rolf, *Betriebserfahrungen mit dem MTU Hot Module,* In: Stationäre Brennstoffzellenanlagen, Markteinführung, VDI Berichte, Nr. 1596, VDI Verlag, Düsseldorf, 2001, 49–57.

[12]   W.E. Schiesser, *The Numerical Method of Lines,* Academic Press, New York, London, 1991.

[13]   K. Schittkowski, *NLPQL: A FORTRAN Subroutine Solving Constrained Nonlinear Programming Problems,* Annals of Operations Research, **5** (1985/86), 485–500.

[14]   K. Schittkowski, *PDEFIT: A FORTRAN code for parameter estimation in partial differential equations,* Optimization Methods and Software, **10** (1999), 539–582.

[15]   K. Schittkowski, *Numerical Data Fitting in Dynamical Systems - A Practical Introduction with Applications and Software,* Kluwer Academic Publishers, Dordrecht, Boston, London, 2002.

[16]   K. Schittkowski, *EASY-FIT: A Software System for Data Fitting in Dynamic Systems,* Structural and Multidisciplinary Optimization, **23** (2002), 153–169.

[17]   K. Schittkowski, *PCOMP: A Modeling Language for Nonlinear Programs with Automatic Differentiation,* in: Modeling Languages in Mathematical Optimization, J. Kallrath ed., Kluwer Academic Publishers, 2004, 349–367.

[18]   K. Schittkowski, *Data Fitting in Partial Differential Algebraic Equations: Some Academic and Industrial Applications,* Journal of Computational and Applied Mathematics **163** (2004), 29–57.

[19]   W. Winkler, *Brennstoffzellenanlagen,* Springer, Berlin, 2002.

Rigid Registration of Medical Images by Maximization of Mutual Information

Rainer Lachner

BrainLAB AG, Kapellenstr. 12, 85622 Feldkirchen, Germany
`rainer.lachner@brainlab.com`

Summary. Multi-modal medical image registration is an important capability for surgical applications. The objective of registration is to obtain a spatial transformation from one image to another by which a similarity measure is optimized between the images. Recently, new types of solutions to the registration problem have emerged, based on information theory. In particular, the mutual information similarity measure has been used to register multi-modal medical images. In this article, a powerful, fully automated and highly accurate registration method developed at BrainLAB AG is described. Key components of image registration techniques are identified and a summary presented of the information-theoretical background that leads to the mutual information concept. In order to locate the maximum of this measure, a dedicated optimization method is presented. A derivative-free algorithm based on trust regions specially designed for this application is proposed. The resulting registration technique is implemented as part of BrainLAB's commercial planning software packages BrainSCAN^{TM} and iPlan$^{\circledR}$.

1 Introduction

During the last decades many radiology departments have installed imaging equipment for three-dimensional image modalities such as Computed Tomography (CT), Magnetic Resonance Imaging (MRI), Positron Emission Tomography (PET), and Single Photon Emission Computed Tomography (SPECT), to name just a few. These systems have fundamentally changed the way many diseases are diagnosed and even treated. Unfortunately, the full power of these systems has not yet been realized. The software and knowledge needed to take full advantage of the imaging systems has not kept pace with the development of the hardware.

Since information obtained from different images acquired in the clinical track of events is usually of complementary nature, proper integration of useful data gained from the separate image modalities is often desired. A first step in the integration process is to bring the relevant images into spatial alignment, a procedure commonly referred to as *registration*. To register images,

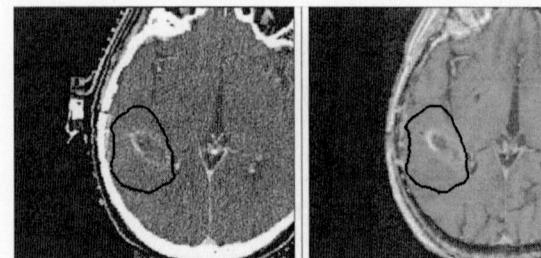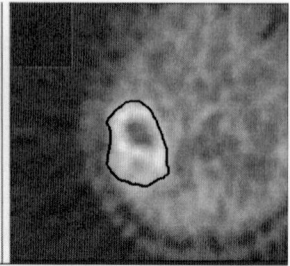

Fig. 1. Malignant tumor (glioblastoma in left temporal lobe) two weeks after surgery. Displayed are corresponding transaxial cuts through the image volumes with region outlined for irradiation. CT (left), MRI (middle), ^{11}C-Methionine-PET (right). Note the poor representation of anatomic structures in the PET image. Using only CT and MRI, residual tumor can not be distinguished from postoperative blood-brain-barrier disturbances. With CT/MRI/PET registration, the residual tumor can be defined with high accuracy

the geometrical relationship between them must be determined. Registration of all the available patient image data allows for better data interpretation, increases diagnostic potential and improves the planning and evaluation of surgery and therapy.

Radiotherapy treatment planning is an example where the registration of different modalities greatly enhances therapy planning. Currently, CT is often used exclusively. However, the combined use of MRI and CT would be beneficial, as the former is better suited for delineation of tumor tissue (and has in general better soft tissue contrast), while the latter is needed for accurate computation of the radiation dose. The usefulness can be enhanced even further when taking PET into account. PET highlights metabolic processes in the tumor and is invaluable if a clear tumor boundary can not be found from CT or MRI alone (see Fig. 1).

2 Medical Imaging and Registration Characteristics

Pereant qui ante nos nostra dixerunt. (To the devil with those who published before us) - Aelius Donatus

Medical images arise from image sensors, e.g., CT- or MRI scanners, which detect properties of the patient being imaged. To create an image, values of the detected properties are mapped into intensity, or alternatively color, scales.

An intensity (i.e. greyscale) image can be interpreted as a function $f(x)$, where x are the spatial coordinates, and the value of f at the point x is the greyscale value of the image at that point. Since we are only concerned with digital images, x and f are discrete quantities. Volumetric (3D) images are composed of a finite number of volume elements, or *voxels*, located at equidistant points x in the image array. The discrete voxel grid V_f is a subset of the (usually cuboid) image domain Ω_f. The range of f is the range of allowable intensity values in the image. Medical images are commonly encoded at 8 or 16 bits per voxel.

Suppose that there are two images, A and B of an object O. The aim of image registration is to find the spatial transformation T, which best aligns the position of features in image A to the position of the corresponding features in image B. The three images in Fig. 1 are registered to each other. Note how corresponding anatomical structures are aligned.

Since medical images typically have different fields of view, their domains Ω_A and Ω_B are different. For an object O, imaged by both A and B, a position x is mapped to x_A by image A and to x_B by image B. Registration finds the transformation T, which maps x_A to x_B over the region of overlap $\Omega_{A,B}^T$, defined as

$$\Omega_{A,B}^T := \{x_A \in \Omega_A : T(x_A) \in \Omega_B\} . \tag{1}$$

Solving the registration problem is a surprisingly difficult task, especially when high accuracy and robustness with minimal user interaction is required. It boils down to an optimization problem of a given correspondence criterion in a transformation space.

Numerous registration methods have been proposed in the literature. They can be roughly divided into the three classes *point based, segmentation based,* and *grey value based.* For a more thorough survey of medical image registration techniques, see, e.g., [5]. Nevertheless, nearly all image registration methods have the following key components in common.

- *images*
  The images A and B involved in the registration process are often called *reference image* and *floating image*, respectively. Technically, they are not equivalent since to compare them, the floating image must be transformed to the voxel grid of the reference image. In Fig. 1, the floating images MRI and PET are transformed and resampled with respect to CT coordinates.
- *search space*
  The most fundamental characteristic of any image registration technique is the type of spatial transformation T needed to properly align the images. The search space contains all admissible geometric transformations T applied to the floating image. If images of the same person are to be registered, rigid transformations are almost always sufficient.
- *feature space*
  The feature space is the representation of the data that will be used for registration. The choice of features determines what is matched. The set of features may consist of the images itself, but other common feature spaces include: edges, contours, surfaces, other salient features such as corners, line intersections, and points of high curvature, statistical features such as moment invariants, and higher level structural descriptions.

- *similarity measure*
  The similarity measure determines how matches are rated. This is closely related with the selection of the feature space. But, while the feature space is precomputed on the images before matching, the similarity measure is calculated using these features for particular transformations T from the search space. A typical similarity measure is *mutual information* which will be introduced in the next section. Given the search space of possible transformations, the similarity measure is used to find the parameters of the final registration transformation.
- *search method*
  For most feature spaces, finding the transformation which yields the best value of the similarity measure requires an iterative search method. This is almost always implemented as a more-or-less sophisticated optimization algorithm.

3 Information Theory and Statistics

> *He uses statistics as a drunken man uses lamp-posts – for support rather than illumination.* - Andrew Lang

Communication theory inspired the desire for a measure of information of a message broadcasted from a sender to a receiver. Shannon [11] focused on characterizing information for communication systems by finding ways of measuring data based on the uncertainty or randomness present in a given system. He proved that for events e_i occurring with probabilities p_i,

$$ H := \sum_i p_i \log \frac{1}{p_i} $$

is the only functional form that satisfies all the conditions that a measure of uncertainty should satisfy. He named this quantity *entropy* since it shares the same form as the entropy of statistical mechanics. The term $\log \frac{1}{p_i}$ signifies that the amount of information gained from an event with probability p_i is inversely related to the probability that the event takes place. The rarer an event, the more meaning is assigned to its occurrence. The information per event is weighted by the probability of occurrence. The resulting entropy term is the average amount of information to be gained from a certain set of events.

3.1 Entropy and Images

In digital imaging, an intensity image X is represented as an array of intensity values. For an image encoded at n bits per voxel, the intensity values are the discrete greyscale values $\mathcal{X} = \{0, 1, \ldots, N - 1\}$, with $N = 2^n$.

In order to apply information theory to digital imaging, we must consider an image as a collection of independent observations of a *random variable*.

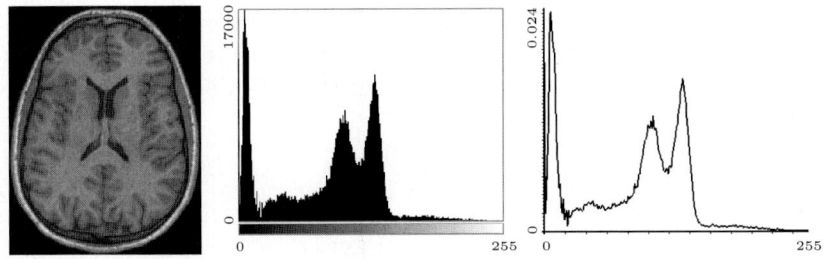

Fig. 2. A 2D image (left), the associated histogram (middle) and probability density estimate (right). Note the peaks in the histogram for the dark background and for the brain tissue types "cerebrospinal fluid", "grey matter", and "white matter"

Given an image X, we use p_X to denote the corresponding intensity value probability distribution function, where $p_X(x) := P(X_i = x)$, for $x \in \mathcal{X}$ and X_i the greyscale value of voxel i in image X. Note that X represents both the image and the random variable that determines the image. Furthermore, since the sample space $\mathcal{X}$ contains only discrete quantities, the random variable X is discrete. An image composed of M voxels is therefore an array of the elements $x \in \mathcal{X}$ as determined by M independent observations of the discrete random variable X.

Definition 1. *For a discrete random variable X with probability density function p_X, the entropy $H(X)$ is defined as*[1]

$$H(X) = H(p_X) := - \sum_{x \in \mathcal{X}} p_X(x) \log p_X(x) \ .$$

In order to actually calculate the entropy of a given image X, we must guess the probability density function p_X of the underlying random variable. Since p_X is usually unknown, we make use of the image data itself to estimate a maximum-likelihood probability density function. This estimation is based on the image histogram[2] and is obtained by dividing each histogram entry by the total number of voxels.

3.2 Joint Entropy and Mutual Information

Following the notation of the last section, we now consider two images (of identical size) to be observations of two discrete random variables, X and Y, with probability distributions p_X and p_Y, and sample spaces $\mathcal{X}$ and $\mathcal{Y}$, respectively.

---

[1] Here, log means $\log_2$ so that entropy is measured in bits. For reasons of continuity, we define $0 \log 0 = 0$.

[2] A histogram can be constructed from an image by counting the frequency of voxels in each intensity bin.

Joint entropy is a measure of the combined randomness of the discrete random variables X and Y. It is a simple extension of entropy since the pair (X, Y) of random variables may be considered a single vector-valued random variable.

Definition 2. *For discrete random variables X and Y with joint probability distribution p_{XY}, the joint entropy $H(X, Y)$ is defined as*

$$H(X, Y) = H(p_{XY}) := -\sum_{x \in \mathcal{X}} \sum_{y \in \mathcal{Y}} p_{XY}(x, y) \log p_{XY}(x, y) .$$

According to

$$\sum_{x \in \mathcal{X}} p_{XY}(x, y) = p_Y(y) \quad \text{and} \quad \sum_{y \in \mathcal{Y}} p_{XY}(x, y) = p_X(x) , \tag{2}$$

the image distributions p_X and p_Y are related to the joint distribution p_{XY}, and in this respect are termed the *marginals* of the joint distribution.

Similar to the one-dimensional case, the joint probability distribution can be estimated from the *joint histogram* (see Fig. 3). The joint histogram is a two-dimensional array of size $\text{card}(\mathcal{X}) \times \text{card}(\mathcal{Y})$, where the value at position (x, y) is the number of voxel pairs with intensity x in image X and intensity y at the same location in image Y.

Lemma 1. *For two independent random variables X and Y, the joint entropy simplifies to $H(X, Y) = H(X) + H(Y)$.*

Proof. This can be directly deduced from Def. 2 and (2), using $p_{XY}(x, y) = p_X(x) p_Y(y)$ for independent random variables.

Relative entropy, or Kullback-Leibler distance [4], is a measure of the distance between one probability distribution and another. Typically, it measures the error of using an estimated distribution q over the true distribution p.

Definition 3. *The relative entropy $D(p||q)$ of two probability distribution functions p and q over $\mathcal{X}$ is defined as*

$$D(p||q) := \sum_{x \in \mathcal{X}} p(x) \log \frac{p(x)}{q(x)} .$$

A special case of relative entropy is *mutual information*. It measures the amount of information shared between two random variables and can be regarded as the distance between their joint probability distribution and the distribution assuming complete independence.

Definition 4. *Let X and Y be two discrete random variables with probability distributions p_X and p_Y, respectively, and joint probability distribution p_{XY}. Mutual information $I(X, Y)$ is the Kullback-Leibler distance between p_{XY} and the product distribution $p_X p_Y$. That is,*

$$I(X, Y) := D(p_{XY}||p_X p_Y) = \sum_{x \in \mathcal{X}} \sum_{y \in \mathcal{Y}} p_{XY}(x, y) \log \frac{p_{XY}(x, y)}{p_X(x) p_Y(y)} .$$

3.3 Properties of Mutual Information

In order to get a better understanding of mutual information, some of its properties are presented below. Additional to the non-negativeness of entropy, it can also be shown that relative entropy is a non-negative quantity[3].

Lemma 2. *Let p and q be two probability distribution functions over $\mathcal{X}$, then $D(p\|q) \geq 0$ and equality holds if and only if $p(x) = q(x)$ for all $x \in \mathcal{X}$.*

Proof. Since $\log_2 a = \ln a / \ln 2$ it is sufficient to prove the statement using the natural logarithm. Let $A = \{x \in \mathcal{X} : p(x) > 0\}$ be the support of p. Then

$$-D(p\|q) = -\sum_{x \in A} p(x) \ln \frac{p(x)}{q(x)} = \sum_{x \in A} p(x) \ln \frac{q(x)}{p(x)} \leq \sum_{x \in A} p(x) \left(\frac{q(x)}{p(x)} - 1 \right)$$

$$= \sum_{x \in A} q(x) - \sum_{x \in A} p(x) \leq \sum_{x \in \mathcal{X}} q(x) - \sum_{x \in A} p(x) = 1 - 1 = 0$$

where we have used $\ln a \leq a - 1$ and equality holds only if $a = 1$, i.e., $\frac{q(x)}{p(x)} = 1$ for all $x \in A$. In this case, equality continues to hold for the second $\leq$-sign.

Lemma 3. *Mutual Information satisfies the following properties:*

(a) $I(X, Y) = I(Y, X)$
   Mutual information is symmetric.
(b) $I(X, X) = H(X)$
   The information image X contains about itself is equal to the information (entropy) of image X.
(c) $I(X, Y) \leq \min(H(X), H(Y))$
   The information the images contain about each other can never be greater than the information in the images themselves.
(d) $I(X, Y) \geq 0$
   The uncertainty about X can not be increased by learning about Y.
(e) $I(X, Y) = 0$ *if and only if X and Y are independent.*
   When X and Y are not related in any way, no knowledge is gained about one image when the other is given.
(f) $I(X, Y) = H(X) - H(X|Y)$
   Mutual information measures the reduction in the uncertainty of X due to the knowledge of Y.
(g) $I(X, Y) = H(X) + H(Y) - H(X, Y)$
   Mutual information equals information of the individual images minus joint information.
(h) $I(X, X) = I(X, g(X))$ *if $g : \mathcal{X} \mapsto \mathcal{X}$ is an injective mapping.*
   This is the justification of mutual information as a similarity measure.

---

[3] Note that although relative entropy is a distance measure, it is not a metric since it is not symmetric and does not satisfy the triangle inequality.

The term "mutual information" is best explained by (f). Here, $H(X|Y) = -\sum_x \sum_y p_{XY}(x,y) \log p_X(x|y)$ denotes the conditional entropy which is based on the conditional probabilities $p_X(x|y) = p_{XY}(x,y)/p_Y(y)$, the chance of observing grey value x in image X given that the corresponding voxel in Y has grey value y. When interpreting entropy as a measure of uncertainty, (f) translates to "the amount of uncertainty about image X minus the uncertainty about X when Y is known". In other words, mutual information is the amount by which the uncertainty about X decreases when Y is given: the amount of information Y contains about X. Because X and Y can be interchanged, see (a), $I(X,Y)$ is also the information X contains about Y. Hence it is *mutual* information.

Form (g) is often used as a definition. It contains the term $-H(X,Y)$, which means that maximizing mutual information is related to minimizing joint entropy. The advantage of mutual information over joint entropy is that it includes the entropies of the individual images. Mutual information and joint entropy are computed for the overlapping part of the images and the measures are therefore sensitive to the size and contents of overlap. A problem that can occur when using joint entropy on its own as similarity measure, is that low values (normally associated with a high degree of alignment) can be found for complete misregistrations.

Why is mutual information at maximum when two images are aligned? When the images show the same object and are acquired by the same sensor, the answer is easy to see. From (b) and (c), it is obvious that $I(X,X)$ is an upper bound for mutual information. Hence, when an image is aligned with itself, mutual information is at a maximum. If the mapping between image intensities is injective, property (h) reveals that mutual information will still peak at alignment. So, the closer the image intensities are related by an injective mapping, the better mutual information will serve.

4 Rigid Registration Based on Mutual Information

The goal is to transform data into information, and information into insight. - Carly Fiorina

The introduction of mutual information as a similarity measure (see the pioneering articles [1] and [14]) has revolutionized image registration and has especially influenced the development of intensity-based image registration techniques. All these methods are based on the same four steps: superposition of reference image and floating image, estimation of their probability distribution function, calculation of mutual information, and optimization of mutual information. See [7] for a survey of the methods using the mutual information concept.

In this section we consider the mutual information between the reference image A and the transformed floating image B, computed in their region of

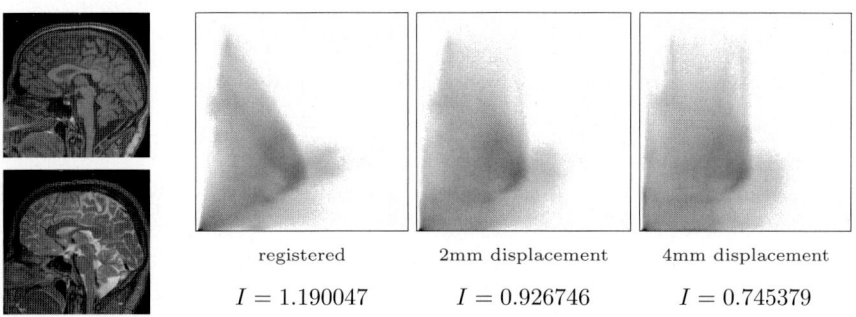

registered	2mm displacement	4mm displacement
$I = 1.190047$	$I = 0.926746$	$I = 0.745379$

Fig. 3. Joint histograms of an image pair displaced relative to each other. Left: Images involved. (The central slice of the image volumes is depicted. Note how different MRI pulse sequences highlight different tissue types.) Right: Joint Histograms and mutual information values for different displacements

overlap (see (1)). As we have seen, there are good reasons to assume that the mutual information value reaches a maximum if the images are correctly registered. Hence, mutual information based registration methods determine the transformation T^* from a class of admissible transformations $\mathcal{T}$ that makes mutual information maximal,

$$T^* = \arg\max_{T \in \mathcal{T}} I(A, B \circ T) \,. \tag{3}$$

The joint histogram of A and $B \circ T$, a key component for evaluation of I, responds very sensitively to different transformations. Figure 3 depicts the joint histogram of two images acquired by different MRI pulse sequences. For correct registration, the joint histogram reveals relatively sharp clusters with a high mutual information value. For increasing misregistration, the structures in the joint histogram get blurry and I decreases.

4.1 Rigid Transformations

Since we are concerned mainly with registration problems involving images of the human brain and since the shape of the brain changes very little with head movement, we use *rigid body transformations* to model different head positions of the same subject. In our implementation, the rigid body model is parameterized in terms of rotations around and translations along each of the three major coordinate axes,

$$\boldsymbol{x}_B = Q(\boldsymbol{x}_A) := R \cdot \boldsymbol{x}_A + \boldsymbol{t} \,. \tag{4}$$

Here, R denotes an orthogonal 3×3 rotation matrix, parameterized by the Euler angles ϕ, θ, and ψ (pitch, roll, yaw); and $\boldsymbol{t} = (t_x, t_y, t_z)^\top$ denotes a translation vector. Altogether, rigid transformations can be described by six parameters or degrees of freedom, combined to a parameter vector $\boldsymbol{\tau}$,

$$\tau := (\phi, \theta, \psi, t_x, t_y, t_z)^\top .$$

In the framework of rigid transformations, (3) translates to

$$\tau^* = \arg\max_{\tau} I\big(A(\boldsymbol{x}_A), B(R \cdot \boldsymbol{x}_A + \boldsymbol{t})\big) , \qquad \boldsymbol{x}_A \in V_A \cap \Omega_{A,B}^Q . \tag{5}$$

The computation of mutual information is restricted to voxel positions $\boldsymbol{x}_A$ in the region of overlap $\Omega_{A,B}^Q$. As the images have finite extent, this region will be of different size for different transformations, introducing mutual information discontinuities due to varying voxel numbers in $\Omega_{A,B}^Q$. In our implementation, we increase the domain Ω_B of the floating image to $\mathbb{R}^3$, either by adding voxels of the background intensity around the image, or by utilizing some intensity replication technique. Hence, due to $\Omega_{A,B}^Q = \Omega_A$, all the voxels in the reference image contribute to I, regardless of the transformation. This avoids discontinuities induced by non-constant voxel numbers.

4.2 Image Interpolation

To apply the concepts of the previous chapter, reference image and transformed floating image must be of identical size. Hence, we must resample B according to the voxel grid of A. For a given transformation Q, the transformed position $\boldsymbol{x}_B = R \cdot \boldsymbol{x}_A + \boldsymbol{t}$ in (5) will most likely not coincide with any of the discrete voxel positions in V_B. Therefore it is necessary to estimate the image intensity at $\boldsymbol{x}_B$ from the intensities at surrounding voxel positions. This process is called *image interpolation*.

Image interpolation reconstructs a continuous 3D signal $s(x, y, z)$ from its discrete samples $s_{i,j,k}$ with $s, x, y, z \in \mathbb{R}$ and $i, j, k \in \mathbb{N}$. Formally, this can be described as the convolution of a coefficient array with a continuous response function $h_{3D}(x, y, z)$,

$$s(x, y, z) = \sum_i \sum_j \sum_k c_{i,j,k} \cdot h_{3D}(x - i, y - j, z - k) . \tag{6}$$

The data-dependent discrete interpolation coefficients $c_{i,j,k}$ are derived from the image samples $s_{i,j,k}$ such that the resulting continuous signal interpolates the discrete one. To reduce the computational complexity, symmetric and separable interpolation kernels are used,

$$h_{3D}(x, y, z) = h(x) \cdot h(y) \cdot h(z) . \tag{7}$$

While numerous interpolation methods[4] have been proposed in the literature, three methods are most popular. These methods are *nearest neighbor interpolation*, *trilinear interpolation*, and *B-spline interpolation*. See Fig. 4 for a plot of their interpolation kernel functions h.

---

[4] For band-limited signals, it is known that the *sinc*-kernel provides the perfect interpolation method. As medical images are not band-limited, a perfect interpolation is not possible here.

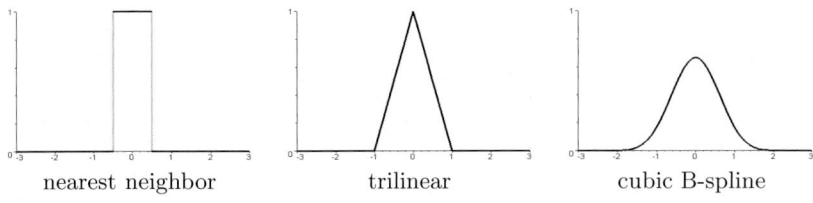

Fig. 4. Popular interpolation kernel functions h

Nearest neighbor is a rapid interpolation technique that requires no computations since each interpolated output voxel is directly assigned the intensity value at the nearest sample point in the input image. It has the advantage that no new image intensities are introduced. The process does not blur an image, but it may produce aliasing effects.

Trilinear interpolation provides a good trade-off between computational cost and interpolation quality. The interpolated intensity is a weighted sum of the intensities at eight surrounding voxels. For trilinear interpolation, the interpolation coefficients $c_{i,j,k}$ in (6) are equal to the image samples $s_{i,j,k}$, so no preprocessing is necessary.

Cubic B-spline interpolation offers a high-quality result for a moderate computational cost. The interpolated intensity is a weighted sum of the coefficients $c_{i,j,k}$ at 64 surrounding voxels. These coefficients must be calculated in advance, either by solving sparse systems of linear equations, or (as in our implementation) by recursive filtering of the array of image intensities. This elegant approach is described in [13]. It is more efficient and considerably faster that solving systems of linear equations.

Figure 5 depicts the performance of the three interpolation methods. A 2D image is repeatedly rotated by 6 degrees around the image center. After each rotation, the resulting interpolated image serves as input image for the next rotation step. After 60 interpolations a full turn is obtained, so ideally, the last interpolated image should be identical to the original input image. However, the results for the three interpolation methods are remarkably different. Nearest neighbor interpolation produces an unacceptable amount of artifacts. Bilinear interpolation (2D analogue of trilinear interpolation) preserves the image features better but exhibits substantial blurring. Cubic B-spline interpolation yields a superior result with minimal blurring.

4.3 Mutual Information Artifacts

One of the key aspects of mutual information based registration is the estimation of the joint probability density function of the reference image and the transformed floating image (see Def. 4). Usually, this estimate is based on the the joint histogram $\#h_{A,B}$,

$$p_{A,B}(i_A, i_B) = \frac{\#h_{A,B}(i_A, i_B)}{\sum_{i_A} \sum_{i_B} \#h_{A,B}(i_A, i_B)},$$

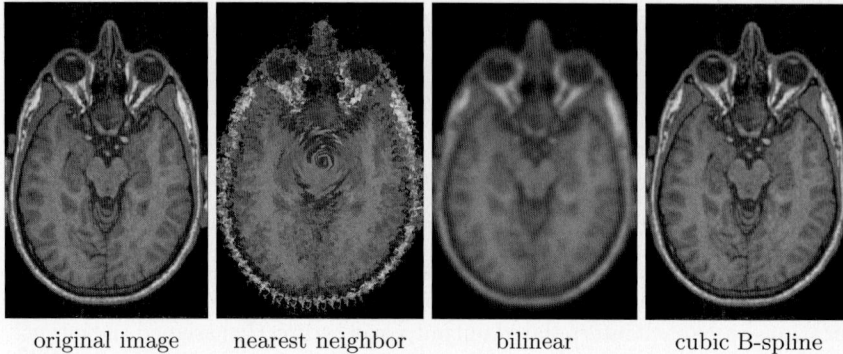

original image   nearest neighbor   bilinear   cubic B-spline

Fig. 5. A 2D image (left) and results of repeated interpolation with three different interpolation methods. See text for description

where i_A and i_B denote intensity levels in the reference image and in the transformed floating image, respectively.

It has been pointed out in [8] that when two images have equal sample spacing (voxel size) in one or more dimensions, existing joint probability estimation algorithms may result in certain types of spurious artifact patterns in a mutual information based registration function. These artifacts, the so-called *grid effect*, interfere with the "ideal" mutual information values. They may seriously inhibit numerical optimization of I and may have fatal impacts on registration accuracy. Hence, an analysis of the smoothness of mutual information is very important in an optimization context. Note that the grid effect is less pronounced for incommensurable voxel sizes and substantial rotations.

All the examples presented in this section are calculated from the MRI images depicted in Fig. 3. These images have identical voxel size (1.100 mm × 1.094 mm × 1.094 mm). The behavior of mutual information as a function of translation along the transverse axis is investigated, and depicted in Fig. 6a–e for various interpolation methods and joint histogram update schemes. For each scheme, mutual information values between A and translated B have been computed and plotted both far away and and in the vicinity of the maximum. Ideally, the curves should be smooth, monotonically increasing (far away) and bell-shaped (around the maximum), respectively.

- *nearest neighbor interpolation*
  Obviously, the nearest neighbor interpolation scheme produces a discontinuous, piecewise constant mutual information function, see Fig. 6a. Both the interpolated intensity and I have jumps at half-integer shifts values. As is known, optimization of a discontinuous function is not an easy task. Hence, nearest neighbor interpolation is rarely the interpolation method of choice, especially when high registration accuracy is desired.
- *trilinear interpolation*
  Both trilinear interpolation and and B-spline interpolation map to $\mathbb{R}$ rather

than to the discrete intensity level set of image B. Explicit calculation of the transformed floating image requires a rounding operation on the interpolated intensities. Using this image for joint histogram computation propagates the discontinuities to mutual information. Figure 6b depicts the situation for trilinear interpolation. Unlike the nearest neighbor case, discontinuities can occur anywhere. The result is a highly jagged mutual information function with many discontinuities and local maxima, a nightmare for any optimization method.

- *partial volume interpolation*
  The authors of [6] propose a very different approach to avoid rounding discontinuities that they call *partial volume interpolation*. It is specifically designed to update the joint histogram of two images, while no explicit transformed image needs to be built. Like trilinear interpolation, partial volume interpolation uses the eight neighbors of the transformed point x_B. However, instead of calculating a weighted average of the neighboring intensity values and incrementing only one joint histogram entry, partial volume interpolation updates several histogram entries at once: For each neighbor v with intensity i_v, the histogram entry $\#h(.,i_v)$ is incremented by its corresponding trilinear weight. This policy does not introduce spurious intensity values, ensures a gradual change in the joint histogram, and makes mutual information continuous. However, due to the additional histogram dispersion (or rather, due to the absence of this dispersion for grid-aligning transformations), a typical pattern of spiky wing-like artifacts emerges whose numerous local maxima at integer shifts might severely hamper the optimization process, especially far away from the optimum (see Fig. 6c).

- *Parzen windowing*
  Another method of updating the joint histogram is based on *Parzen windows*. Let $w : \mathbb{R} \mapsto \mathbb{R}$ be continuous and nonnegative with $\int_{-\infty}^{\infty} w(\xi)\,d\xi = 1$. Then we compute the discrete Parzen joint histogram as

$$\#h(i_A, i_B) = \sum_{x_A \in V_A} w(A(x_A) - i_A) \cdot w(B(T(x_A)) - i_B) . \qquad (8)$$

Basically, this approach was also used in [14], where the estimated probability density function is a superposition of Gaussian densities. The largest drawback is the infinite extension of the Gaussian which makes updating the joint histogram extremely costly. We prefer Parzen window functions w with finite support, e.g., the cubic B-spline kernel (see Fig. 4). The Parzen window method suffers less from grid effect induced artifacts than partial volume interpolation, nevertheless they are not completely suppressed. Figure 6d shows the resulting I for trilinear interpolation and Parzen windowing with a cubic B-spline kernel. The grid effect manifests itself as ridges in I.

- *random sampling*
  In our implementation, we use additional *random sampling* of the refer-

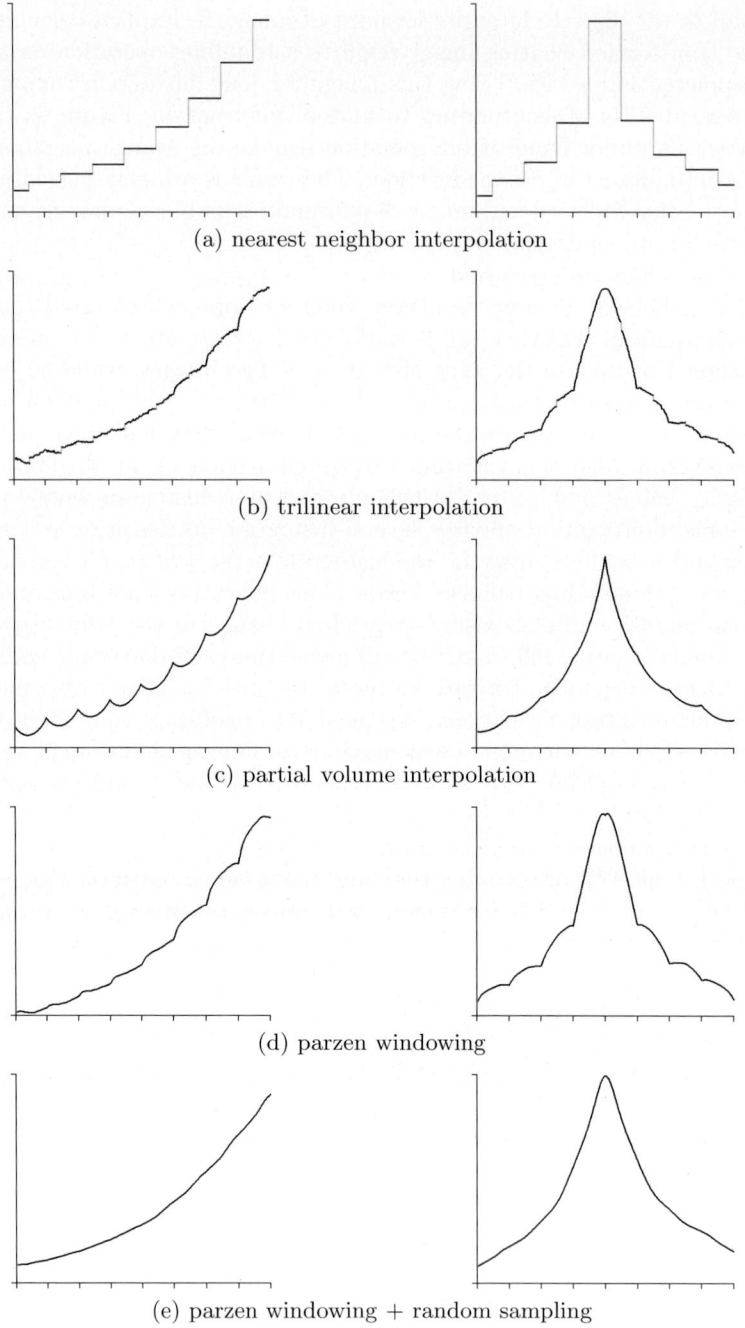

(a) nearest neighbor interpolation

(b) trilinear interpolation

(c) partial volume interpolation

(d) parzen windowing

(e) parzen windowing + random sampling

Fig. 6. Mutual Information as a function of translational misregistration for different interpolation and joint histogram update schemes, far away (left) and in the vicinity of the maximum (right)

ence image to get rid of the interpolation artifacts. Instead of calculating the joint histogram from the intensity values at voxel positions $x_A \in V_A$, we first perturb each x_A by a small random vector and interpolate the intensity at the perturbed position. The random perturbation and interpolation need only be performed once and is reused for each evaluation of mutual information. This effectively breaks the grid effect and frees mutual information almost completely from interpolation artifacts, see Fig. 6e.

5 Derivative-free Optimization

Optimization hinders evolution. - Anonymous

Solving the registration problem (5) requires the localization of the mutual information maximum. The non-differentiability of mutual information for, e.g., trilinear interpolation or partial volume interpolation prevents an application of conjugate-gradient or quasi-Newton optimization methods. Therefore, a derivative-free approach using only function evaluations seems natural.

In the context of image registration, Powell's method [9] is regularly proposed in the literature. Nevertheless, there are more efficient derivative-free algorithms. We implemented a *trust region method* based on the construction of quadratic models of an objective function F. Each model is defined by $\frac{1}{2}(n+1)(n+2)$ values of F, where n is the number of variables, six in our case. This model is assumed to approximate the objective function sufficiently well in a region called trust region. A typical iteration of the algorithm generates a new vector of variables x, either by minimizing the quadratic model subject to a trust region bound, or by improving the accuracy of the model. Then $F(x)$ is evaluated and one of the interpolation points is replaced by x. Explicit usage of Lagrange polynomials assists the procedure by improving the model and providing an error estimate of the quadratic approximation to F, which allows the algorithm to achieve a fast rate of convergence.

The user has to provide an initial vector of variables x_0, and initial and final values of the trust region radius ϱ. Typical distances between successive points at which F is evaluated are of magnitude ϱ, and ϱ is decreased if the objective function stops decreasing for such changes to the variables. Derivative estimates are given by a quadratic model

$$S(x) = \frac{1}{2}(x - x_0)^\top G\,(x - x_0) + g^\top (x - x_0) + c\,, \qquad x \in \mathbb{R}^n\,, \quad (9)$$

whose parameters are the $n \times n$ symmetric matrix G, the n-vector g, and the real number c. Hence the linear space of quadratic polynomials over $\mathbb{R}^n$ has dimension $m = \frac{1}{2}(n+1)(n+2)$ and the parameters of S are defined by a system of interpolation conditions of the from

$$S(x_i) = F(x_i)\,, \qquad i = 1, \ldots, m\,. \quad (10)$$

The model points $\boldsymbol{x}_i, i = 1, \ldots, m$ are generated automatically by the algorithm by taking into account the additional property that no nonzero quadratic polynomial vanishes at all of them.

The quadratic model is used in a sequence of constrained quadratic programming problems. Specifically, $\boldsymbol{d}^* \in \mathbb{R}^n$ is set to the vector of variables that solves the optimization problem

$$\boldsymbol{d}^* = \arg\max_{d} S(\boldsymbol{x}_k + \boldsymbol{d}) \quad \text{subject to} \quad ||\boldsymbol{d}|| \leq \varrho , \tag{11}$$

where $\boldsymbol{x}_k$ is the point in $\{\boldsymbol{x}_1, \ldots, \boldsymbol{x}_m\}$ such that $F(\boldsymbol{x}_k)$ is the greatest of the values $F(\boldsymbol{x}_i)$, $i = 1, \ldots, m$. Depending on its F-value, $\boldsymbol{x}_k + \boldsymbol{d}^*$ replaces one of the current model points.

Also required are the Lagrange polynomials of the interpolation problem (10). For $j = 1, \ldots, m$, the j-th Lagrange polynomial l_j is a quadratic polynomial on $\mathbb{R}^n$ such that

$$l_j(\boldsymbol{x}_i) = \delta_{ij} , \quad i = 1, \ldots, m .$$

Our implementation makes explicit use of first and second derivatives of all the Lagrange polynomials. They are also used to construct the model parameters G and $\boldsymbol{g}$ in (9).

The main benefit of the Lagrange polynomials is their ability to assess and improve the quality of the quadratic model. If one or more of the distances $||\boldsymbol{x}_k - \boldsymbol{x}_j||$, $j = 1, \ldots, m$ is greater than 2ϱ, where $\boldsymbol{x}_k$ is defined as in (11), we move the interpolation point $\boldsymbol{x}_j$ into the neighborhood of $\boldsymbol{x}_k$. This is accomplished by solving the auxiliary optimization problem

$$\boldsymbol{d}^* = \arg\max_{d} |l_j(\boldsymbol{x}_k + \boldsymbol{d})| \quad \text{subject to} \quad ||\boldsymbol{d}|| \leq \varrho , \tag{12}$$

and replacing $\boldsymbol{x}_j$ by $\boldsymbol{x}_k + \boldsymbol{d}^*$. This choice maintains and assists the nonsingularity of the interpolation problem (10). After the initial quadratic model has been formed, each vector of variables for a new value of F is generated by one of the two trust region subproblems in (11) and (12).

The ways of choosing between these alternatives, the details of adjustments to the trust region radii, and analysis of convergence are beyond the scope of this article. Descriptions of trust region methods can be found, e.g., in [2]. Our implementation is inspired by the methods described in [10].

6 Implementation Details, Results and Extensions

> *If I had eight hours to chop down a tree, I'd spend six hours sharpening my axe.* - Abraham Lincoln

A mutual information based image registration algorithm has been implemented in the commercial BrainLAB planning software packages iPlan and

BrainSCAN. These packages are routinely used in numerous hospitals around the world and assist the physician in the creation of treatment plans both for radiotherapy and neurosurgery. Multimodal rigid image registration, a key feature of both iPlan and BrainSCAN, is performed by our algorithm in a matter of seconds. The average registration accuracy has been evaluated and is below 1 mm within the intracranial cavity for typical CT-MRI registrations and around 2 mm for CT-PET registrations, as reported in [3]. The success of the method is partially based on the following implementation details that increase accuracy, robustness, and speed.

- *region of interest*
  Mutual information is only calculated with respect to those parts of the reference image containing the patient's head. Prior to registration, large background regions and potential non-anatomical objects like stereotactic frames are automatically detected and excluded. Additionally, the user has the possibility to exclude regions if he is interested in increased accuracy within a certain region of interest.
- *multi-resolution*
  The basic idea behind multi-resolution is to first solve the registration problem at a coarse scale where the images have fewer voxels. The solution is propagated to initialize the next step, which finds an optimal transformation for larger, more refined images. The process iterates until it reaches the finest scale. In order to not introduce artificial maxima to mutual information, the images must be lowpass-filtered (smoothed) before subsampling. On a coarse scale, the images have no fine details. Hence, mutual information should have a broad maximum at the correct transformation at that scale and the optimization method should converge to this maximum from a large set of initial states. When applied correctly, multi-resolution increases the robustness of the registration method by tremendously increasing the basin of attraction of the final optimal transformation. A positive side effect of multi-resolution is a reduced computation time since most optimizer iterations are performed at the coarse levels where evaluation of mutual information is inexpensive.
- *intensity rebinning*
  The number of grey levels (intensity bins) in the images determines the joint histogram's accuracy. The crucial task is to determine the number of bins that provides an optimal trade-off between histogram variance and bias. A simple rule for the number of bins is Sturge's rule [12]. Fortunately, the algorithm's performance is not too sensitive to the number of bins. An investigative study revealed that 64 bins is a good choice for typical registration problems, so we use this number of for most modality combinations.
- *parameter space scaling*
  Translations are measured in millimeters in our implementation, rotations are measured in radians. Due to this inconsistency, we rescale the param-

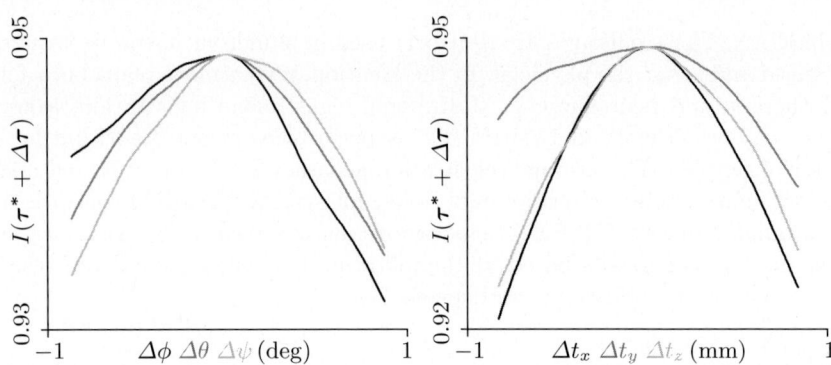

Fig. 7. Mutual Information around its maximum

eter space prior to optimization to enforce comparable changes in I for unit changes in the parameters, both for rotations and translations. This ensures good numerical properties for the resulting optimization problem.

Exemplarily, the result of a typical CT-MRI registration is presented. The CT scan consists of 90 slices with a resolution of 512×512, encoded at 12 bits per voxel, the MRI has 80 slices with a resolution of 256×256, encoded at 8 bits per voxel. This translates to a voxel size of $0.8\,\mathrm{mm} \times 0.8\,\mathrm{mm} \times 2.0\,\mathrm{mm}$ for the MRI. The CT scan has varying voxel size ($0.7\,\mathrm{mm} \times 0.7\,\mathrm{mm} \times 2.1\,\mathrm{mm}$ in the upper part and $0.7\,\mathrm{mm} \times 0.7\,\mathrm{mm} \times 1.5\,\mathrm{mm}$ in the lower part).

The colored screenshot in Fig. A.9 on page 329 depicts the registration result and the graphical user interface of the iPlan software. Solving the optimization problem (5) requires 98 function evaluations. Figure 7 shows mutual information I around the optimal transformation parameters τ^*. Note the smoothness of I, even in the small range of $\pm 1\,\mathrm{mm}$ and $\pm 1\,\mathrm{deg}$. around the optimum. Integrated images obtained by means of image registration are the base for a variety of other applications, e.g, image guided surgery, see Fig. A.10 on page 330.

6.1 Mutual Information Derivative

Some MRI pulse sequences yield images that are nonlinearly distorted due to inhomogeneities in the main magnetic field and nonlinearities in the gradient field. If one of the images is distorted, rigid registration will perform poorly. Furthermore, for extracranial applications, e.g., chest or thorax registration, rigid transformations are also insufficient. In these cases, spatial transformations with a high number of parameters are required to "warp" the floating image in order to align it properly with the reference image. Derivative-free algorithms are not suited for such high-dimensional search spaces.

Fortunately, the Parzen window-based technique provides a differentiable variant of mutual information whose derivative can be calculated analytically. According to (8), the estimated probability distribution is given by

$$p_{A,B}(i_A, i_B) = \alpha \sum_{\boldsymbol{x}_A \in V_A} w(A(\boldsymbol{x}_A) - i_A) w(B(T(\boldsymbol{x}_A)) - i_B) , \qquad (13)$$

where the normalization factor α ensures that $\sum_{i_A} \sum_{i_B} p_{A,B}(i_A, i_B) = 1$. Let $T : \mathbb{R}^3 \mapsto \mathbb{R}^3$ denote a diffeomorphism parameterized by a set of transformation parameters. Formal differentiation of I (see Def. 4) with respect to such a parameter u yields

$$\frac{\partial}{\partial u} I = \sum_{i_A} \sum_{i_B} \frac{\partial p_{A,B}(i_A, i_B)}{\partial u} \left(1 + \log \frac{p_{A,B}(i_A, i_B)}{p_A(i_A) p_B(i_B)} \right)$$
$$- p_{A,B}(i_A, i_B) \left(\frac{\frac{\partial p_A(i_A)}{\partial u}}{p_A(i_A)} + \frac{\frac{\partial p_B(i_B)}{\partial u}}{p_B(i_B)} \right) .$$

The derivatives of the probability distribution estimate exist if both the Parzen window w and the interpolation kernel h (see (7)) are differentiable. In our implementation, we use Parzen windows with the additional "partition of unity" property,

$$\sum_{i \in \mathbb{Z}} w(\xi - i) = 1 \quad \forall \xi \in \mathbb{R} ,$$

e.g., the cubic B-spline kernel. For such w, the normalization factor α in (13) does not depend on the transformation and is equal to $1/\mathrm{card}(V_A)$. Furthermore, the marginal probability p_A does not depend on the transformation and the derivative of mutual information simplifies to

$$\frac{\partial}{\partial u} I = \sum_{i_A} \sum_{i_B} \frac{\partial p_{A,B}(i_A, i_B)}{\partial u} \log \frac{p_{A,B}(i_A, i_B)}{p_B(i_B)} . \qquad (14)$$

Numerical computation of $\frac{\partial}{\partial u} I$ requires the derivatives of the probability distribution estimates, given by

$$\frac{\partial p_{A,B}(i_A, i_B)}{\partial u} = \alpha \sum_{\boldsymbol{x}_A} w(A(\boldsymbol{x}_A) - i_A) \, w'(B(T(\boldsymbol{x}_A)) - i_B) \left(\nabla B|_{T(\boldsymbol{x}_A)} \cdot \frac{\partial T}{\partial u}|_{\boldsymbol{x}_A} \right) .$$

The expression in parentheses denotes the scalar product of the image gradient ∇B (which exists for differentiable interpolation schemes and is calculated by means of (6)) and the derivatives of the transformation with respect to the parameter u.

This approach unfolds its full potential only for transformation spaces of high dimensions (up to several thousands for highly nonlinear warps), and in combination with an efficient gradient-based optimization scheme. Interestingly, using this method for rigid transformations yields a considerable reduction in the number of function evaluations during optimization. However, the total computation time increases due to the costly computation of $\frac{\partial}{\partial u} I$. Therefore, for rigid registration, we do not compute derivatives and stick to the faster and comparably accurate derivative-free optimization scheme.

Acknowledgement. The author is indebted to Dr. Anca-Ligia Grosu, Dept. of Radiation Oncology, TU Munich for providing medical data and to his colleague Dr. Rowena Thomson for helpful comments.

References

[1]   Collignon, A., Maes, F., Delaere, D., Vandermeulen, D., Suetens, P., Marchal, G.: Automated multi-modality image registration based on information theory. In: Bizais, Y., Barrilot, C., Di Paola, R. (eds) Information Processing in Medical Imaging. Kluver Academic Publishers, Dordrecht (1995)

[2]   Conn, A.R., Gould, N.I.M., Toint, Ph.L.: Trust-region methods. SIAM (2000)

[3]   Grosu, A.-L., Lachner, R., Wiedenmann, N., Stärk, S., Thamm, R., Kneschaurek, P., Schwaiger, M., Molls, M., Weber, W.A.: Validation of a method for automatic image fusion (BrainLAB system) of CT data and ^{11}C-methionine-PET data for stereotactic radiotherapy using a LINAC: First clinical experience. Int. J. Rad. Oncol. Phys. vol. 56(5) (2003)

[4]   Kullback, S., Leibler, R.A.: On information and sufficiency. Annals of Mathematical Statistics, vol. 22(1) (1951)

[5]   Maintz, J.B.A., Viergever, M.A.: A survey of medical image registration. Medical Image Analysis, vol. 2(1) (1998)

[6]   Maes, F., Collignon, A., Vandermeulen, D., Marchal, G., Suetens, P.: Multimodality image registration by maximization of mutual information. IEEE Transactions on medical imaging, vol. 16 (1997)

[7]   Maes, F., Vandermeulen, D., Suetens, P.: Medical image registration using mutual information. Proceedings of the IEEE, vol. 91(10) (2003)

[8]   Pluim, J.P.W., Maintz, J.B.A, Viergever, M.A.: Interpolation artefacts in mutual information-based image registration. Computer Vision and Image Understanding, Vol. 77 (2000)

[9]   Powell, M.J.D.: An efficient method for finding the minimum of a function of several variables without calculating derivatives. Computer Journal, vol. (7) (1964)

[10] Powell, M.J.D.: On trust region methods for unconstrained minimization without derivatives. Math. Programming, vol. (97) (2003)

[11] Shannon, C.E.: A mathematical theory of communication. Bell System Technical Journal, vol. 27 (1948)

[12] Sturges, H.: The choice of a class-interval. J. Amer. Statist. Assoc. vol. 21 (1926)

[13] Unser, M., Aldroubi, A., Eden, M.: B-spline signal processing: Part I – Theory. IEEE transactions on signal processing, vol. 41(2) (1993)

[14] Viola, P., Wells III, W.M.: Alignment by maximization of mutual information. In: Grimson, E., Shafer, S., Blake, A., Sugihara, K. (eds) International Conference on Computer Vision. IEEE Computer Society Press, Los Alamitos, CA (1995)

Early Delay
with Hopf Bifurcation

R. Seydel

Mathematisches Institut der Universität zu Köln
Weyertal 86-90, D-50931 Köln
seydel@math.uni-koeln.de

The collection of case studies [5], which owes a lot to Bulirsch's inspiration, was based on rather elementary mathematics. The prerequisite to work on the exercises was the kind of calculus that prevailed mathematical education until 20 years ago. That is, no computer support was needed. Today relevant software packages are in general use, which has opened access to a much broader range of possible cases to study. The immense pedagogical value of the classic analytical approach remains untouched, but computational tools with additional applications must be added. In some way, modern simulation tools make it much easier to bridge the gap between elementary mathematics and current research. This holds in particular for nonlinear dynamics, where computational methods enable rapid experimental approaches into exciting phenomena. This is the broader context of the present paper.

The scope of the paper is threefold. First, two case studies from biology and economy are discussed. Secondly, we attempt to introduce Hopf bifurcation for delay differential equations in an elementary way. Finally we point at early papers by Polish scientists, which have not found the attention they deserve, partly because their essential papers were written in Polish. In the spirit of Bulirsch's work, we attempt to combine fundamental applications and constructive numerical approaches, doing justice to historical roots, all with a strong educational motivation.

1 Delay Differential Equations

Let $\mathbf{y}$ be a vector function depending on time t. Further assume a "law" defined by some vector function $\mathbf{f}$. For numerous applications of dynamical systems the prototype equation $\dot{\mathbf{y}} = \mathbf{f}(\mathbf{y})$ is not adequate. Rather, the time derivative $\dot{\mathbf{y}}$ may depend on the state $\mathbf{y}(t - \tau)$ at a earlier time instant $t - \tau$. The past time is specified by the delay τ. The *delay differential equation* is then $\dot{\mathbf{y}}(t) = \mathbf{f}(\mathbf{y}(t - \tau))$, or more general,

$$\dot{\mathbf{y}}(t) = \mathbf{f}(t, \mathbf{y}(t), \mathbf{y}(t - \tau)) \ .$$

Delay equations where solutions for $t \geq t_0$ are investigated, require an initial *function* ϕ on the interval $t_0 - \tau \leq t \leq t_0$,

$$\mathbf{y}(t) = \phi(t) \quad \text{for } t \in [t_0 - \tau, t_0] \ .$$

Delay equations are used to model numerous systems, in particular in engineering or biology, and in economy.

2 Production of Blood Cells

A nice example of a system where delay occurs naturally is the production of red blood cells in the bone marrow. It takes about four days for a new blood cell to mature. When $b(t)$ represents the number of red blood cells, then their death rate is proportional to $-b(t)$, whereas the reproduction rate depends on $b(t - \tau)$ with a delay τ of about four days. A pioneering model is given in the 1977 paper [3]. Here we analyze the model of [10], an earlier and less known paper by Polish scientists.

For positive δ, p, ν, τ the scalar delay differential equation

$$\dot{b}(t) = -\delta\, b(t) + p\, e^{-\nu b(t-\tau)} \tag{1}$$

serves as model for production of blood cells in the bone marrow. The first term on the right-hand side is the decay, with $0 < \delta < 1$ because δ is the probability of death. The second term describes the production of new cells with a nonlinear rate depending on the earlier state $b(t - \tau)$. The equation has an equilibrium or stationary state b^s, given by

$$\delta\, b^s = p \exp(-\nu b^s) \ .$$

A Newton iteration for the parameter values

$$\delta = 0.5\,, \ p = 2\,, \ \nu = 3.694$$

gives the equilibrium value $b^s = 0.5414$. This model exhibits a Hopf bifurcation off the stationary solution b^s, which will be explained in the following sections. We postpone a discussion of the stability of b^s, and start with a transformation to a framework familiar from elementary calculus: For an experimental investigation we create an approximating *map*, based on a discretization by forward Euler. (For Euler's method, see [7].)

3 A Simple Discretization

We stay with the above example of Eq. (1). A transformation to normal delay by means of the normalized time $\tilde{t} := t/\tau$ and $u(\tilde{t}) := b(\tau \tilde{t})$ leads to

$$b(t - \tau) = b(\tau(\tilde{t} - 1)) = u(\tilde{t} - 1) \quad \text{and} \quad \frac{du}{d\tilde{t}} = \dot{b} \cdot \tau \,,$$

so

$$\frac{du}{d\tilde{t}} = -\delta\tau u + p\tau e^{-\nu u(\tilde{t}-1)} \,.$$

Let us introduce a t-grid with step size $h = \frac{1}{m}$ for some $m \in \mathbf{N}$, and denote the approximation to $u(jh)$ by u_j. Then the forward Euler approximation is

$$u_{j+1} = u_j - \delta h \tau u_j + p h \tau \exp(-\nu u_{j-m}) \,. \tag{2}$$

The equation has the equilibrium $u^{\mathrm{s}}(= b^{\mathrm{s}})$, given by $\delta u^{\mathrm{s}} = p \exp(-\nu u^{\mathrm{s}})$. With $\mathbf{y}^{(j)} := (u_j, u_{j-1}, \ldots, u_{j-m})^{tr}$ the iteration is written in vector form

$$\mathbf{y}^{(j+1)} := \mathbf{P}(\mathbf{y}^{(j)}, \tau) \,, \tag{3}$$

with $\mathbf{P}$ defined by (2). This discretization approach has replaced the scalar delay equation (1) by an m-dimensional vector iteration (3). The stationary solution b^{s} corresponds to a fixed point $\mathbf{y}^{\mathrm{s}}$.

4 Experiments

By a computer experiment based on (2) we study the dynamical behavior depending on the delay parameter τ. Simulations for $\tau = 2$ and $\tau = 3$ show different qualitative behavior, see Fig. 1. For $\tau = 2$ an initial oscillation caused by the discrepancy between b^{s} and the chosen initial ϕ, decays gradually: The equilibrium is stable. For $\tau = 3$ the discrepancy has another effect: The trajectory approaches a stable periodic oscillation. For this value of the parameter τ the equilibrium b^{s} is unstable.

The simple experiment has revealed two different scenarios: A stable equilibrium exists for $\tau = 2$, and for $\tau = 3$ an unstable equilibrium is surrounded by oscillation. It turns out that in the parameter interval $2 < \tau < 3$ there is a threshold value τ_{H} which separates the two qualitatively different states. For τ_{H} the oscillatory realm is born (as seen from smaller values $\tau < \tau_{\mathrm{H}}$). And seen from the other side ($\tau > \tau_{\mathrm{H}}$), the oscillation dies out. This mechanism at τ_{H} is called *Hopf bifurcation*. (For an exposition into the field, see for instance [6].) An analysis below will show that also the delay differential equation (1) has a Hopf bifurcation, with the parameter value $\tau_{\mathrm{H}} = 2.4184$.

5 Stability Analysis of the Example

The discussion of the stability of the stationary state and of the fixed point follow different lines. We still postpone the analysis of the scalar delay equation and first discuss the stability of the fixed point of the discretized problem (3). With the matrix of size $(m + 1) \times (m + 1)$

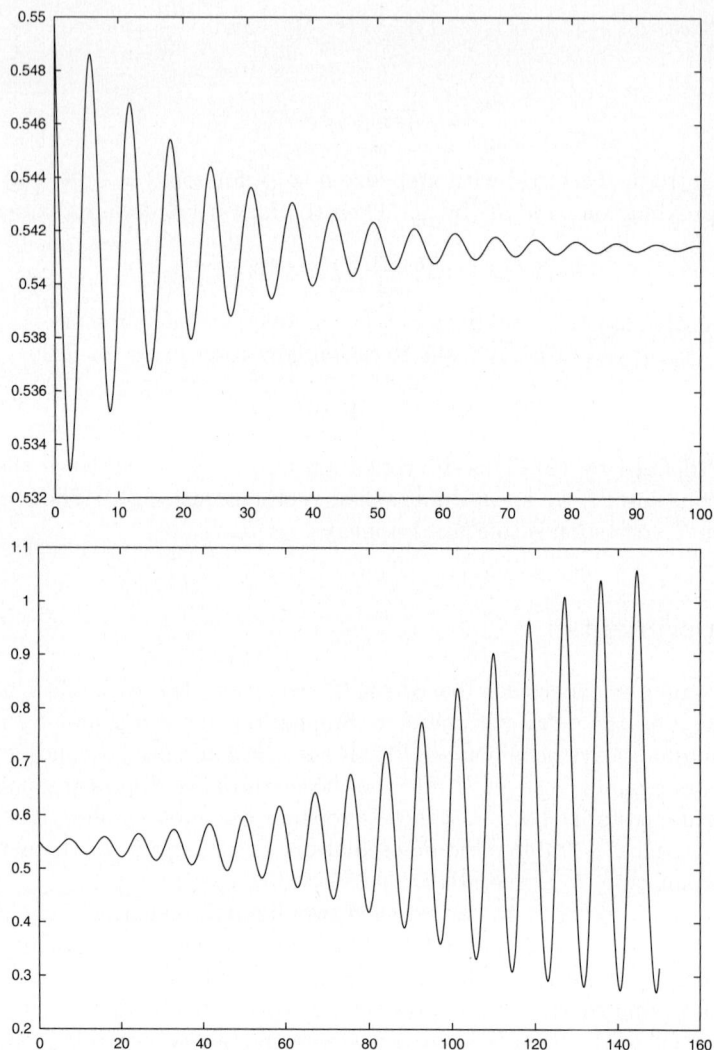

Fig. 1. Blood production model (1), approximation to $b(t)$; $m = 100$, initial state $\phi \equiv 0.55$; top figure: delay $\tau = 2$, bottom: $\tau = 3$

$$\mathbf{A} := \begin{pmatrix} (1 - \delta h\tau) & 0 & & 0 & 0 \\ 1 & 0 & & & \vdots \\ 0 & 1 & \ddots & & \vdots \\ \vdots & & \ddots & \ddots & 0 & 0 \\ 0 & 0 & 0 & 1 & 0 \end{pmatrix}$$

the iteration is written

$$\mathbf{y}^{(j+1)} = \mathbf{A}\mathbf{y}^{(j)} + \begin{pmatrix} ph\tau \exp(-\nu u_{j-m}) \\ 0 \\ \vdots \\ 0 \end{pmatrix}$$

The linearization about the equilibrium u^{s} (the fixed point $\mathbf{y}^{\mathrm{s}}$) leads to a Jacobian matrix $\mathbf{P_y}$ which essentially consists of $\mathbf{A}$ with the additional top-right element

$$\gamma := -\nu ph\tau \exp(-\nu u^{\mathrm{s}}) = -\nu h\tau \delta u^{\mathrm{s}} .$$

By inspecting the structure of the Jacobian matrix we realize that its eigenvalues z are the zeroes of

$$(z - 1 + \delta h\tau)z^{m} + \nu h\delta \mu u^{\mathrm{s}} = 0 . \tag{4}$$

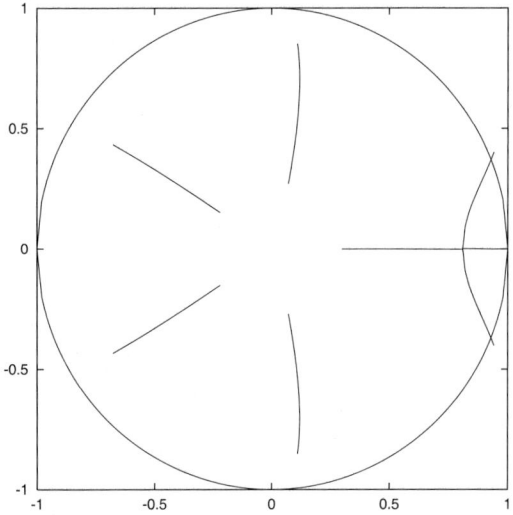

Fig. 2. Blood production model (1)/(2), $m = 6$; paths of the m eigenvalues in the complex plane, for $0.01 \leq \tau \leq 3$

It is interesting to trace the paths of the eigenvalues of the Jacobian in the complex plane, as they vary with the delay parameter τ, see Fig. 2. For $\tau = 0$ the eigenvalue equation (4) simplifies to $(z - 1)z^{m} = 0$. Then $z = 0$ is an m-fold root and $z = 1$ a simple root. For increasing τ, the eigenvalues are initially all inside the unit circle, representing stability of the equilibrium u^{s} for $0 < \tau < \tau_{\mathrm{H}}$, see also [12]. For example, for $m = 6$, two real eigenvalues merge at $\tau = 0.338$ and become a complex-conjugate pair. Eventually, for growing τ, all eigenvalues approach the unit circle (Fig. 2). For the Hopf value τ_{H} one pair passes the circle, indicating instability of the equilibrium.

6 General Stability Analysis of a Stationary Solution

For an analysis of delay differential equations of a fairly general type we study the autonomous system

$$\dot{\mathbf{y}}(t) = \mathbf{f}(\mathbf{y}(t), \mathbf{y}(t - \tau))$$

and assume a stationary solution $\mathbf{y}^s$,

$$\mathbf{f}(\mathbf{y}^s, \mathbf{y}^s) = 0 \ . \tag{5}$$

This equation (5) is obviously independent of the delay. We stay with the general vector case although—as seen with the above example—Hopf bifurcation for delay equations is already possible in the scalar case. The analysis of the stability follows the lines of the ODE situation. Small deviations from $\mathbf{y}^s$ are approximated by the solution $\mathbf{h}$ of the linearized system

$$\dot{\mathbf{h}}(t) = \frac{\partial \mathbf{f}(\mathbf{y}^s, \mathbf{y}^s)}{\partial \mathbf{y}(t)} \, \mathbf{h}(t) + \frac{\partial \mathbf{f}(\mathbf{y}^s, \mathbf{y}^s)}{\partial \mathbf{y}(t - \tau)} \, \mathbf{h}(t - \tau) \ . \tag{6}$$

To find solutions of (6), we start from an ansatz $\mathbf{h}(t) = \mathrm{e}^{\mu t}\mathbf{w}$, which yields the nonlinear eigenvalue problem

$$\mu \mathbf{w} = (\mathbf{A}_1 + \mathrm{e}^{-\mu \tau} \mathbf{A}_2)\mathbf{w} \ , \tag{7}$$

where $\mathbf{A}_1$ and $\mathbf{A}_2$ denote the matrices of the first-order partial derivatives in (6). The corresponding characteristic equation has infinitely many solutions μ in the complex plane. But stability can be characterized in an analogous way as with ODEs: In case $\mathrm{Re}(\mu) < 0$ for all μ, then $\mathbf{y}^s$ is locally stable. And in case one or more μ are in the "positive" side of the complex plane ($\mathrm{Re}(\mu) > 0$), then $\mathbf{y}^s$ is locally unstable.

As in the above example, we take the delay parameter as bifurcation parameter, and denote the real part and imaginary part of μ by α and β, so that

$$\mu(\tau) = \alpha(\tau) + \mathrm{i}\beta(\tau) \ .$$

Analogously as with ODEs, Hopf bifurcation occurs for a critical delay parameter τ_{H} that is characterized by

$$\alpha(\tau_{\mathrm{H}}) = 0 \ , \quad \beta(\tau_{\mathrm{H}}) \neq 0 \ , \quad \frac{\mathrm{d}\alpha(\tau_{\mathrm{H}})}{\mathrm{d}\tau} \neq 0 \ . \tag{8}$$

Of course, the analysis is more complicated in the delay case than in the ODE case because of the infinite number of eigenvalues.

7 Business Cycles

After a decision to invest money in a business there is a time delay ("gestation period") until new capital is installed and investment goods are delivered. Following the famous Polish scientist Kalecki (already 1935, see [2]), a net level of investment J can be modeled by the scalar delay equation

$$\dot{J}(t) = aJ(t) - bJ(t - \tau) \tag{9}$$

for nonnegative constants a, b, which are lumped parameters. Although this equation is linear, it can be used to study the essential mechanism of Hopf bifurcation; just see J in the role of $\mathbf{h}$, a for $\mathbf{A}_1$ and $-b$ for $\mathbf{A}_2$. The eigenvalue equation (7) for Kalecki's scalar model reduces to

$$\mu = a - be^{-\mu\tau} .$$

Separating this equation for the complex variable $\mu = \alpha + i\beta$ into real and imaginary parts results in the pair of real equations

$$\alpha = a - be^{-\alpha\tau} \cos \beta\tau$$
$$\beta = be^{-\alpha\tau} \sin \beta\tau .$$

To check for Hopf bifurcation according to (8), we ask whether there is a solution $\beta \neq 0$ and $\alpha = 0$ for some τ ($\beta > 0$ without loss of generality). This amounts to the system

$$\frac{a}{b} = \cos \beta\tau \tag{10a}$$

$$\frac{\beta}{b} = \sin \beta\tau , \tag{10b}$$

the solution of which requires $a \leq b$ and $\beta \leq b$. Squaring the equation and adding gives

$$a^2 + \beta^2 = b^2$$

or $\beta = \sqrt{b^2 - a^2}$. For $a < b$ and this value of β the real part α vanishes. For a sufficiently small, a solution $\tau = \tau_H$ to (10) exists. By inspecting Eq. (10b) more closely (e.g. graphically) we realize that

$$\tau_H > 1/b .$$

For a further discussion of the Hopf bifurcation in this example (9), see [8]. With such a delay model Kalecki explained the occurrence of business cycles. Using his numbers $a = 1.5833$ and $b = 1.7043$, we obtain $\beta = 0.6307$, and hence a time period $\frac{2\pi}{\beta} = 9.962$. This corresponds to the duration of a business cycle of 10 years, as calculated and predicted by Kalecki. For more general nonlinear versions and two-dimensional economic growth models with delay see [9], [11].

We leave it as an exercise to establish the value $\tau_H = 2.4184$ for the blood cell model (1).

Let us finally comment on numerical algorithms for delay equations. Some "unusual" behavior (as seen from the ODE viewpoint) of solutions of delay equations makes their numerical solution tricky. For the purpose of integrating delay equations there are specifically designed methods. A popular approach is to approximate and replace the delayed term $\mathbf{y}(t - \tau)$ by an interpolating function [4]. In this way the delay equation can be reduced to an ODE. Another approach is Bellman's method of steps, which subdivides the t-axis into intervals of length τ. For each subinterval $[t_0 + (k-1)\tau,\ t_0 + k\tau]$ $(k = 2, 3, \ldots)$ a new system is added, and $\mathbf{y}(t-\tau)$ is always given by an ODE subsystem. For a general exposition see [1]. (Each course on ODEs should include a chapter on delay equations.) We have chosen the simple numerical approach of Sect. 3 because it is elementary and does not require specific knowledge on numerical analysis or delay equations.

References

[1]   Bellen, A., Zennaro, M.: Numerical Methods for Delay Differential Equations. Oxford (2003)
[2]   Kalecki, M.: A marcrodynamic theory of business cycles. Econometrica **3**, 327-344 (1935)
[3]   Mackey, M.C., Glass, L.: Oscillation and chaos in physiological control systems. Science **197**, 287-289 (1977)
[4]   Oberle, H.J., Pesch, H.J.: Numerical treatment of delay differential equations by Hermite interpolation. Numer.Math. **37**, 235-255 (1981)
[5]   Seydel, R., Bulirsch, R.: Vom Regenbogen zum Farbfernsehen. Höhere Mathematik in Fallstudien aus Natur und Technik. Springer, Berlin (1986)
[6]   Seydel, R.: Practical Bifurcation and Stability Analysis. From Equilibrium to Chaos. Second Edition. Springer Interdisciplinary Applied Mathematics Vol. 5. Springer, New York (1994)
[7]   Stoer, J., Bulirsch, R.: Introduction to Numerical Analysis. Springer, New York (1980) (Third Edition 2002)
[8]   Szydlowski, M.: Time to build in dynamics of economic models I: Kalecki's model. Chaos Solitions & Fractals **14**, 697-703 (2002)
[9]   Szydlowski, M.: Time-to-build in dynamics of economic models II: models of economic growth. Chaos Solitions & Fractals **18**, 355-364 (2003)
[10]   Wazewska-Czyzewska, M., Lasota, A.: Mathematical problems of the dynamics of a system of red blood cells (in Polish). Ann.Polish Math.Soc.Ser. III, Appl.Math. **17**, 23-40 (1976)
[11]   Zhang, C., Wei, J.: Stability and bifurcation analysis in a kind of business cycle model with delay. Chaos Solitions & Fractals **22**, 883-896 (2004)
[12]   Zhang, C., Zu, Y., Zheng, B.: Stability and bifurcation of a discrete red blood cell survival model. Chaos Solitions & Fractals **28**, 386-394 (2006)

A Singular Value Based Probability Algorithm for Protein Cleavage

T. Stolte and P. Rentrop

M2 – Zentrum Mathematik, Technische Universität München, Boltzmannstr.3, 85748 Garching, Germany, stolte@ma.tum.de

Summary. Protein degradation by the proteasome is one of the key steps in cell biology and in immune biology. For degradation a mathematical model can be based on a binomial distribution, where the wanted cleavage probabilities are characterized by a minimization problem. The proposed method analyzes the singular values of the Jacobian, thus giving a significant parameter reduction. The computation time is 50 fold reduced compared to the unreduced model, which in practice means hours instead of days for a run. The implementation is based on MATLAB.

1 The Biochemical Problem

Proteasomal protein degradation plays an important role for antigen presentation. A mathematical model should allow statistical analysis and identification of resulting fragments. To study the proteasomal function, a protein, e. g. enolase, is used as a model substrate. After incubation together with proteasomes in vitro, the resulting degradation products of enolase are separated by high pressure liquid chromatography (HPLC) and analyzed by mass spectrometry and Edman sequencing as described else where [7], [13]. The obtained data evolve from two processes, firstly from the cleaving process by proteasome and secondly from the experimental detection process which causes a loss of information. Our investigations are based on the dataset 3039_1 from [10], [13], which describes the start- and end-position and the frequency of the detected fragments from the protein enolase. In [7] and [5] the same data set is treated via a neural net approach. And the second dataset is from [12] who used the protein prion.

The paper is organized as follows. In Sect. 2 a statistical model for the degradation and for the experimental detection process is derived. This model could be reduced by the properties of the binomial distribution. The singular value decomposition (SVD), see [4], is applied in a minimization method, see [1] and [11], in Sect. 3. Based on first numerical experiments, in Sect. 4 a refined model for a reduced parameter set is derived and its performance is discussed.

2 Mathematical Model

For the derivation of the model, independence of the degradation process and the experimental detection process is assumed, each with two possibilities for to cut or not to cut, to detect or not to detect. This setting allows to use a binomial distribution, see [6].

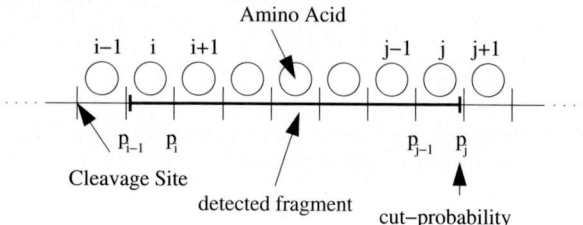

Fig. 1. Fragments of amino acids and probabilities

Definition 1.

a) *Let the* splitting probability *of a cleavage site* in a protein after the *i*-th amino acid be p_i, with $p_i \in I := \{p \mid 0 \le p \le 1\}$. $[i,j]$ *stands for the* frag-ment *between position i and j of the substrate. Let the* splitting probability of a fragment *ranging from the i-th to the j-th amino acid be* p_{ij}, *see Fig. 1*
b) *Let the* detection probability *be* $q(l,h)$, *depending on the length l and the* frequency h *of this fragment type, see discussion follows.*
c) $[i,j] = \nu$ *means that the fragment* $[i,j]$ *was detected with frequency* ν.
d) *n is the maximum number of available enolase molecules.*

For the fragment cleavage probability p_{ij} holds

$$p_{ij} := p_{i-1} \prod_{k=i}^{j-1}(1 - p_k)\, p_j \quad i < j \,. \tag{1}$$

The detection probability is denoted by $q(l,h)$ and depends on the length l and the frequency h of this fragment type. In order to keep the model simple, we assume the independence of the length dependent detection probability $q_l(l)$ and the frequency dependent detection probability $q_h(h)$, i.e.

$$q(l,h) = q_l(l)q_h(h).$$

The functions q_l (respectively q_h) are assumed to be continuous, 0 below a length l_1 (a frequency h_1) and 1 above l_2 (h_2). While q_l is piecewise linear, q_h is chosen to be square of a piecewise linear function shown in Fig. 2. For more detail see [9]. We used the parameters

$$l_1 = 2,\, l_2 = 6,\, h_1 = 15,\, h_2 = 50.$$

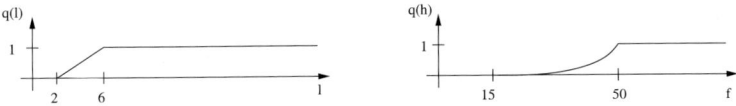

Fig. 2. The functions $q_l(l)$ and $q_h(h)$

Now the probability P of one special fragment $[i, j]$ with measured frequency ν is equal to the sum of all combinations where m fragments, with frequency ν detected, exists.

A binomial distribution allows a product splitting of both probabilities, then

$$P([i,j] = \nu) = \sum_{m=\nu}^{n} \underbrace{\binom{m}{\nu} q(l,h)^{\nu} (1 - q(l,h))^{m-\nu}}_{\text{detection probability}} \underbrace{\binom{n}{m} p_{ij}^{m} (1 - p_{ij})^{n-m}}_{\text{splitting probability}}.$$

$$(2)$$

Remark 1. n is in the range up to 4000 and ν in range of 100. This means that for every fragment more than 3000 terms have to be computed.

With the abbreviation

$$\text{bin}(\nu, n, p) = \binom{n}{\nu} p^{\nu} (1 - p)^{n-\nu}$$

the probability of a detected fragment is, see (2),

$$P([i,j] = \nu) = \sum_{m=\nu}^{n} \text{bin}(\nu, m, q(l,h)) \, \text{bin}(m, n, p_{ij}) \,. \tag{3}$$

This expression can be reduced by Lemma 1.

Lemma 1 (Model Reduction). *The probability of a fragment with detection frequency ν is*

$$P([i,j] = \nu) = \text{bin}(\nu, n, p_{ij}) q(l,h)^{\nu} \left(\frac{1 - q(l,h)p_{ij}}{1 - p_{ij}} \right)^{n-\nu}. \tag{4}$$

Proof.

$$P([i,j] = \nu) = \sum_{m=\nu}^{n} \text{bin}(\nu, m, q(l, h)) \, \text{bin}(m, n, p_{ij})$$

$$= \sum_{m=\nu}^{n} \binom{m}{\nu} q(l, h)^{\nu} (1 - q(l, h))^{m-\nu} \binom{n}{m} p_{ij}^{m} (1 - p_{ij})^{n-m}$$

$$= p_{ij}^{\nu} q(l, h)^{\nu} \sum_{m=0}^{n-\nu} \binom{m+\nu}{\nu} \binom{n}{m+\nu} ((1 - q(l, h)) p_{ij})^{m}$$

$$(1 - p_{ij})^{(n-\nu)-m}$$

$$= \binom{n}{\nu} p_{ij}^{\nu} q(l, h)^{\nu} \sum_{m=0}^{n-\nu} \binom{n-\nu}{m} ((1 - q(l, h)) p_{ij})^{m}$$

$$(1 - p_{ij})^{(n-\nu)-m}$$

$$= \binom{n}{\nu} p_{ij}^{\nu} q(l, h)^{\nu} (1 - q(l, h) p_{ij})^{n-\nu}$$

$$= \binom{n}{\nu} p_{ij}^{\nu} (1 - p_{ij})^{n-\nu} q(l, h)^{\nu} \left(\frac{1 - q(l, h) p_{ij}}{1 - p_{ij}} \right)^{n-\nu}$$

$$= \text{bin}(\nu, n, p_{ij}) q(l, h)^{\nu} \left(\frac{1 - q(l, h) p_{ij}}{1 - p_{ij}} \right)^{n-\nu}$$

$\square$

Now independent of n and ν, only one term for every fragment has to be computed. Now let k the number of possible fragments which will be discussed later on. There are 435 parameters in enolase to be identified for the probabilities p_i of a cut after the i-th amino acid.

Definition 2.

(a) The nonlinear function

$$M(p) = \prod_{r=1}^{k} P([i_r, j_r] = \nu_r) \tag{5}$$

with p is the vector of splitting probabilities, see Def. 1

$$p := (p_1, \ldots, p_k)^T$$

defines the likelihood function.
(b) optimization problem
   For a given detection probability $q(l,h)$ find the splitting probability $\hat{p}$ of maximum likelihood, where $\hat{p}$ is defined as

$$M(\hat{p}) = \max_{p} M(p). \tag{6}$$

Remark 2. Instead of (5) usually the minimum of $-\log(M(p))$ is investigated.

Defining

$$F(p) := \begin{pmatrix} -\log(P([i_1, j_1] = \nu_1)) \\ \vdots \\ -\log(P([i_k, j_k] = \nu_k)) \end{pmatrix} \tag{7}$$

the maximum likelihood problem (5) can be written as the equivalent nonlinear least square problem

$$\min_p ||F(p)||_2 \tag{8}$$

Definition 3 (nonlinear least square problem). *For a given detection probability $q(l,h)$ find the splitting probability $\hat{p}$, where $\hat{p}$ is defined as*

$$||F(\hat{p})||_2 = \min_{p \in I^{435}} ||F(p)||_2 . \tag{9}$$

In a first step every possible fragment, i.e. 436 amino acids from enolase, is considered, so the number of possible fragments k in (5) results to

$$k = \sum_{r=1}^{436} r = (1 + 436)\frac{436}{2} = 95266$$

(For the prion with 260 amino acids, we get 32896.)

Tackling this problem is not meaningful. It does not make sense to consider an optimization problem with 90000 fragments where experiments tell that only 100–200 fragments with detectable frequency occur. The possible cleavage sites in enolase (prion) include 114 (74) start or end positions and the available data describe 135 (133) different fragments only. A possible way for reduction of the original problem uses the singular value decomposition of the Jacobian. Therefore fragments and splitting probabilities are identified via a singular value decomposition of the Jacobian of $F(p)$:

$$DF(p) = \begin{pmatrix} * & \cdots & * \\ \vdots & \ddots & \vdots \\ * & \cdots & * \end{pmatrix} \iff DF(p) = U \cdot S \cdot V^T . \tag{10}$$

3 Application of the Singular Value Decomposition

Any real $m \times n$ matrix A can be decomposed

$$A = U \cdot S \cdot V^T$$

where U is an $m \times m$ orthogonal matrix, V is an $n \times n$ orthogonal matrix, and S is an unique $m \times n$ matrix with real non-negative elements σ_i, $i = 1, .., \min(m, n)$, the singular values. These are arranged in descending order $\sigma_1 \geq \sigma_2 \geq \ldots \geq \sigma_{\min(m,n)} \geq 0$. S has the form

$$\begin{bmatrix} \Sigma \\ 0 \end{bmatrix} \text{ if } m \geq n \text{ and } \begin{bmatrix} \Sigma\ 0 \end{bmatrix} \text{ if } m < n,$$

where Σ is a diagonal matrix with the diagonal elements σ_i. Thus the matrix S can be used for rank determination.

It is well known that the SVD provides a numerically robust solution to the linear least-squares problem,

$$x = (A^T \cdot A)^{-1} \cdot A^T \cdot b \,.$$

Replacing $A = U \cdot S \cdot V^T$ gives

$$x = V \cdot [\Sigma^{-1}] \cdot U^T \cdot b \,.$$

For more details see [3]. To compute the singular value decomposition the function SVD in [14], [8] is used, which is equivalent to the LAPACK routine $DGESVD$, see [2]. For the minimization problem the function $LSQNONLIN$ from the Optimization Toolbox in [8] is applied.

To check the behaviour of the algorithm three different typical sets of initial values are tested, denoted by initial1 to initial3.

- initial1:

$$p_i \equiv 0, \, \forall i$$

- initial2:

$$p_i \equiv 0.1, \, \forall i$$

- initial3:

$$p_i = \begin{cases} 0 & i \text{ is never} \\ 0.1 & i \text{ is} \end{cases} \text{ cleavage site}$$

The algorithm is insensitive to the choice of the initial values. Initial values equal to 0.1 or 0.5 or 0.7 lead to the same solution. Only the initial data set 1, when all initial values are set to zero terminates immediately without any changes.

In Fig. 3 (left) and in Table 1 the singular values (SV) for enolase are presented. For initial sets initial2 and initial3: 321 SV from 435 tend to 0, fixing the number of the non-cleavage sites. This indicates that many splitting probabilities are redundant and many fragments imply no information.

The SV of initial3 include further information. The singular values larger than 10^{-6} (total number 135) fix the number of detected fragments. The intermediate singular values (second column of the table) influence both, the cleavage sites and the detected fragments. This is nicely shown in Fig. 3 (left) see the vertical marks and Table 1 (for the prion see Fig. 3 (right)).

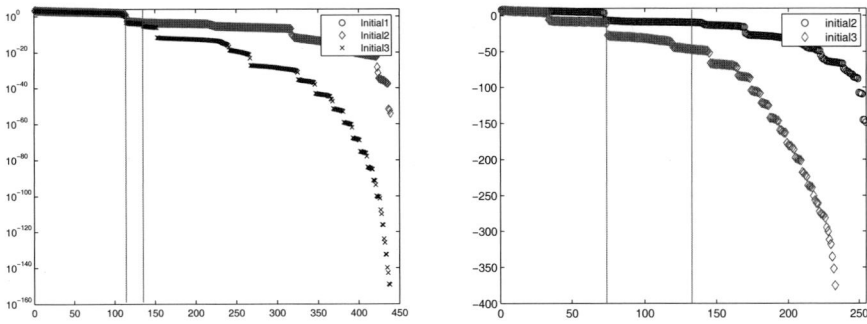

Fig. 3. Singular values for different p values (logarithmic scale) in test example enolase (left) and prion (right)

Table 1. Singular Value Distribution for enolase

p Value	$SV \geq 0.5$	$0.5 > SV > 10^{-6}$	$10^{-6} \geq SV$
initial1	0	0	435
initial2	114	106	215
initial3	114	21	300

4 Derivation of the Reduced Model

Inner extremal values are characterized by the zeros of the first derivative $DF(p)$, see (7). The SVD of DF allows a parameter reduction. SV equal to zero or near to zero characterize the terms without any or small influence on the solution, thus reducing the complexity.

Lemma 2 (Parameter Reduction). *For the r-th fragment given by $[i, j]$ and the splitting probability p_k with $i \leq k \leq j$ follows*

$$\frac{\partial F_r}{\partial p_k} = \begin{cases} > 0 & \text{if } k \text{ is cleavage site} \\ < 0 & \text{if } k \text{ is no cleavage site.} \end{cases} \tag{11}$$

Proof. For $F_r = -\log(P([i_r, j_r] = \nu_r))$ see (7), let $[i_r, j_r] = [i, j]$ and $\nu_r = \nu$. The first derivative

$$\frac{\partial P([i, j] = \nu)}{\partial p_{ij}} = \binom{n}{\nu} q(l, h)^\nu p_{ij}^{\nu-1} (1 - q(l, h)p_{ij})^{n-\nu-1} [\nu - nq(l, h)p_{ij}]$$

possesses the zeros

$$\text{first: } p_{ij} = 0, \text{ second: } p_{ij} = \frac{\nu}{nq(l, h)}, \text{ third: } p_{ij} = \frac{1}{q(l, h)}$$

The second candidate p_{ij} is a zero if $\frac{\nu}{n} \leq q(l,h)$ and the third candidate p_{ij} is a zero if $q(l,h) = 1 \Rightarrow p_{ij} = 1$. The second derivative gives

$$\frac{\partial^2 P([i,j] = \nu)}{\partial^2 p_{ij}} = \binom{n}{\nu} q(l,h)^\nu p_{ij}^{\nu-2}(1 - q(l,h)p_{ij})^{n-\nu-2}$$
$$(\nu^2 - \nu + 2\nu q(l,h)p_{ij} - 2\nu n q(l,h)p_{ij}$$
$$-nq(l,h)^2 p_{ij}^2 + n^2 q(l,h)^2 p_{ij}^2)$$

If the second zero $p_{ij} = \frac{\nu}{nq(l,h)}$ of the first derivative is inserted one gets

$$\frac{\partial^2 P([i,j]=\nu)}{\partial^2 p_{ij}} = \binom{n}{\nu} q(l,h)^2 \left(\frac{\nu}{n}\right)^{\nu-2} \left(1 - \frac{\nu}{n}\right)^{n-\nu-2}\left(-\nu + \frac{\nu^2}{n}\right)$$

In this term every factor is positive except the last. Therefore

$$-\nu + \frac{\nu^2}{n} \leq 0 \quad \Rightarrow \quad \frac{\nu^2}{n} \leq \nu \quad \Rightarrow \quad \frac{\nu}{n} \leq 1 \,.$$

That means that $P([i,j] = \nu)$ reaches a maximum at $p_{ij} = \frac{\nu}{nq(l,h)}$ if $\frac{\nu}{n} \leq q(l,h)$. Otherwise for $\nu = 0$, $q(l,h) = 0$, or for $\frac{\nu}{n} > q(l,h)$ $q(l,h) = 1$ holds. This explains the convergence of the non-cut position to zero because of

$$\frac{\partial F_l}{\partial p_k} = -\frac{1}{P([i,j] = \nu)} \frac{\partial P([i,j] = \nu)}{\partial p_{ij}} \frac{\partial p_{ij}}{\partial p_k}$$

with (consider $\nu = 0$)

$$\frac{\partial P([i,j] = \nu)}{\partial p_{ij}} = (1 - q(l,h)p_{ij})^{n-1}(-n)q(l,h)$$

and

$$\frac{\partial p_{ij}}{\partial p_k} = \begin{cases} i) \quad k = i - 1 & \frac{\partial p_{ij}}{\partial p_{i-1}} = \prod_{l=i}^{j-1}(1 - p_l)p_j \\ ii) \quad k = j & \frac{\partial p_{ij}}{\partial p_j} = p_{i-1}\prod_{l=i}^{j-1}(1 - p_l) \\ iii) \quad k \neq i,j & \frac{\partial p_{ij}}{\partial p_k} = -p_{i-1}\prod_{l=i}^{k-1}(1 - p_l)\prod_{l=k+1}^{j-1}(1 - p_l)p_j \end{cases}$$

This gives

$$\frac{\partial F_r}{\partial p_k} = \begin{cases} > 0 & \text{if } k \text{ is cleavage site} \\ < 0 & \text{if } k \text{ is no cleavage site} \end{cases}$$

and justifies the two cases

- i or j are never start or end position of a fragment with $\nu > 0$ $\Rightarrow p_{i-1} = 0, p_j = 0$ thus $p_{ij} = 0$. This creates the zero-entries in the first derivative.

$$DF = \begin{pmatrix} * & \cdots & * \\ \vdots & \ddots & \vdots \\ * & \cdots & * \\ 0 & \cdots & 0 \\ * & \cdots & * \\ \vdots & \ddots & \vdots \\ * & \cdots & * \end{pmatrix}$$

- Position k is never a cut position $\Rightarrow p_k = 0$. This creates the zero-entries in the first derivative.

$$DF = \begin{pmatrix} * & \cdots & * & 0 & * & \cdots & * \\ \vdots & & \vdots & \vdots & \vdots & & \vdots \\ * & \cdots & * & 0 & * & \cdots & * \end{pmatrix}.$$

□

Remark 3. As a consequence of these results the first derivative has the structure below and thus the corresponding singular value is equal to 0.

$$DF = \begin{pmatrix} * & \cdots & * & 0 & * & \cdots & * \\ \vdots & \ddots & \vdots & \vdots & \vdots & \ddots & \vdots \\ * & \cdots & * & & * & \cdots & * \\ 0 & \cdots & 0 & & \cdots & 0 \\ * & \cdots & * & & * & \cdots & * \\ \vdots & \ddots & \vdots & \vdots & \vdots & \ddots & \vdots \\ * & \cdots & * & 0 & * & \cdots & * \end{pmatrix}.$$

This means that fragments denoted by $[i, j]$ are never in a cut-position, i.e. one has to show $\max P([i, j] = 0) = 1$. Both cases from above gives $p_{i-1} = 0$ or $p_j = 0$, thus

$$\Rightarrow P([i, j] = 0) = \sum_{m=0}^{N} \binom{m}{0} q(l, h)^0 (1 - q(l, h))^m \underbrace{\binom{N}{m} 0^m 1^{N-m}}_{=0 \text{ for } m \neq 0}$$

$$= \binom{0}{0} q(l, h)^0 (1 - q(l, h))^0 \binom{N}{0} 0^0 1^N = 1$$

The same arguments hold for fragments where only one side is never a cut position, i.e. only p_{i-1} or p_j is equal to zero.

These considerations reduce the complexity of the model drastically. For an optimization only those fragments have to be considered with both sides define possible cut positions. In our case 114 cut positions are found and therewith for k

$$k = \text{Number of possible fragments} = \sum_{r=1}^{114} r = (1 + 114)\frac{114}{2} = 6555$$

and nontrivial parameters

$$dim(p) = 114.$$

With Def. 1, Def. 3 and (7) we have the two problems:
The non-reduced minimization problem

$$\min_{p} \|F(p)\|_2$$

with

$$p \in I^{435} \text{ and } F(p) : I^{435} \to \mathbb{R}^{95266} .$$

The reduced minimization problem, denoted by $\sim$

$$\min_{\tilde{p}} \left\|\tilde{F}(\tilde{p})\right\|_2$$

with

$$\tilde{p} \in I^{114} \text{ and } \tilde{F}(\tilde{p}) : I^{114} \to \mathbb{R}^{6555} .$$

The overall work to compute the first derivative DF is reduced from $435 \cdot 95266 = 41440710$ to $114 \cdot 6555 = 747270$ function evaluations. For the prion it is reduced from $255 \cdot 32896 = 8388480$ to $74 \cdot 8411 = 659414$ function evaluations. The acceleration factor for the reduced model is larger than 50.

The solutions of the original and the reduced model are given in Fig. 4 by plotting $(p_{original}, p_{reduced})$, where only seven positions differ significantly. A more detailed database analysis revealed that little information is available for these parameters, i.e. in a special cut position only one fragment with low frequency is found. On the other hand the gain in computing time is significant. Whereas the original problem uses several days on a 1,6 GHz Centrino the reduced model is solved in 3–4 hours.

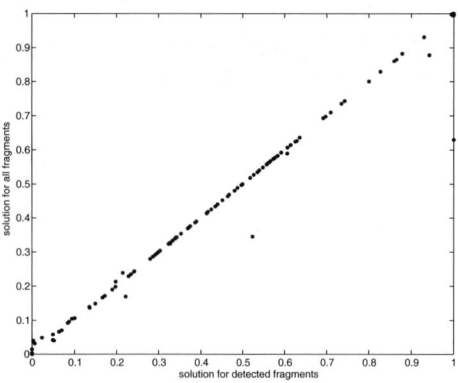

Fig. 4. Comparison of the solutions from the model and the reduced model

The presented approach can be summarized as a Probability Method for Protein Cleavages whereby the last step corresponds to the original problem and the acceleration is reached by the dimension reduction.
Given:

data-set: Set of N different data entries with same structure

$$data\_set = \{d_r, r = 1, \ldots, N\}$$

initial sets: different initial sets $p_{init,j}$
model-function: Function depending on m parameters and a data set entry d_r,

$$F_r(p) := F_{d_r}(p) : I^m \to \mathbb{R}, \quad F(p) := (F_1(p), \ldots, F_N(p))^T$$

minimization problem: see (7)

$$\min_p \|F(p)\|_2$$

Singular-Value based Method

1. Model Reduction (Lem. 1): Simplify $F_r(p)$ (in example use the properties of the binomial distribution) $\Rightarrow \tilde{F}_r(p)$
2. Parameter Reduction (Lem. 2):
   a) Compute singular value decomposition for initial value $p_{init,j}$ from $DF(p_{init,j})$

   $$DF(p_{init,j}) = U_{init,j} \cdot S_{init,j} \cdot V_{init,j}^T$$

   b) Analyse the singular values
   c) Identify with help of $U_{init,j}$ and $V_{init,j}$ unnecessary data set entries and parameters $\Rightarrow \tilde{p}$ and $\tilde{F}(\tilde{p})$ (Fig. A.11 on page 331)
3. Parameter Identification: Solve the reduced model

   $$\min_{\tilde{p}} \left\| \tilde{F}(\tilde{p}) \right\|_2$$

As a last result in Fig. A.11 the entries (with values bigger than 10^{-5}) of the matrix U (in extracts) and the matrix V are plotted. The matrices are from the SVD of the Jacobian of the model function for the data set Enolase and for the initial values initial3. The positions 114, 135 and 435 are highlighted as they correspond to the stages within the singular values. With the interrelations

$$u_{i,j} = \text{part from } i\text{-th fragment of } j\text{-th singular value}$$
$$v_{i,j} = \text{part from } i\text{-th parameter of } j\text{-th singular value}$$

from step 2 (c) obviously, these figure allow the identification of the essential input data (fragments) and parameters (probabilities).

Acknowledgement. The authors are indebted to B. Schönfisch from University of Tübingen and J. Müller from GSF Munich for the first mathematical model of proteasome [9] and A.K. Nussbaum for the experimental data.

References

[1]   Alt, W.: Nichtlineare Optimierung. Vieweg, Braunschweig (2002)
[2]   Anderson, E., Bai, Z., Bischof, C., Blackford, S., Demmel, J., Dongarra, J., Du Croz, J., Greenbaum, A., Hammarling, S., McKenney, A. and Sorensen, D.: LAPACK User's Guide, Third Edition. SIAM, Philadelphia (1999)
[3]   Bock, R. K., Krischer, W.: The Data Analysis BriefBook. Springer, New York (1998)
[4]   Golub, G. H., van Loan, C. F.: Matrix Computations. The Johns Hopkins University Press, Baltimore London (1996)
[5]   Hadeler, K.P., Kuttler, C., Nussbaum, A. K.: Cleaving proteins for the immune system. Mathematical Biosciences, **188**, 63–79 (2004)
[6]   Higham, D. J.: Nine ways to implement the binomial method for option valuation in MATLAB. SIAM Review, **44**, 661–677 (2002)
[7]   Kuttler, C., Nussbaum, A. K., Dick, T. P., Rammensee, H.-G., Schild, H., Hadeler, K.-P.: An algorithm for the prediction of proteasomal cleavages. J. Mol. Biol., **298**, 417–429 (2000)
[8]   MATLAB 6 R13 Documentation: MATLAB functions and Optimization Toolbox, (2003)
[9]   Müller, J., Schönfisch, B., Nussbaum, A. K.: Modeling Proteasomal Protein Degradation, Preprint GSF PP02-033 (2002)
[10]   Nussbaum, A. K., Dick, T. P., Keilholz, W., Schirle, M., Stevanovic, S., Dietz, K., Heinemeyer, W., Groll, M., Wolf, D. H., Huber, R., Rammensee, H.-G., Schild, H.: Cleavage motifs of the yeast 20S proteasome β subunits deduced from digests of enolase. Proc. Natl. Acad. Sci. USA, **95**, 12504–12509 (1998)
[11]   Spellucci, P.: Numerische Verfahren der nichtlinearen Optimierung. Birkhäuser, Berlin (1993)
[12]   Tenzer S, Stoltze L, Schönfisch B, Dengjel J, Müller M, Stevanovic S, Rammensee HG, Schild H.: Quantitative analysis of prion-protein degradation by constitutive and immuno-20S proteasomes indicates differences correlated with disease susceptibility. J Immunol., **172(2)**, 1083–1091 (2004)
[13]   Toes, R. E. M., Nussbaum, A. K., Degermann, S., Schirle, M., Emmerich, N., Kraft, M., Laplace, C., Zwinderman, A., Dick, T., Müller, J., Schönfisch, B., Schmid, C., Fehling, H.-J., Stevanovic, S., Rammensee, H.-G., Schild, H.: Discrete cleavage motifs of constitutive and immunoproteasomes revealed by quantitative analysis of cleavage products. J. Exp. Med., **194**, 1–12 (2001)
[14]   van Loan C. F.: Introduction to scientific computing. MATLAB Curriculum Series, New Jersey (1997)

Calculation of Magnetic Fields with Finite Elements

G. Wimmer[1], M. Clemens[1] and J. Lang[2]

[1] Chair for Theory of Electrical Engineering and Computational Electromagnetics, Helmut-Schmidt-University, University of the Federal Armed Forces Hamburg, Holstenhofweg 85, 22043 Hamburg, Germany, {g.wimmer,m.clemens}@hsu-hh.de

[2] Numerical Analysis and Scientific Computing, Darmstadt University of Technology, Department of Mathematics, Schlossgartenstr. 7, 64289 Darmstadt, Germany, lang@mathematik.tu-darmstadt.de

Summary. The discretization of transient and static magnetic field problems with the Whitney Finite Element Method results in differential-algebraic systems of equations of index 1 and nonlinear systems of equations. Hence, a series of nonlinear equations have to be solved. This involves e.g. the solution of the linear(-ized) equations in each time step where the solution process of the iterative preconditioned conjugate gradient method reuses and recycles spectral information of previous linear systems. Additionally, in order to resolve induced eddy current layers sufficiently and regions of ferromagnetic saturation that may appear or vanish depending on the external current excitation a combination of an error controlled spatial adaptivity and an error controlled implicit Runge-Kutta scheme is used to reduce the number of unknowns for the algebraic problems effectively and to avoid unnecessary fine grid resolutions both in space and time. Numerical results are presented for 2D and 3D nonlinear magneto-dynamic problems.

1 Magneto-Quasistatic Fields

1.1 Continuous Formulation in 3D

Electro-magnetic phenomena are described by a set of partial differential equations known as Maxwell's equations [4]. These are given by Faraday's law

$$\frac{\partial \mathbf{B}}{\partial t} + \operatorname{curl} \mathbf{E} = 0 \quad \text{in} \quad \mathbb{R}^3, \tag{1}$$

the nonexistence of magnetic monopoles

$$\operatorname{div} \mathbf{B} = 0 \quad \text{in} \quad \mathbb{R}^3, \tag{2}$$

Ampère's law

$$\operatorname{curl} \mathbf{H} = \mathbf{J} + \frac{\partial \mathbf{D}}{\partial t} \quad \text{in} \quad \mathbb{R}^3 \tag{3}$$

and Gauss' law

$$\operatorname{div} \mathbf{D} = \rho \quad \text{in} \quad \mathbb{R}^3 \tag{4}$$

with the electric field $\mathbf{E}$, the electric displacement $\mathbf{D}$, the current density $\mathbf{J}$, the magnetic field $\mathbf{H}$, the magnetic flux density $\mathbf{B}$ and the magnetization $\mathbf{M}$. These equations are joined by the material relations

$$\mathbf{B} = \mu(\mathbf{H} + \mathbf{M}), \qquad \mathbf{J} = \sigma \mathbf{E} + \mathbf{J_e}, \qquad \mathbf{D} = \varepsilon \mathbf{E}, \tag{5}$$

where $\mathbf{J_e}$ denotes the impressed current density, μ, σ and ε the magnetic permeability, the electric conductivity and the electric permittivity.

The solenoidal character of the magnetic flux density $\mathbf{B}$ according to (2) implies the existence of a magnetic vector potential $\mathbf{A}$ such that

$$\mathbf{B} = \operatorname{curl} \mathbf{A}. \tag{6}$$

Hence, Ampère's law (3) together with the material relation (5) yields

$$\operatorname{curl}(\mu^{-1} \operatorname{curl} \mathbf{A}) = \mathbf{J} + \operatorname{curl} \mathbf{M} + \frac{\partial \mathbf{D}}{\partial t}. \tag{7}$$

Faraday's law (1) can be rewritten as

$$\operatorname{curl}(\mathbf{E} + \frac{\partial \mathbf{A}}{\partial t}) = 0 \tag{8}$$

from which the existence of an electric scalar potential ϕ can be postulated with

$$\mathbf{E} = -\frac{\partial \mathbf{A}}{\partial t} - \operatorname{grad} \phi. \tag{9}$$

The eddy current equation arises from Maxwell's equations as the magneto-quasistatic approximation by neglecting the displacement current $\partial \mathbf{D}/\partial t$. This is reasonable for low-frequency and high-conductivity applications. Finally, Eq. (7) with the additional gauging $\operatorname{grad} \phi = 0$ results in

$$\sigma \frac{\partial \mathbf{A}}{\partial t} + \operatorname{curl}(\mu^{-1} \operatorname{curl} \mathbf{A}) = \mathbf{J_e} + \operatorname{curl} \mathbf{M}. \tag{10}$$

In order to solve the eddy current equation (10) numerically a bounded domain $\Omega \subset \mathbb{R}^3$ as well as an artificial boundary $\Gamma := \partial \Omega$ sufficiently removed from the region of interest are introduced. The parabolic initial-boundary value problem for the magnetic vector potential $\mathbf{A}(r, t)$ can be formulated as

$$\sigma \frac{\partial \mathbf{A}}{\partial t} + \operatorname{curl}(\mu^{-1} \operatorname{curl} \mathbf{A}) = \mathbf{J_e} + \operatorname{curl} \mathbf{M} \quad \text{in} \quad \Omega, \tag{11}$$

$$\mathbf{A} \times \mathbf{n} = 0 \quad \text{on} \quad \Gamma, \tag{12}$$

$$\mathbf{A}(r, 0) = \mathbf{A_0} \quad \text{in} \quad \Omega, \tag{13}$$

where $\mathbf{A_0}$ describes the initial state and $\mathbf{n}$ represents the outward unit vector of Γ. If only magneto-static problems are considered (11)-(13) reduce to

$$\operatorname{curl}(\mu^{-1}\operatorname{curl}\mathbf{A}) = \mathbf{J_e} + \operatorname{curl}\mathbf{M} \quad \text{in} \quad \Omega, \tag{14}$$

$$\mathbf{A} \times \mathbf{n} = 0 \quad \text{on} \quad \Gamma. \tag{15}$$

Another formulation for magneto-static problems can be derived if the source current $\mathbf{J_e}$ vanishes. Hence, from Ampère's law $\operatorname{curl}\mathbf{H} = 0$ (3) the existence of a magnetic scalar potential ψ_m can be derived with

$$\mathbf{H} = -\operatorname{grad}\psi_m. \tag{16}$$

Equation (2) and the material relation (5) yield

$$-\operatorname{div}(\mu\operatorname{grad}\psi_m) = -\operatorname{div}\mu\mathbf{M} \quad \text{in} \quad \Omega, \tag{17}$$

$$\psi_m = 0 \quad \text{on} \quad \Gamma, \tag{18}$$

where the homogeneous Dirichlet boundary condition (18) is introduced to simulate the decay of the magnetic field. An example of a homogeneously magnetized sphere is shown in Fig. A.14 on page 332.

1.2 Continuous Formulation in 2D

If only planar problems with a current

$$\mathbf{J_e} = (0, 0, J_z)^T \tag{19}$$

perpendicular to the xy-plane of a Cartesian coordinate system and a magnetization of the form

$$\mathbf{M} = (M_x, M_y, 0)^T \tag{20}$$

are considered the magnetic vector potential

$$\mathbf{A} = (0, 0, A_z)^T \tag{21}$$

is shown to have only a component A_z in z-direction. Hence, Eq. (10) yields

$$\sigma\frac{\partial}{\partial t}A_z - \operatorname{div}\mu^{-1}\operatorname{grad}A_z = J_z + \operatorname{curl}^{2D}(M_x, M_y) \quad \text{in} \quad \mathbb{R}^2, \tag{22}$$

with

$$\operatorname{curl}^{2D}(M_x, M_y) := \left[\frac{\partial M_y}{\partial x} - \frac{\partial M_x}{\partial y}\right]. \tag{23}$$

Again, the problem is posed in a bounded domain $\Omega \subset \mathbb{R}^2$ and an artificial boundary $\Gamma := \partial\Omega$ is introduced. The parabolic initial-boundary value problem for the magnetic scalar potential $A_z(x, y; t)$ can be formulated as

$$\sigma \frac{\partial}{\partial t} A_z - \mathrm{div}\mu^{-1}\mathrm{grad}A_z = J_z + \mathrm{curl}^{2\mathrm{D}}(M_x, M_y) \qquad \mathrm{in} \quad \Omega, \qquad (24)$$

$$A_z = 0 \qquad \mathrm{on} \quad \Gamma, \qquad (25)$$

$$A_z(x, y; 0) = A_z^0 \qquad \mathrm{in} \quad \Omega. \qquad (26)$$

The magnetic flux density B vanishes in z-direction and can be computed from A_z by

$$\mathbf{B} = \begin{pmatrix} B_x \\ B_y \\ 0 \end{pmatrix} = \begin{pmatrix} \partial A_z/\partial y \\ -\partial A_z/\partial x \\ 0 \end{pmatrix}. \qquad (27)$$

In materials with nonlinear behaviour μ is a function of the absolute value of magnetic flux density $\mathbf{B}$ and can be written as

$$\mu(|\mathbf{B}|) = \mu(|\mathrm{grad}A_z|). \qquad (28)$$

The magneto-static problem with $\sigma = 0$ can be stated as

$$-\mathrm{div}\mu^{-1}\mathrm{grad}A_z = J_z + \mathrm{curl}^{2\mathrm{D}}(M_x, M_y) \qquad \mathrm{in} \quad \Omega, \qquad (29)$$

$$A_z = 0 \qquad \mathrm{on} \quad \Gamma. \qquad (30)$$

2 The Finite Element Method

2.1 Time Dependent Eddy Current Equation

A numerical solution of (24)-(26) is sought in a space of continuous functions, which are not differentiable according to the classical definition. Since the strong solution of (24) needs to be at least twice differentiable an appropriate formulation which is called the weak formulation has to be found. For this, the Sobolev spaces $H^1(\Omega)$ and $H_0^1(\Omega)$ are introduced as

$$H^1(\Omega) := \{v \in L_2(\Omega) : v' \in L_2(\Omega)\}, \qquad (31)$$

$$H_0^1(\Omega) := \{v \in H^1(\Omega) : v = 0 \quad \mathrm{on} \quad \Gamma\}, \qquad (32)$$

$$V(T, \Omega) := \{v \in H_0^1(\Omega) \text{ for nearly all } t \in (0, T)\}. \qquad (33)$$

Under the assumption $J_z \in L_2(\Omega)$ and $\mathrm{curl}^{2\mathrm{D}}(M_x, M_y) \in L_2(\Omega)$ Eq. (24) is multiplied by $u \in H_0^1(\Omega)$ and Green's formula

$$-\int_\Omega \triangle u v \, \mathrm{d}\Omega = \int_\Omega \mathrm{grad}u \, \mathrm{grad}v \, \mathrm{d}\Omega - \int_\Gamma \frac{\partial u}{\partial \mathbf{n}} v \, \mathrm{d}s \qquad \forall u, v \in H^1(\Omega) \quad (34)$$

is applied which yields

$$\sigma \int_\Omega \frac{\partial}{\partial t} A_z v \, d\Omega + \int_\Omega (\mathrm{grad}\, A_z)^T M_\mu \mathrm{grad}\, v \, d\Omega - \int_\Gamma \frac{\partial A_z}{\partial \mathbf{n}} v \, ds \qquad (35)$$

$$= \int_\Omega J_z v \, d\Omega + \int_\Omega \mathrm{curl}^{2\mathrm{D}} (M_x, M_y) \, v \, d\Omega, \qquad (36)$$

where

$$M_\mu := \begin{pmatrix} \mu^{-1} & 0 \\ 0 & \mu^{-1} \end{pmatrix} \in \mathbb{R}^{2\times 2}. \qquad (37)$$

Since $v \in H_0^1(\Omega)$ the line integral in (35) vanishes and the last term in the right hand side of (35) can be written as

$$\int_\Omega \mathrm{curl}^{2\mathrm{D}} (M_x, M_y) \, v \, d\Omega = \int_\Omega M_x \frac{\partial v}{\partial y} - M_y \frac{\partial v}{\partial x} \, d\Omega. \qquad (38)$$

Hence, the variational or weak formulation can be statet as: Find $A_z(x, y; t) \in V(T, \Omega)$ such that

$$b(A_z, v; t) + a(A_z, v; t) = l(v; t) \qquad \forall v \in H_0^1(\Omega) \qquad (39)$$

for nearly all $t \in (0, T)$ and the initial value condition

$$g(A_z(x, y; 0), v) = g(A_z^0, v) \qquad \forall v \in H_0^1(\Omega) \qquad (40)$$

is fulfilled with

$$b(A_z, v; t) := = \sigma \int_\Omega \frac{\partial}{\partial t} A_z v \, d\Omega, \qquad (41)$$

$$a(A_z, v; t) := \int_\Omega (\mathrm{grad}\, A_z)^T M_\mu \mathrm{grad}\, v \, d\Omega, \qquad (42)$$

$$l(v; t) := \int_\Omega J_z v \, d\Omega + \int_\Omega M_x \frac{\partial v}{\partial y} - M_y \frac{\partial v}{\partial x} \, d\Omega, \qquad (43)$$

$$g(u, v) := \int_\Omega uv \, d\Omega. \qquad (44)$$

$T \in \mathbb{R}$ denotes the end point of the simulation. In order to obtain a discretization of (39) and (40) Lagrangian nodal finite elements $N_i(x, y) \in H_0^1(\Omega_h)$ for every node which does not belong to the boundary Γ are introduced with a triangulation of the computational domain Ω_h which is generally an approximation of Ω (see Fig. 1). In order to state the discrete variational formulation the function spaces

$$V_{0h}(\Omega_h) := \{v_h(x, y) : v_h(x, y) = \sum_{i=1}^n v_i N_i(x, y)\}, \qquad (45)$$

$$V_h(T, \Omega_h) := \{v_h(x, y; t) : v_h(x, y; t) = \sum_{i=1}^n v_i(t) N_i(x, y)\} \qquad (46)$$

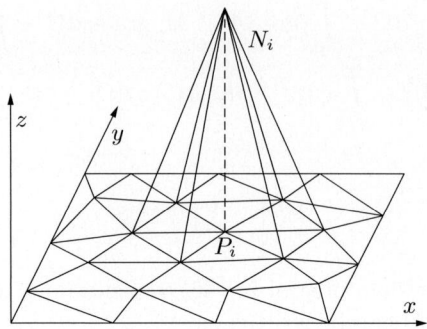

Fig. 1. Ansatz function N_i with $N_i(P_i) = 1$.

are introduced, where n denotes the number of nodes which are not contained in the boundary of Ω_h. Hence, the variational or weak formulation in the finite dimensional function spaces can be statet as: Find $A_{zh}(x, y; t) \in V_h(T, \Omega)$ such that

$$b(A_{zh}, v_h; t) + a(A_{zh}, v_h; t) = l(v_h; t) \qquad \forall v_h \in V_{0h}(\Omega_h) \qquad (47)$$

for nearly all $t \in (0, T)$ and the initial value condition

$$g(A_{zh}(x, y; 0), v_h) = g(A_z^0, v_h) \qquad \forall v \in V_{0h}(\Omega_h). \qquad (48)$$

is fulfilled. If the expansion $A_{zh}(x, y; t) = \sum_{i=1}^{n} v_i(t) N_i(x, y)$ and $v_h = N_j(x, y)$ for all $j = 1, \ldots, n$ is applied to (47) and (48) the differential equation for $v_h(t) := (v_1(t), \ldots, v_n(t))^T$

$$M_h \frac{d}{dt} v_h(t) + K_h(t) v_h(t) = f_h(t) \qquad \text{for nearly all } t \in (0, T), \qquad (49)$$

$$M_h v_h(0) = g_h, \qquad (50)$$

is obtained, where matrix and vector entries in row $i \in \{1, \ldots, n\}$ and column $j \in \{1, \ldots, n\}$ of M_h, $K_h(t)$, $f_h(t)$ and g_h are defined as

$$M_h^{(ij)} := \sigma g(N_i, N_j), \qquad (51)$$

$$K_h^{(ij)}(t) := a(N_i, N_j; t), \qquad (52)$$

$$f_h^{(i)} := l(N_i; t), \qquad (53)$$

$$g_h^{(i)} := g(A_z^0, N_i). \qquad (54)$$

Since the conductivity σ vanishes in non-conducting regions, M_h is singular and positive semi-definite. Therefore, (49) and (50) represents a differential algebraic equation of index 1 [8]. If the material behaviour is nonlinear the stiffness matrix $K_h(t) = K_h(v_h; t)$ is additionally dependent on v_h.

2.2 Magneto-Static Equation

In magneto-static problems with $\sigma = 0$ the weak formulation of (29)-(30) according to the derivation of (39) can be stated as: Find $A_z \in H_0^1(\Omega)$ such that

$$a(A_z, v) = l(v) \qquad \forall v \in H_0^1(\Omega), \tag{55}$$

where

$$a(A_z, v) := \int_\Omega (\text{grad} A_z)^T M_\mu \text{grad} v \, \mathrm{d}\Omega, \tag{56}$$

$$l(v) := \int_\Omega J_z v \, \mathrm{d}\Omega + \int_\Omega M_x \frac{\partial v}{\partial y} - M_y \frac{\partial v}{\partial x} \, \mathrm{d}\Omega. \tag{57}$$

is defined. Discretization with $A_{zh}(x,y) = \sum_{i=1}^n v_i N_i(x,y)$ and $v_h = N_j(x,y)$ for all $j = 1, \ldots, n$ yields a system of equations for $\mathrm{v}_h := (v_1, \ldots, v_n)^T$

$$K_h \mathrm{v}_h = f_h, \tag{58}$$

where matrix and vector entries in row $i \in \{1, \ldots, n\}$ and column $j \in \{1, \ldots, n\}$ of K_h and f_h are defined as

$$K_h^{(ij)} := a(N_i, N_j), \tag{59}$$

$$f_h^{(i)} := l(N_i). \tag{60}$$

If nonlinear materials are considered (58) is replaced by the nonlinear system of equations

$$K_h(\mathrm{v}_h)\mathrm{v}_h = f_h. \tag{61}$$

3 Discrete Differential Forms

In the last section the magnetic scalar potential A_{zh} was approximated in the space $V_{0h}(\Omega_h)$ of continuous functions which are in the range of the nodal Lagrangian basis functions. This approach is reasonable because the magnetic scalar potential is continuous. Since other electromagnetic quantities such as electric field or magnetic flux density possess different mathematical properties (continuity, differentiability, etc.) adjusted basis functions have to be used [3]. Therefore Discrete Differential Forms (DDF) which are finite element basis functions defined on a mesh have been devised. DDF exist for 2D elements and 3D elements such as triangle, quadrilateral, tetrahedron, hexahedron and prism. In the C++ library FEMSTER [1] also higher-order DDF basis functions for 3D elements are implemented. Higher-order DDF basis functions for 2D elements have been added by the authors. Classic nodal basis functions

are used for continuous scalar quantities such as magnetic scalar potentials (0-form field), H(curl) basis functions for quanities which are tangential continous across different elements for the magnetic vector potential (1-form), H(div) basis functions for quantities which have a continuous normal component such as the magnetic flux density (2-form) and piecewise constant functions are used for the charge density (3-form). Table 1 shows different electromagnetic quantities (field) the corresponding degree of the form (form), the topological quantity to which the form is associated (integral), the continuity property of the form (continuity) and the derivative operator that can be applied to the form (derivative). The magnetic flux density $\mathbf{B}$ for example can be integrated

Table 1. Different electromagnetic quantities and the properties of the associated form

form	field	integral	continuity	derivative
0-form	scalar magnetic potential ψ_m, A_z electric potential	point	total	gradient
1-form	electric field $\mathbf{E}$ magnetic field $\mathbf{H}$ magnetization $\mathbf{M}$ magnetic vector potential $\mathbf{A}$	edge	tangential	curl
2-form	magnetic flux density $\mathbf{B}$ electric displacement $\mathbf{D}$ current density $\mathbf{J}$	surface	normal	div
3-form	charge density ρ	volume	-	none

over a surface giving the value of flux flowing through this surface, the normal component of $\mathbf{B}$ is continuous across elements and the divergence of $\mathbf{B}$ is well defined. In order to derive a variational formulation of (14)-(15) the Sobolev spaces H(curl, Ω), H$_0$(curl, Ω) and H(div, Ω) are introduced as

$$H(\text{curl}, \Omega) := \{\mathbf{v} \in L_2(\Omega) : \text{curl}\,\mathbf{v} \in L_2(\Omega)\}, \tag{62}$$

$$H_0(\text{curl}, \Omega) := \{\mathbf{v} \in H(\text{curl}, \Omega) : \mathbf{v} \times \mathbf{n} = 0 \quad \text{on} \quad \Gamma\}, \tag{63}$$

$$H(\text{div}, \Omega) := \{\mathbf{v} \in L_2(\Omega) : \text{div}\,\mathbf{v} \in L_2(\Omega)\}. \tag{64}$$

Equation (14) is multiplied by a test function $\mathbf{v} \in H_0(\text{curl}, \Omega)$ and integrated over the domain Ω which results in

$$\int_\Omega \text{curl}\left(\mu^{-1}\text{curl}\,\mathbf{A} - \mathbf{M}\right)\mathbf{v}\,d\Omega = \int_\Omega \mathbf{J_e}\mathbf{v}\,d\Omega. \tag{65}$$

Finally, Greens formula for vector valued functions yields

$$\int_\Omega (\text{curl}\,\mathbf{A})^T M_\mu\,\text{curl}\,\mathbf{v}\,d\Omega = \int_\Omega \mathbf{J_e}\mathbf{v}\,d\Omega + \int_\Omega \mathbf{M}\,\text{curl}\,\mathbf{v}\,d\Omega, \tag{66}$$

where

$$M_\mu := \begin{pmatrix} \mu^{-1} & 0 & 0 \\ 0 & \mu^{-1} & 0 \\ 0 & 0 & \mu^{-1} \end{pmatrix} \in \mathbb{R}^{3\times 3}. \tag{67}$$

The variational formulation amounts to the computation of $\mathbf{A} \in H_0(\text{curl}, \Omega)$ such that

$$a(\mathbf{A}, \mathbf{v}) = l(\mathbf{v}) \ \forall \ \mathbf{v} \in H_0(\text{curl}, \Omega). \tag{68}$$

Here, the bilinear form $a : H_0(\text{curl}, \Omega) \times H_0(\text{curl}, \Omega) \to \mathbb{R}$ and the linear form $l : H_0(\text{curl}, \Omega) \to \mathbb{R}$ are given by

$$a(\mathbf{A}, \mathbf{v}) = \int_\Omega (\text{curl}\,\mathbf{A})^T M_\mu \,\text{curl}\,\mathbf{v}\,\mathrm{d}\Omega, \tag{69}$$

$$l(\mathbf{v}) = \int_\Omega \mathbf{J_e v}\,\mathrm{d}\Omega + \int_\Omega \mathbf{M}\,\text{curl}\,\mathbf{v}\,\mathrm{d}\Omega. \tag{70}$$

The variational equation (66) is discretized by lowest order edge elements. The basis functions are associated with the edges of the finite element mesh and can be written in the form

$$\mathbf{w}_{ij}^{(1)} := N_i \text{grad} N_j - N_j \text{grad} N_i, \tag{71}$$

where N_i and N_j are the nodal basis functions associated with node i and j. The edge basis function has the property that the line integral of $\mathbf{w}_{ij}^{(1)}$ along the edge E_{ij} connecting the nodes i and j is one and along other edges it is zero, that means

$$\int_{E_{ij}} \mathbf{t}_{ij} \cdot \mathbf{w}_{ij}^{(1)} \mathrm{d}s = 1, \tag{72}$$

$$\int_{E_{kl}} \mathbf{t}_{kl} \cdot \mathbf{w}_{ij}^{(1)} \mathrm{d}s = 0 \text{ for } E_{kl} \neq E_{ij}. \tag{73}$$

Here, $\mathbf{t}_{ij}$ denotes the unit vector pointing from node i to node j. The magnetic vector potential in (66) is expressed as a linear combination of edge elements

$$\mathbf{A} = \sum_{i=1}^{n_e} a_i \mathbf{w}_{ei}^{(1)}, \tag{74}$$

where n_e denotes the number of edges in the mesh and $\mathbf{w}_{ei}^{(1)}$ the basis function associated to edge i. As test functions $\mathbf{v}$ the basis functions $\mathbf{w}_{ej}^{(1)}, j = 1, \ldots, n_e$ are chosen. Equation (68) results in a linear equation for $a_h := (a_1, \ldots, a_{n_e})^T$

$$K_h a_h = f_h, \tag{75}$$

where matrix and vector entries in row $i \in \{1, \ldots, n\}$ and column $j \in \{1, \ldots, n\}$ of K_h and f_h are defined as

$$K_h^{(ij)} := a(\mathbf{w}_{ei}^{(1)}, \mathbf{w}_{ej}^{(1)}), \tag{76}$$

$$f_h^{(i)} := l(\mathbf{w}_{ei}^{(1)}). \tag{77}$$

The structure of K_h is investigated in more detail. The magnetic flux density $\mathbf{B}$ is obtained from $\mathbf{A}$ by $\mathbf{B} = \operatorname{curl} \mathbf{A}$ and is an element of $H(\operatorname{div}, \Omega)$. The lowest order basis functions of the discretized space are known as Raviart-Thomas elements and are associated to the surfaces. A basis is given by $\{\mathbf{w}_{f1}^{(1)}), \ldots, \mathbf{w}_{fn_f}^{(1)})\}$. The number of surfaces is denoted by n_f. Since $\operatorname{curl} H(\operatorname{curl}, \Omega) \subset H(\operatorname{div}, \Omega)$ and this relation also for the discretized spaces of edge elements and Raviart-Thomas elements holds, the curl of $\mathbf{w}_{ei}^{(1)}$ can be expressed as a linear combination

$$\operatorname{curl} \mathbf{w}_{ei}^{(1)} = \sum_{k=1}^{n_f} c_{ki} \mathbf{w}_{fk}^{(1)}. \tag{78}$$

The matrix element $K_h^{(ij)}$ of K_h can be expressed as

$$
\begin{aligned}
a(\mathbf{w}_{ei}^{(1)}, \mathbf{w}_{ej}^{(1)}) &= \int_{\Omega} \left(\operatorname{curl} \mathbf{w}_{ei}^{(1)} \right)^T M_\mu \operatorname{curl} \mathbf{w}_{ej}^{(1)} \, d\Omega \\
&= \int_{\Omega} \sum_{k=1}^{n_f} c_{ki} \left(\mathbf{w}_{fi}^{(1)} \right)^T M_\mu \sum_{l=1}^{n_f} c_{lj} \mathbf{w}_{fl}^{(1)} \, d\Omega.
\end{aligned} \tag{79}
$$

It follows that K_h can be written in the form $K_h = C^T N_h C$ with the curl matrix $C \in \mathbb{R}^{n_f \times n_e}$ and the stiffness matrix $N_h \in \mathbb{R}^{n_f \times n_f}$. The matrix entries $N_h^{(ij)}$ in row $k \in \{1, \ldots, n_f\}$ and column $l \in \{1, \ldots, n_f\}$ of N_h are defined as

$$N_h^{(kl)} := \int_{\Omega} \left(\mathbf{w}_{fk}^{(1)} \right)^T M_\mu \mathbf{w}_{fl}^{(1)} \, d\Omega. \tag{80}$$

Since $\operatorname{grad} H^1(\Omega) \subset H(\operatorname{curl}, \Omega)$ the matrix entry $K_h^{(ij)}$ of (58) and (59) can be written as

$$
\begin{aligned}
K_h^{(ij)} &= \int_{\Omega} (\operatorname{grad} N_i)^T M_\mu \operatorname{grad} N_j \, d\Omega \\
&= \int_{\Omega} \sum_{k=1}^{n_e} g_{ki} \left(\mathbf{w}_{ek}^{(1)} \right)^T M_\mu \sum_{l=1}^{n_e} g_{lj} \mathbf{w}_{el}^{(1)} \, d\Omega,
\end{aligned} \tag{81}
$$

for some coefficients $g_{ki} \in \mathbb{R}$. Here, N_i and $\mathbf{w}_{ek}^{(1)}$ denote the lowest order nodal and edge basis functions related to a two dimensional mesh. It follows that K_h in (58) can be written as $G^T T_h G$ with the gradient matrix $G \in \mathbb{R}^{n_e \times n}$ and the matrix $T_h \in \mathbb{R}^{n_e \times n_e}$ whose components are defined by

$$T_h^{(kl)} := \int_\Omega \left(\mathbf{w}_{ek}^{(1)} \right)^T M_\mu \mathbf{w}_{el}^{(1)} \, d\Omega. \tag{82}$$

The structure of the matrices $C^T N_h C$ and $G^T T_h G$ is identical to the structure that is obtained by discretizing the Maxwell equation by the finite integration theory (FIT) [2]. Scalar quantities are directly associated to nodes, edges, surfaces and volumes on a primal and a dual grid. The Maxwell equations are transfered to the Maxwell grid equations (MGE). The equations on the primal and dual grid are then connected by the material relations. Lowest order finite element theory is closely related to the FIT.

4 Subspace Recycling Methods

The application of Newton- or Quasi-Newton methods on Eqs. (49), (58) and (75) results in a sequence of linear systems of equations

$$A^l x^l = b^l, \qquad l = 1, \ldots, m \tag{83}$$

with similar positive or positive semi definite system matrices and different right-hand sides. For these multiple right-hand side (mrhs) problems usually preconditioned conjugate gradient or algebraic multigrid methods are applied, which do not reuse information from previous solution processes. In [5] it is shown that cg methods can also applied to singular systems which are positive semi definite. If (83) is solved with the conjugate gradient (cg) method the smallest eigenvalues in A^l slow down the convergence. Consequently, if the smallest eigenvalues could be "removed" in some way the convergence is expected to be improved. This removal is generally achieved by a deflated cg method (see [7], [9]) where search directions are orthogonal to the eigenvectors belonging to the extremal eigenvalues. Hence, the cg method does not "see" these eigenvalues. The effectivity of the method depends mainly on the closeness of the approximation of the subspace that is spanned by the eigenvectors belonging to the extremal eigenvalues. If a series of linear equations (83) have to be solved the Krylov subspace $Q = [z_0, \ldots, (M^{-1}A)^k z_0]$, $z_0 := M^{-1}(b - Ax_0)$ that is created when solving the first linear system with a cg method will be useful when solving all subsequent equations. M is a preconditioning matrix, x_0 the start vector and k the number of cg iterations. The i-th Krylov subspace is augmented by the range of Q and the augmented preconditioned cg method is characterized by

$$x_i \in \{x_0\} + \bar{\mathcal{K}}_i(M^{-1}A, z_0, Q) \quad \text{with} \quad r_i \perp \bar{\mathcal{K}}_i(M^{-1}A, z_0, Q), \tag{84}$$

$$\bar{\mathcal{K}}_i(M^{-1}A, z_0, Q) := \mathcal{K}_i(M^{-1}A, z_0) \oplus \text{Range}(Q). \tag{85}$$

The algorithm can be considered as a continuation of the classical cg method with constructed Krylov subspace Q. This leads to the algorithm (PPCG-1). The dependence on the index l is omitted:

```
1: initial value x₀₀ from previous linear system
2: i = 0
```
1: initial value x_{00} from previous linear system
2: $i = 0$
3: $x_0 = Q(Q^T A Q)^{-1} Q^T b + P_Q x_{00}$
4: $r_0 = b - A x_0)$
5: **while** $\|r_i\| \neq 0$ **do**
6:     $z_i = P_Q M^{-1} r_i$
7:     **if** $i = 0$ **then**
8:         $p_i = z_i$
9:     **else**
10:        $\beta_i = r_i^T z_i / r_{i-1}^T z_{i-1}$
11:        $p_i = \beta_i p_{i-1} + z_i$
12:    **end if**
13:    $\alpha_i = r_i^T z_i / p_i^T A p_i$
14:    $x_{i+1} = x_i + \alpha_i p_i$
15:    $r_{i+1} = r_i - \alpha_i A p_i$
16:    $i = i + 1$
17: **end while**

5 Numerical Examples

5.1 Magneto-Static Motor Simulation

A nonlinear magneto-static simulation is shown in Fig. A.12 on page 332. The motor consists of rotor, stator and two exciting coils and is surrounded by air. The different algorithms are compared with the classical cg method (PCG)

Table 2. Comparison of different cg methods for a nonlinear magneto-static simulation with 2804 nodes.

solver type	# newton iterations	# cg iterations	cpu in s
PCG	18	490	24.4
PPCG-1	18	330	17.9

Table 3. Comparison of different cg methods for a nonlinear magneto-static simulation with 25945 nodes.

solver type	# newton iterations	# cg iterations	cpu in s
PCG	16	798	367.3
PPCG-1	16	514	278.7

concerning the number of cg iterations and computing time for a mesh with 2804 nodes, $\dim(P_Q)$=64 (Table 2) and 25945 nodes, $\dim(P_Q)$=106 (Table 3). The preconditioned augmented cg method PPCG-1 reduces the number of iterations and the computing time considerably.

5.2 3D Nonlinear Electro Motor Simulation

A 3D model of an electro motor design (Fig. A.13 on page 332) with ferromagnetic, nonlinear material characteristic and two current coils was discretized resulting in 379313 degrees of freedom (dof). The results in Table 4 show a significant decrease in the time required for the nonlinear iteration even for a rather high dimension of 108 vectors in the augmented space Q.

Table 4. 3D electro motor: nonlinear magneto-static model (379313 dof).

solver type	# newton iterations	# cg iterations	cpu in s
PCG	11	680	17409.0
PPCG-1	11	347	8834.0

5.3 Sinusoidal Excited C-Magnet

A c-magnet structure which consists of an iron core surrounded by air is investigated. The current density of the exciting coils at the long side of the core has a sinusoidal form over 60 ms. The magnetic potential is set to zero at the boundaries to approximate infinity. Figure 2 depicts the geometry with the coarse mesh of level 0. The initial coarse mesh consists of 96 nodes and 165 triangles. A combination of simultaneous error controlled adaptivity for the space and time discretization is presented. The time integration is performed using a Singly-Diagonal-Implicit-Runge-Kutta method (SDIRK3(2))

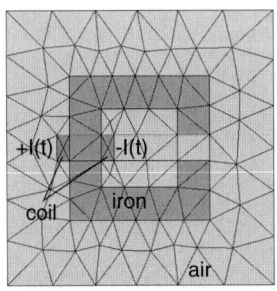 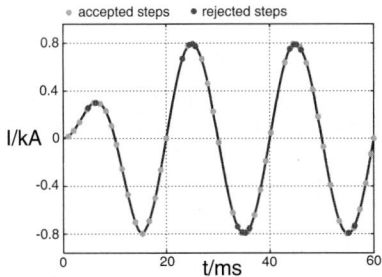

Fig. 2. Geometry with initial coarse mesh and excitation.

of order 3 with embedded order 2 solution for error control. This approach features a time evolving mesh where an error controlled refinement of the mesh resolution takes into account appearing and disappearing local transient saturation effects. In the field of parabolic partial differential equations related to advection-diffusion type computational fluid dynamics a similar combination has been investigated by Lang [6]. The error tolerances for the space $\delta_H = 10^{-8}$ and time $\delta_T = 10^{-5}$ are used. The simulation over the three periods needs 46 time steps where 17 steps are rejected at a specific grid level. Figure 2 shows the current excitiation in the coils together with the accepted and rejected steps.

Acknowledgement. The authors would like to thank Prof. R. Hiptmair from the Swiss Federal Institute of Technology for helpful discussions on discrete differential forms at the 19th Chemnitz FEM Symposium 2006 and Prof. S. Kurz from ETAS in Stuttgart for discussions on the treatment of permanent magnets with finite elements.

References

[1] P. Castillo, J. Koning, R. Rieben, M. Stowell, and D. White. *Discrete Differential Forms: A Novel Methodology for Robust Computational Electromagnetics.* Lawrence Livermore National Laboratory, 2003.

[2] M. Clemens and T. Weiland. Discrete electromagnetism with the Finite Integration Technique. In F. L. Teixeira, editor, *Geometric Methods for Computational Electromagnetics*, number 32 in PIER, pages 65–87. EMW Publishing, Cambridge, Massachusetts, USA, 2001.

[3] R. Hiptmair. Finite elements in computational electromagnetism. *Acta Numerica*, pages 237–339, 2002.

[4] J. Jin. *The Finite Element Method in Electromagnetics.* John Wiley and Sons, second edition, 2002.

[5] E. F. Kaasschieter. Preconditioned conjugate gradients for solving singular systems. *Journal of Computational and Applied Mathematics*, 24:265–275, 1988.

[6] J. Lang. *Adaptive Multilevel Solution of Nonlinear Parabolic PDE Systems: Theory, Algorithm and Application.* Springer-Verlag, Berlin, Heidelberg, New York, 2001.

[7] R. Nabben and C. Vuik. A comparison of deflation and coarse grid correction applied to porous media flow. Report 03-10, TU Delft, Department of Applied Mathematical Analysis, 2003.

[8] A. Nicolet and F. Delincé. Implicit Runge Kutta methods for transient magnetic field computation. *IEEE Transactions on Magnetics*, 32:1405–1408, May 1996.

[9] Y. Saad, M. Yeung, J. Erhel, and F. Guyomarch. A deflated version of the conjugate gradient algorithm. *SIAM Journal on Scientific Compututing*, 21(5):1909–1926, 2000.

Mathematics and Applications in Macroscale

Smooth Approximation and Rendering of Large Scattered Data Sets

Jörg Haber[1], Frank Zeilfelder[1], Oleg Davydov[2], and Hans-Peter Seidel[1]

[1] Max-Planck-Institut für Informatik, Saarbrücken, Germany
[2] Justus-Liebig-Universität Giessen, Germany

Summary. We present an efficient method to automatically compute a smooth approximation of large functional scattered data sets given over arbitrarily shaped planar domains. Our approach is based on the construction of a C^1-continuous bivariate cubic spline and our method offers optimal approximation order. Both local variation and non-uniform distribution of the data are taken into account by using local polynomial least squares approximations of varying degree. Since we only need to solve small linear systems and no triangulation of the scattered data points is required, the overall complexity of the algorithm is linear in the total number of points. Numerical examples dealing with several real world scattered data sets with up to millions of points demonstrate the efficiency of our method. The resulting spline surface is of high visual quality and can be efficiently evaluated for rendering and modeling. In our implementation we achieve real-time frame rates for typical fly-through sequences and interactive frame rates for recomputing and rendering a locally modified spline surface.

1 Introduction

The problem of scattered data fitting is to efficiently compute a suitable surface model that approximates a given large set of arbitrarily distributed discrete data samples. This problem arises in many scientific areas and fields of application, for instance, in chemistry, engineering, geology, medical imaging, meteorology, physics, and terrain modeling. Moreover, scattered data methods play an important role in scientific visualization to get a better understanding of a given set of scattered data points and for subsequent treatment needed in many applications.

In this paper we concentrate on the problem of functional scattered data fitting, which can be described as follows: Given a finite set of points $(x_i, y_i) \in \Omega$, $i = 1, \ldots, N$, where Ω is a bounded domain in the plane, and corresponding values z_i, $i = 1, \ldots, N$, find a method to construct a surface $s : \Omega \mapsto \mathbb{R}$ that meets as many as possible of the following goals:

- **Approximation**: s should approximate the data, i.e. $s(x_i, y_i) \approx z_i$ $(i = 1, \ldots, N)$, while offering optimal approximation order
- **Quality**: s should be of high visual quality (i.e., for "smooth data", s should be at least C^1-continuous) and have convenient properties for further processing.
- **Usability**: Large real world data sets, where N is typically at least of order 10^6, should be manageable.
- **Efficiency**: Both the computation and the evaluation of s should be fast and efficient.
- **Stability**: The computation of s should be numerically stable, i.e., the method should work for any distribution of scattered points.
- **Adaptiveness**: The local variation and distribution of the data should be taken into account.
- **Simplicity**: The method should be easy to implement.

Although many approaches have been developed in the last 30 years, the literature shows that it is a difficult task to meet all of the above goals by using one single method. In fact, the algorithms proposed in the literature typically have at least one of the following drawbacks: limitations in approximation quality, severe restrictions on the number of points, limited visual quality of the resulting surface, high computation times, and restrictions on the domain and distribution of the data.

In this paper, we develop a new approach to scattered data fitting which is based on differentiable bivariate cubic splines. We decided to construct a smooth surface since such surfaces look more pleasant and have nice properties for further processing and rendering. The method we propose belongs to the class of so-called two-stage methods [40]: In the first step of the algorithm, we compute discrete least squares polynomial pieces for local parts of the spline s by using only a small number of nearby points. Then, in the second step, the remaining polynomial pieces of s are obtained directly by using C^1 smoothness conditions. Our approach uniquely combines the following advantages: The data need not be triangulated, the domain Ω can be of arbitrary shape, no estimations of derivatives need to be computed, and we do not perform any global computations. As a result, we obtain a fast method that is applicable to large real world data, local data variations do not have an undesirable global effect, and the differentiable approximating spline is of high visual quality. Thus, we have a fully automatic method which is stable, easy to implement, and the local distribution and variation of the data are taken into account.

The spline representation of our surface model allows to employ Bernstein-Bézier techniques efficiently for evaluation and rendering of the spline. In contrast to previous methods for terrain visualization, we render smooth surfaces and thus do not need to decimate or (re-)triangulate the scattered data. Moreover, we have fast and robust algorithms for view frustum culling and computation of surface points, true surface normals, and texture coordinates.

2 Previous Work

In this section we give an overview on previous and related work in the fields of scattered data fitting and rendering of large terrain data sets. There are many different approaches to scattered data fitting, see for instance the surveys and overview in [19, 26, 30, 35, 40]. A very active area of research are radial basis methods [2, 20, 35, 38]. However, these methods usually require solving large, ill-conditioned linear systems. Therefore, sophisticated iterative techniques are needed for the computation of the radial function interpolants [2]. An approach based on regularization, local approximation, and extrapolation has been proposed in [1].

Motivated by some classic results from approximation theory, various methods based on bivariate splines were proposed. There are several types of splines that can be used. The simplest approach is to consider tensor product splines [13, 18, 22, 23] and their generalizations to NURBS surfaces [37], which have important applications, e.g., in modeling and designing surfaces. These spaces are essentially restricted to rectangular domains. In general, tensor product methods are straightforward to apply only for data given on a grid. If the data points are irregularly distributed, there is no guarantee that the interpolation problem has a solution. Also, global least squares approximation and related methods have to deal with the problem of rank deficiency of the observation matrix. Alternatively, lower dimensional spaces and/or adaptive refinement combined with precomputation in those areas where the approximation error is too high can be employed [39, 28, 45]. In [39], parametric bicubic splines possessing G^1 geometric continuity are adaptively subdivided to approximate 3D points with a regular quadmesh structure. Multilevel B-splines are used in [28] to approximate functional scattered data.

Other spline methods are based on box splines [9, 24], simplex splines [46], or splines of finite-element type. The simplest example of finite-element splines are continuous piecewise linear functions with respect to a suitable triangulation of the planar domain. It is well-known that the approaches based on piecewise linear functions can not exceed approximation order 2. To achieve higher smoothness and approximation order, polynomial patches of greater degree have to be considered. In particular, there are scattered data methods based on classical smooth finite elements such as Bell quintic element, Frajies de Veubecke-Sander and Clough-Tocher cubic elements, and Powell-Sabin quadratic element, see the above-mentioned surveys and more recent papers [10, 14, 31]. In general, these methods are local, and it is obvious that these splines possess much more flexibility than tensor product splines. Many methods require that the vertices of the triangulation include all data points or a suitable subset of these points obtained by a thinning procedure. Such a triangulation is not very expensive to obtain (computational cost $\mathcal{O}(N \log N)$), but the arising spline spaces can become difficult to deal with if the triangulation is complicated and its triangles are not well-shaped. In addition, the above-mentioned methods based on finite elements require accurate estimates of derivatives at the data points, which is a nontrivial task by itself assuming the data points are irregularly distributed. To overcome these difficulties, global least squares approximation and other global methods were considered, e.g., in [14, 21, 34, 46].

The basic idea of our method is related to the interpolation scheme of [33, 32]. Essential differences are, however, that we neither triangulate (quadrangulate) the data points, nor make use of any interpolation scheme as for instance in [10, 31, 32]. In particular, we do not need any estimates of z-values at points different from the given data points. Instead, we compute local least squares approximations directly in the Bernstein-Bézier form, and then settle the remaining degrees of freedom by using the standard C^1 smoothness conditions [16], which results in very short computation times. Since our method does not even require a triangulation of the data points, it is very well suited for extremely large datasets. In addition, our method allows (local) reproduction of cubic polynomials and hence offers optimal approximation order. Theoretical aspects of the method are treated in [11].

A large number of techniques for efficient rendering of terrain data has been proposed in the literature. However, these techniques usually operate on piecewise linear surface representations only. Moreover, many of these methods are restricted to data that are regularly sampled on a rectangular grid. This kind of data is commonly referred to as a *digital elevation map* (DEM) or, in the case of additional color values associated to each data point, as a *digital terrain map* (DTM).

Among the early methods based on DEM / DTM data we mention techniques such as clipping surface cells against the view frustum for ray tracing [8], extracting feature points from the data set by curvature analysis and constructing a Delaunay triangulation [41], and using quadtree data structures to accelerate ray casting [6]. More recent approaches achieve interactive frame rates for rather low resolution DTM by determining visible and occluded terrain regions [7] or by exploiting vertical ray coherence for ray casting [27]. Several methods incorporate level-of-detail (LOD) techniques to speed up rendering. Such LOD techniques can be embedded in multiresolution BSP trees [43] or combined with hierarchical visibility for culling occluded regions [42]. However, changing the LOD during an animation might result in visual discontinuities. This problem has led to the development of continuous LOD techniques [17, 29, 15]. In [29], a continuous LOD rendering is obtained through on-line simplification of the original mesh data while maintaining user-specified screen-space error bounds. A related approach is proposed in [15], where additional optimizations such as flexible view-dependent error metrics, incremental triangle stripping, and predefined triangle counts are introduced.

The methods that do not require data to be sampled on a uniform grid typically construct a *triangulated irregular network* (TIN) from the data. A multiresolution representation of arbitrary terrain data is presented in [5], where every resolution level consists of a TIN that is obtained through incremental Delaunay triangulation. The approximation error can be chosen to be constant or variable over the data domain. The approach in [25] is similar to the one before, but here the triangulation is computed on-the-fly, avoiding the storage requirements of the hierarchy.

3 Construction of the Spline

3.1 Overview and Basic Idea

Many of the methods mentioned in the previous section are based on global least squares approximation and related procedures, thus facing the problem of rank deficiency. One possibility to overcome this is by applying well known numerical techniques for rank deficient matrices, such as singular value decomposition. However, this procedure is very expensive for large coefficient matrices arising from the global methods and destroys their sparse structure. In contrast, our method only relies on *local* least squares computations, which allows us to employ the singular value decomposition efficiently.

Given a finite set of points (x_i, y_i), $i = 1, \ldots, N$, in a bounded domain $\Omega \subset \mathbb{R}^2$ and corresponding values z_i, $i = 1, \ldots, N$, we first determine a suitable space $\mathcal{S}$ consisting of smooth bivariate splines of degree 3 such that the total number of degrees of freedom (i.e., the dimension of $\mathcal{S}$) is approximately N. We construct a quadrilateral mesh covering the domain Ω, see Fig. 1, and define $\mathcal{S}$ to be the space of C^1-continuous piecewise cubics with respect to the uniform triangle mesh Δ obtained by adding both diagonals to every quadrilateral. We call the union of these quadrilaterals the *spline domain $\mathcal{Q}$*. The basic idea of the method is to choose a subset $\mathcal{T}$ of triangles in Δ with the following properties:

(i) the triangles of $\mathcal{T}$ are uniformly distributed in Δ;
(ii) the polynomial patches $s|_T$ $(T \in \mathcal{T})$ can be chosen freely and independently from each other;
(iii) if a spline $s \in \mathcal{S}$ is known on all triangles in $\mathcal{T}$, then s is also completely and uniquely determined on all other triangles that cover the domain;
(iv) each patch $s|_T$, where $T \in \Delta \backslash \mathcal{T}$ has a non-empty intersection with Ω, can be computed using only a small number of nearby patches corresponding to triangles in $\mathcal{T}$.

The approximating spline is constructed in two steps. First, we compute for every triangle $T \in \mathcal{T}$ a least squares polynomial p_T^q in its Bernstein-Bézier form by using singular value decomposition (SVD) and taking into account only points in T and several adjacent triangles. The degree q of p_T^q may vary from triangle to triangle, though not exceeding 3, and is chosen adaptively in accordance with the local density of the data points. We set $s = p_T^q$ on each $T \in \mathcal{T}$. Then, in the second step, the remaining polynomial pieces of s on the triangles $T \in \Delta \setminus \mathcal{T}$ are computed by using Bernstein-Bézier smoothness conditions. In order to guarantee property (iii), it is necessary to add some auxiliary *border cells* containing both diagonals to Δ as shown in Fig. 1.

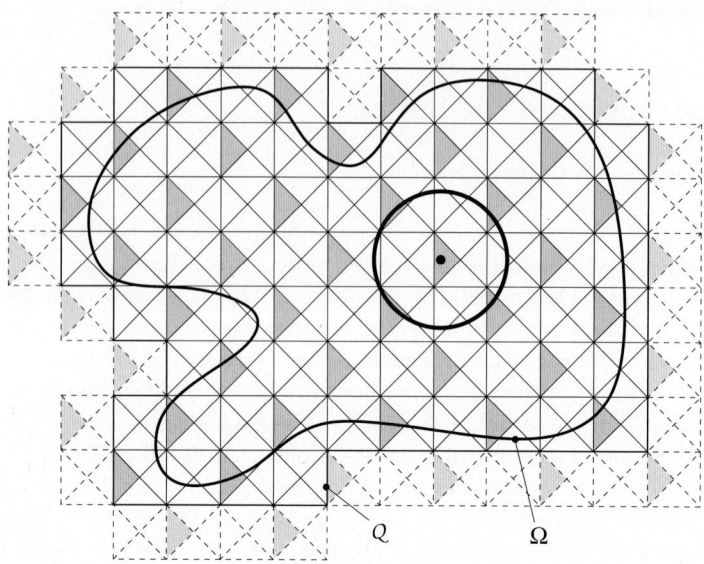

Fig. 1. Layout of the Bézier triangle patches for an arbitrarily shaped domain Ω of scattered data points. In addition to the Bernstein-Bézier coefficients of the grey triangles inside the spline domain $\mathcal{Q}$, the coefficients of the light grey triangles in the dashed border cells are needed to determine all remaining triangle patches in $\mathcal{Q}$ by smoothness conditions. The bold circle shows the position of one of the circles C_T.

3.2 The Spline Space

Our method is implemented for arbitrary domains as shown in Fig. A.17 on page 334. For a detailed description of the spline space and its Bernstein-Bézier representation it is sufficient to consider the square domain $\Omega = [0, 1]^2$.

For given scattered points $(x_i, y_i) \in \Omega$, $i = 1, \ldots, N$, we set $n = \lfloor \sqrt{N/5} \rfloor$ and we cover the domain Ω with squares Q_{ij}, $i, j = 1, \ldots, n$ of edge length $h = 1/n$. This choice of n ensures that the dimension of the spline space approximately coincides with the number of scattered data points. In addition, a ring of square *border cells* surrounding the union $\mathcal{Q} = \bigcup Q_{ij}$ is needed to completely determine the approximating spline on $\mathcal{Q}$. A uniform triangle mesh Δ (a so-called Δ^2 partition) is obtained by adding both diagonals to every square Q_{ij} as shown in Fig. 1.

We consider the space of bivariate splines $\mathcal{S}$ consisting of all cubic C^1 piecewise polynomials with respect to Δ. It is well-known that the dimension of the spline space $\mathcal{S}$ is $5n^2 + 8n + 3 \approx N$ [3]. Moreover, this space combines a number of attractive features. First, the splines of this kind are highly flexible in contrast to, e.g., tensor product splines [13]. Second, the comparatively low degree allows fast and efficient computation of the approximating splines. Third, the approximation order of the space $\mathcal{S}$ is optimal [3], i.e., any sufficiently smooth function can be approximated by a spline $s \in \mathcal{S}$ with the error $\mathcal{O}(h^4)$, which is the best possible approximation order for piecewise cubics.

For computing the cubic patches of the approximating spline, we use the well-known Bernstein-Bézier representation of a cubic polynomial p_T^3 defined on a triangle $T = [\mathbf{t}_0, \mathbf{t}_1, \mathbf{t}_2] \in \Delta$:

$$p_T^3(\mathbf{u}) = \sum_{|\alpha|=\alpha_0+\alpha_1+\alpha_2=3} B_T^{\alpha,3}(\mathbf{u}) \, b_\alpha, \quad \mathbf{u} \in T. \tag{1}$$

Here,

$$B_T^{\alpha,3}(\mathbf{u}) := \frac{3!}{\alpha_0!\alpha_1!\alpha_2!} \lambda_0^{\alpha_0}(\mathbf{u}) \, \lambda_1^{\alpha_1}(\mathbf{u}) \, \lambda_2^{\alpha_2}(\mathbf{u})$$

are the ten Bernstein basis polynomials of degree 3, where $\lambda_\nu(\mathbf{u})$, $\nu = 0, 1, 2$, are the barycentric coordinates of $\mathbf{u}$ with respect to T. The $b_\alpha \in \mathbb{R}$ are called Bernstein-Bézier coefficients of p and represent the local degrees of freedom of the polynomial patch.

The advantages of the Bernstein-Bézier techniques include the stability of the Bernstein-Bézier basis, easily implementable smoothness conditions (see Sect. 3.4 below), and an efficient algorithm for the evaluation of the spline and its corresponding normal (de Casteljau algorithm).

3.3 Adaptive Least Squares Approximation

We start our algorithm by choosing a subset $\mathcal{T}$ of the triangles in Δ as in Fig. 1 and initial circles C_T, $T \in \mathcal{T}$, with radius $\frac{5}{4}h$ and midpoint at the barycenter of T. This choice of the circles ensures that the domain Ω is completely covered by the union of the circles C_T, $T \in \mathcal{T}$. In the first step of our algorithm we determine the polynomial pieces of the approximating spline s on the triangles of $\mathcal{T}$. To this end we compute $n^2/2$ different local discrete least squares approximations for bivariate polynomials by using singular value decomposition.

Since we treat scattered data, there is in principle no restriction on the number m of points within a particular circle C_T. However, both too few and too many points are not desirable. Therefore, we use the initial circle only if m satisfies $M_{\min} \leq m \leq M_{\max}$ with $M_{\min}, M_{\max}$ chosen as described below. Thus, two different situations that require further processing can occur:

1. the number of data points within C_T is too large: $m > M_{\max}$;
2. the number of data points within C_T is too small: $m < M_{\min}$.

In the first case ($m > M_{\max}$), we thin the data inside of C_T down to at most $M_{\max}$ points. This is done by sorting all data points from within C_T

into an auxiliary regular grid, which is constructed in such a way that at most $M_{\max}$ grid cells lie inside of C_T. Then we choose the most central data point from each non-empty grid cell to be taken into account for computing the local polynomial patch. Such thinning of the data is justified by the assumption that unreasonably large sets of local data points carry redundant information. By thinning the data points, we avoid expensive computation of SVD for large matrices.

In the second case ($m < M_{\min}$), we simply increase the radius of C_T until at least $M_{\min}$ scattered points are inside of C_T. The parameter $M_{\min}$ controls the minimal local approximation order of the spline surface, while $M_{\max}$ acts as a balancing parameter between detail reproduction and overall smoothness. In order to locally reproduce at least linear functions in the areas of very sparse data, we choose $M_{\min} = 3$. Since we need at least 10 scattered data points to fit a cubic polynomial, we require $M_{\max} \geq 10$. In our tests, we found a good heuristic choice to be $20 \leq M_{\max} \leq 60$.

The process of finding the data points that lie inside of a given circle C_T is accelerated using an additional uniform grid data structure $\mathcal{G}$ constructed during the initial input of the data. This grid covers the domain Ω and its resolution is chosen such that an average number of K data points lie within each grid cell. By experiment we found that values of $10 \leq K \leq 20$ lead to a reasonable speed up. Every grid cell is assigned a list of the data points inside that cell. Thus, we can reduce the number of point-in-circle tests significantly by considering only those data points, which are associated to the cells of $\mathcal{G}$ partially or completely covered by C_T. These grid cells can be determined efficiently by using pre-computed two-dimensional bit masks that depend on the radius of C_T and the position of its midpoint relative to the grid cells of $\mathcal{G}$.

The above procedure determines for each $T \in \mathcal{T}$ a set of data points $(\tilde{x}_i, \tilde{y}_i)$, $i = 1, \ldots, m$, that is either the set of all scattered points lying in the circle C_T, or a subset of it obtained by thinning. Figure A.18 on page 335 shows some examples of such sets of data points. We now consider the system of linear equations

$$\sum_{|\alpha|=q} B_T^{\alpha,q}(\tilde{x}_i, \tilde{y}_i)\, b_\alpha \;=\; \tilde{z}_i, \quad i = 1, \ldots, m, \tag{2}$$

where the $\tilde{z}_i$ are the z-values at points $(\tilde{x}_i, \tilde{y}_i)$ and q is the local degree of p_T^q ($q \leq 3$). Denote by A_q the matrix of the system (2). The number M_q of unknown coefficients b_α depends on q and is equal to 10, 6, 3, or 1 if q is 3, 2, 1, or 0, respectively. Depending on m, we initially choose q so that $m \geq M_q$. If, for instance, $m \geq 10$, we choose $q = 3$.

To solve (2) in the least squares sense, we compute the singular value decomposition $A_q = UDV^\top$, where $U \in \mathbb{R}^{m,M_q}$ and $V \in \mathbb{R}^{M_q,M_q}$ are (column-) orthogonal matrices and $D = \mathrm{diag}\{\sigma_1, \ldots, \sigma_{M_q}\}$ is a diagonal matrix containing the singular values σ_j of A_q ($j = 1, \ldots, M_q$). We use the algorithm given in [36] for the computation of the SVD. Since the dimension of A_q is at most $M_{\max} \times 10$, we compute the SVD for small matrices only, making this step fast and robust. The least squares solution of (2) can be efficiently computed as $b = VD^{-1}U^\top \tilde{z}$, since the inverse of D is again a diagonal matrix with reciprocal diagonal elements.

In general, the condition number of the system (2) is given by the ratio $\max_j\{\sigma_j\}/\min_j\{\sigma_j\}$. If at least one of the singular values σ_j is smaller than a prescribed bound $\varepsilon_{\text{cond}}$, i.e., the points $(\tilde{x}_i, \tilde{y}_i)$ lie close to an algebraic curve of degree q, we consider the system (2) to be ill-conditioned and drop the degree q of the least squares polynomial. If the initial degree was $q = 3$, this means that we consider the system (2) once again, but this time for quadratic polynomials, i.e., $q = 2$. If the system (2) for $q = 2$ is ill-conditioned again, we drop the degree further down to $q = 1$ or even $q = 0$. The bound $\varepsilon_{\text{cond}}$ is obtained from a user-specified condition number κ: $\varepsilon_{\text{cond}} := \max_j\{\sigma_j\}/\kappa$. Extensive numerical simulations show that the quality of the resulting spline surface is quite sensitive to the choice of κ, see also Fig. A.19 on page 335. If κ is chosen too high, the spline patches constructed over the triangles in $\mathcal{T}$ tend to be of a higher degree. Although this behavior can reduce the average approximation error at the data points, our tests show that individual data points may exhibit a larger approximation error. On the other hand, if κ is chosen too low, more and more spline patches possess a lower degree, thereby decreasing the local approximation order of the spline. In our tests, we successfully use $\kappa \in [80, 200]$.

In this way, for every $T \in \mathcal{T}$ we determine a polynomial $p_T^q = s|_T$ on T. If the degree q of p_T^q is less than three, the cubic Bernstein-Bézier representation (1) of

$$p_T^q(\mathbf{u}) = \sum_{|\alpha|=q} B_T^{\alpha,q}(\mathbf{u}) a_\alpha, \quad \mathbf{u} \in T, \tag{3}$$

is finally obtained by degree raising. The corresponding relations between the coefficients of p_T^q in its two representations (1) and (3) result from the following equation [16]:

$$\left(\sum_{|\alpha|=q} B_T^{\alpha,q} a_\alpha \right) (\lambda_0 + \lambda_1 + \lambda_2)^{3-q} = \sum_{|\alpha|=3} B_T^{\alpha,3} b_\alpha.$$

3.4 C^1-Conditions

In the second step of the algorithm, we determine the approximating spline s on the remaining triangles $T \in \Delta \setminus \mathcal{T}$ that have non-empty intersection with Ω.

For computing the remaining coefficients of s we use the well-known C^1 smoothness conditions for two adjacent triangular Bézier patches [16]. By using these simple formulas, we compute the Bernstein-Bézier coefficients of the polynomial pieces of the smooth spline s on the triangles $T \in \Delta \setminus \mathcal{T}$ that have a non-empty intersection with Ω. In particular, we do not need to perform this step for the remaining triangles in the border cells. This computation works step by step as illustrated in Fig. A.16 on page 334. Here, so-called *domain points* in the part of the domain surrounded by four triangles in $\mathcal{T}$ are shown. Each domain point represents a Bernstein-Bézier coefficient of the spline. In particular, the points shown in black correspond to the Bernstein-Bézier coefficients on the triangles in $\mathcal{T}$. We first use these coefficients and the above C^1 smoothness conditions to compute the coefficients corresponding to the red points. Then we compute successively the coefficients shown in blue, green, and yellow. Note that in the last step there are several possibilities to compute the coefficient shown in yellow. We just choose one of them since the result is unique as follows by a standard argument.

4 Rendering

To efficiently render our spline surface, we adapted several techniques from [8, 17, 29, 44] to smooth Bézier spline surfaces. We start by overlaying the domain Ω with a uniform, axis-aligned *render grid*. The resolution of this grid is adjustable, we typically use about 30×30 grid cells. Each of the grid cells c_{ij} is assigned a minimum and a maximum z-value $z_{ij}^{\min}$ and $z_{ij}^{\max}$, which are taken from the minimum and the maximum Bernstein-Bézier coefficient of all the Bézier triangles covered by c_{ij}, respectively. Since the spline surface is known to lie completely inside the convex hull of its Bernstein-Bézier control points, we can use the boxes b_{ij} defined by the grid cells c_{ij} and their associated z-values $z_{ij}^{\min}$ and $z_{ij}^{\max}$ as bounding boxes for a cheap visibility test. Our rendering algorithm is organized as follows:

```
for each box b_{ij}
    cull b_{ij} against view frustum;
    if b_{ij} not completely culled
        mark b_{ij} as visible;
        compute bounding rectangle of projection
            of b_{ij} into screen space;
compute LOD from max bounding rect extent;
for each visible box b_{ij}
    if LOD ≠ b_{ij}.LOD
        evaluate spline on uniform sub-grid of
            level LOD over c_{ij};
        draw triangle strips into b_{ij}.displayList;
        b_{ij}.LOD = LOD;
    execute b_{ij}.displayList;
```

The culling of the axis-aligned bounding boxes b_{ij} is performed in a conservative but cheap way: Each of the eight vertices of a box is projected into screen space. The box is rejected from visibility if and only if all of the projected vertices lie completely inside the same of one of the four half planes above, below, left, or right of the viewport. An early acceptance of the box is found, if any of the projected vertices lies inside the viewport. We found that this conservative test performs better than a more accurate test that includes 2D line clipping for each of the twelve edges of a box.

For each box b_{ij} that passes the culling test, we compute the bounding rectangle of the projected vertices. The maximum width / height of all these bounding rectangles is used to compute the global level-of-detail (LOD) for the current viewpoint. Obviously, it would be more efficient to have an individual LOD for each box: grid cells that are projected onto only a few pixels need not be subdivided as finely as grid cells that cover a large area of the viewport. However, such an adaptive LOD might result in visible gaps between adjacent grid cells, since the common boundary curve of the spline surface between two grid cells might be represented by a different number of linear segments for each cell. To overcome these problems, we are currently investigating the applicability of view-dependent sampling and triangle mesh generation as suggested for instance in [15].

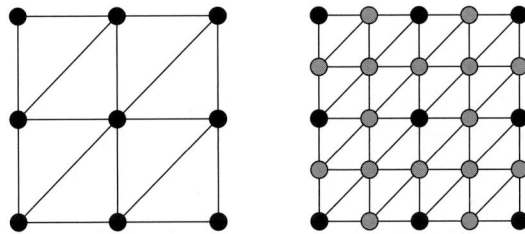

Fig. 2. Sub-grids and resulting triangle meshes corresponding to LOD = 1 (left) and LOD = 2 (right). The spline surface is evaluated at the positions ● for level one. For level two, the spline needs to be evaluated only at the new positions ●.

After the global LOD has been computed, the spline surface is evaluated within every visible box b_{ij} on a uniform sub-grid over the domain of the grid cell c_{ij}. The resolution of the sub-grid is determined by the LOD. In our implementation we use a discrete LOD such that incrementing the LOD by one doubles the resolution of the sub-grid along both of its dimensions (see Fig. 2). This approach has the advantage that all surface points and normals from level j can be re-used for level $j+1$. Thus the spline surface has to be evaluated only at the new positions for a higher level. In the case of going down from level $j+1$ to level j, we do not need to evaluate the spline at all – all the surface points and normals we need have already been computed. Since the number of triangles quadruples when going from level j to level $j+1$, it might seem reasonable to split every triangle into only two triangles, instead. In this case, however, it is not possible to generate optimal triangle strips for efficient rendering: Either the triangle strips are running diagonally, so that additional expensive texture binding calls are needed, or the triangle strips become much shorter than in our current approach.

To avoid possible visual discontinuities ("popping effects") when changing the LOD during an animation, we blend the surfaces S_{new} (corresponding to the new LOD) and S_{old} (corresponding to the previous LOD) over k successive frames using the alpha buffer and a blend function provided by our graphics hardware. Since we achieve real-time frame rates for rendering, we found that $k = 20$ yields visually smooth transitions from S_{old} to S_{new}.

The evaluation of the spline within every visible box b_{ij} is done using the de Casteljau algorithm [12]. In contrast to tensor product patches of the same degree, the de Casteljau algorithm for bivariate triangle patches takes about half as many multiplications and additions to compute a point on the spline surface. If, like in our case for proper shading, the surface normal has to be computed as well, the advantage of bivariate triangle patches becomes even more evident: In addition to the costs of evaluating a surface point, we need only three multiplications and additions each to compute the (unnormalized) exact normal vector in that point.

5 Results

In order to verify the working of our method, we performed numerical tests on several real world scattered data sets which vary widely in size and quality. All simulations were run on an sgi Octane with a 300 MHz R12k processor and 1 GB main memory. The approximation quality of a spline is measured using the one-sided Hausdorff distance $d_H : \mathbb{R}^3 \times \mathcal{S} \rightarrow \mathbb{R}$ between the scattered data points $(x_i, y_i, z_i) \in \mathbb{R}^3$ and the spline $s \in \mathcal{S}$. To facilitate comparison of different data sets, we divide the maximum approximation error by the length δ of the diagonal of the scattered data's bounding-box. The results of our simulations are summarized in Table 1.

The results in Table 1 clearly show that both the runtime and the memory usage of our algorithm is linear in the number N of scattered data points. The approximation quality of the spline depends on the quality of the input data: The more densely the scattered data points lie, the more least squares polynomial patches of higher order will be created in general, thereby increasing the local approximation order of the spline. Our simulations show that large real world scattered data can be approximated well within some seconds: The maximum approximation error for our largest data set (588 km^2 with 360 meters elevation range) is about 14 meters (relative error $5.2 \cdot 10^{-4}$). In particular, in the case of huge and dense data sets, the deviation of the approximating spline to the majority of data points is less than one meter.

Once a spline approximation to a scattered data set has been computed, this spline surface can be efficiently evaluated for rendering using the techniques described in Sect. 4. On our Octane with a Vpro/8 graphics board

Table 1. Results of smooth approximation for different real world scattered data sets. Timings are given for the construction of the spline; see Sect. 5 for details on operating platform and Hausdorff distance d_H. The diameter of the scattered data's bounding-box is denoted by δ.

N	9,704	45,324	736,577	2,958,078
avg. density (data / km^2)	81	278	5,001	5,020
memory usage	1.8 MB	4.2 MB	48 MB	189 MB
CPU time	0.3 s	1.4 s	19.9 s	92.9 s
percentage of patches of degree 3	61.9%	95.6%	77.6%	78.1%
percentage of patches of degree 2	8.4%	1.4%	20.4%	20.6%
percentage of patches of degree 1	13.7%	2.8%	1.4%	0.8%
percentage of patches of degree 0	16.0%	0.2%	0.6%	0.5%
$\max(|d_H|/\delta)$	3.1E-3	2.4E-3	8.5E-4	5.2E-4
percentage of data with $|d_H| > 10\,\mathrm{m}$	3.3%	9.0%	0.002%	0.002%
percentage of data with $|d_H| > 5\,\mathrm{m}$	17.5%	28.5%	0.06%	0.07%
percentage of data with $|d_H| > 1\,\mathrm{m}$	95.3%	84.7%	8.4%	8.5%

we achieve real-time frame rates of 30–60 fps in typical fly-through sequences with up to 100k Gouraud-shaded and textured triangles per frame. Moreover, it is possible to locally modify the data, e.g., for animation or modeling, and render the adapted spline surface at interactive frame rates. We have created an animation sequence of a surface that has been constructed from 736k scattered data points. In an area that covers about 0.4% of the surface, the scattered data points are moved from frame to frame. Due to the locality of the method, we are able to recompute the Bernstein-Bézier coefficients of the polynomial pieces for each frame and render the spline surface at about 10 fps.

An additional area of application for our method is (lossy) geometric compression (see Fig. A.17 on page 334). Storing the $5n^2$ Bernstein-Bézier coefficients that have been computed as described in Sect. 3.3 allows to reconstruct the complete spline due to C^1-conditions. For $n = \lfloor \sqrt{N/5} \rfloor$, this requires storage of approximately N floating point numbers only, since the position of the Bernstein-Bézier coefficients is implicitly defined by our scheme[3]. Since the scattered data require storage of $3N$ floats, we obtain a compression ratio of about $3:1$. By decreasing the resolution n of the spline grid, we can control the compression ratio. Unlike many other compression techniques, our method becomes faster when using higher compression ratios. Compressing three million scattered data points from an area of 588 km^2 with an elevation range of 360 meters by a ratio of $30:1$ takes about 15 seconds and yields a maximum approximation error of 20 meters. Although our method will probably

---

[3] In addition, the number n and the min/max coordinates of the scattered data in the xy-plane have to be stored.

not achieve the same reconstruction quality as state-of-the-art geometric compression or mesh reduction techniques [4], it can compete very well with such techniques concerning run-time complexity. Most of these techniques operate on polygonal meshes. If, however, the input data are given as unorganized scattered data, it takes $\mathcal{O}(N \log N)$ time to compute a triangulation. Since our method computes the Bernstein-Bézier coefficients of a smooth approximation of the scattered data in $\mathcal{O}(N)$ time, it might be the method of choice for very large N.

6 Conclusion and Future Work

We have presented an efficient method to automatically compute smooth approximations of large sets of unorganized scattered data points. The method is based on the construction of a differentiable bivariate spline with respect to a uniform triangle mesh over an arbitrarily shaped planar domain. For a uniformly distributed subset of triangles we compute local polynomial least squares approximations by using singular value decomposition (SVD) of small matrices. The smooth approximating spline is constructed by gluing together these patches using Bernstein-Bézier smoothness conditions. We emphasize the following key features of our method:

- We develop a completely local approach, which means that we do not use any global optimization or other techniques involving computation with large portions of the data set.
- We employ the rank-revealing features of SVD to control the polynomial degree of the initial patches, which allows to take into account the local variation and distribution of the data points.
- The algorithm does not make use of any interpolation scheme. In particular, no estimation of derivatives is needed.
- Our method offers optimal approximation order and the constructed spline is by its nature C^1-continuous. In addition, the spline surface does not have artifacts like, for instance, peaks or flat spots close to the data points.
- The use of a uniform triangle mesh also contributes to great savings in computation time. As a result, the overall complexity of the algorithm is linear in the number of scattered data points.

The numerical examples with millions of scattered points of real world data sets show that the approximating spline can be computed very fast, the approximation error is very small, the resulting surfaces are of high visual quality and can be easily used for further processing in the context of rendering, modeling, and geometric compression.

In addition we note that the most expensive step of our algorithm is the computation of local discrete least squares approximations (more than 95 % of the overall computation time). Since these approximations are computed independently from each other, this step can be easily parallelized, thus leading to savings proportional to the number of processors on a multi-processor machine.

Considering these results, a variety of generalizations and applications of the new method can be thought of. In particular, the local adaptiveness of the method can be increased by designing multilevel spline spaces and adjusting the local dimension of the spline space adaptively to the local distribution of data points. However, because of the additional computational complexity, it is not clear if this approach will increase the overall efficiency of the algorithm. Other natural and important questions that include generalizations of the method to the reconstruction of surfaces of higher smoothness and more general topology or to the reconstruction of trivariate functions using, e.g., Bézier tetrahedra are currently under investigation.

Acknowledgement. The data used for the simulations and images was taken from the digital terrain model, scale 1:5000 (DGM 5) by kind permission of "Landesamt für Kataster-, Vermessungs- und Kartenwesen des Saarlandes" under license numbers G-07/00 (9/26/00) and D 90/2000 (10/17/00).

This paper was published originally in *Proceedings of IEEE Visualization*, October 2001, pages 341–347.

References

[1] E. Arge, M. Dæhlen, and A. Tveito. Approximation of Scattered Data Using Smooth Grid Functions. *J. Computational and Applied Math.*, 59:191–205, 1995.

[2] M. D. Buhmann. Radial Basis Functions. *Acta Numerica*, pp. 1–38, 2000.

[3] C. K. Chui. *Multivariate Splines*. SIAM, 1988.

[4] P. Cignoni, C. Montani, and R. Scopigno. A Comparison of Mesh Simplification Algorithms. *Computers & Graphics*, 22(1):37–54, 1998.

[5] P. Cignoni, E. Puppo, and R. Scopigno. Representation and Visualization of Terrain Surfaces at Variable Resolution. *The Visual Computer*, 13(5):199–217, 1997.

[6] D. Cohen and A. Shaked. Photo-Realistic Imaging of Digital Terrains. In *Computer Graphics Forum (Proc. Eurographics '93)*, volume 12, pp. C363–C373, 1993.

[7] D. Cohen-Or and A. Shaked. Visibility and Dead-Zones in Digital Terrain Maps. In *Computer Graphics Forum (Proc. Eurographics '95)*, volume 14, pp. C171–C180, 1995.

[8] S. Coquillart and M. Gangnet. Shaded Display of Digital Maps. *IEEE Computer Graphics and Applications*, 4(7):35–42, July 1984.

[9] M. Dæhlen and V. Skyth. Modelling Non-rectangular Surfaces using Box-splines. In D. C. Handscomb, *Mathematics of Surfaces III*, pp. 287–300. 1989.

[10] W. A. Dahmen, R. H. J. Gmelig Meyling, and J. H. M. Ursem. Scattered Data Interpolation by Bivariate C^1-piecewise Quadratic Functions. *Approximation Theory and its Applications*, 6:6–29, 1990.

[11] O. Davydov and F. Zeilfelder. Scattered Data Fitting by Direct Extension of Local Polynomials to Bivariate Splines. *Advances in Comp. Math.*, 21:223-271, 2004.

[12] P. de Casteljau. Outillages Méthodes Calcul. Technical report, Andre Citroen Automobiles, Paris, 1959.

[13] P. Dierckx. *Curve and Surface Fitting with Splines.* Oxford University Press, 1993.

[14] P. Dierckx, S. Van Leemput, and T. Vermeire. Algorithms for Surface Fitting using Powell-Sabin Splines. *IMA Journal of Numerical Analysis*, 12(2):271–299, 1992.

[15] M. Duchaineau, M. Wolinsky, D. E. Sigeti, M. C. Miller, C. Aldrich, and M. B. Mineev-Weinstein. ROAMing Terrain: Real-time Optimally Adapting Meshes. In *Proc. IEEE Visualization*, pp. 81–88, 1997.

[16] G. Farin. *Curves and Surfaces for Computer Aided Geometric Design.* Academic Press, 4. edition, 1993.

[17] R. L. Ferguson, R. Economy, W. A. Kelley, and P. P. Ramos. Continuous Terrain Level of Detail for Visual Simulation. In *Proc. Image V Conference 1990*, pp. 145–151, 1990.

[18] D. R. Forsey and R. H. Bartels. Surface Fitting with Hierarchical Splines. *ACM Transactions on Graphics*, 14(2):134–161, April 1995.

[19] R. Franke. Scattered Data Interpolation: Test of Some Methods. *Mathematics of Computation*, 38(157):181–200, January 1982.

[20] R. Franke and H. Hagen. Least Squares Surface Approximation using Multiquadrics and Parameter Domain Distortion. *Computer Aided Geometric Design*, 16(3):177–196, 1999.

[21] R. H. J. Gmelig Meyling and P. R. Pfluger. Smooth Interpolation to Scattered Data by Bivariate Piecewise Polynomials of Odd Degree. *Computer Aided Geometric Design*, 7(5):439–458, August 1990.

[22] B. F. Gregorski, B. Hamann, and K. I. Joy. Reconstruction of B-spline Surfaces from Scattered Data Points. In *Proc. Computer Graphics International 2000*, pp. 163–170, 2000.

[23] G. Greiner and K. Hormann. Interpolating and Approximating Scattered 3D Data with Hierarchical Tensor Product Splines. In A. Le Méhauté, C. Rabut, and L. L. Schumaker, *Surface Fitting and Multiresolution Methods*, pp. 163–172. 1996.

[24] H. Hoppe, T. DeRose, T. Duchamp, M. Halstead, H. Jin, J. McDonald, J. Schweitzer, and W. Stuetzle. Piecewise Smooth Surface Reconstruction. In *Computer Graphics (SIGGRAPH '94 Conf. Proc.)*, pp. 295–302, 1994.

[25] R. Klein, D. Cohen-Or, and T. Hüttner. Incremental View-dependent Multiresolution Triangulation of Terrain. *The Journal of Visualization and Computer Animation*, 9(3):129–143, July–September 1998.

[26] P. Lancaster and K. Šalkauskas. *Curve and Surface Fitting.* Academic Press, 1986.

[27] C.-H. Lee and Y. G. Shin. A Terrain Rendering Method Using Vertical Ray Coherence. *Journal of Visualization and Computer Animation*, 8(2):97–114, 1997.

[28] S. Lee, G. Wolberg, and S. Y. Shin. Scattered Data Interpolation with Multilevel B-Splines. *IEEE Transactions on Visualization and Computer Graphics*, 3(3):228–244, July 1997.

[29] P. Lindstrom, D. Koller, W. Ribarsky, L. F. Hughes, N. Faust, and G. Turner. Real-Time, Continuous Level of Detail Rendering of Height Fields. In *Computer Graphics (SIGGRAPH '96 Conf. Proc.)*, pp. 109–118, 1996.

[30] S. K. Lodha and R. Franke. Scattered Data Techniques for Surfaces. In H. Hagen, G. Nielson, and F. Post, *Proc. Dagstuhl Conf. Scientific Visualization*, pp. 182–222, 1999.

[31] M. Morandi Cecchi, S. De Marchi, and D. Fasoli. A Package for Representing C^1 Interpolating Surfaces: Application to the Lagoon of Venice's Bed. *Numerical Algorithms*, 20(2,3):197–215, 1999.

[32] G. Nürnberger, L. L. Schumaker, and F. Zeilfelder. Local Lagrange Interpolation by Bivariate C^1 Cubic Splines. In *Proc. Conference on Curves and Surfaces*, 2001. in print.

[33] G. Nürnberger and F. Zeilfelder. Local Lagrange Interpolation by Cubic Splines on a Class of Triangulations. In *Proc. Conf. Trends in Approximation Theory 2000*, pp. 341–350, 2001.

[34] R. Pfeifle and H.-P. Seidel. Fitting Triangular B-splines to Functional Scattered Data. In *Proc. Graphics Interface '95*, pp. 80–88, 1995.

[35] M. J. D. Powell. Radial Basis Functions for Multivariable Interpolation. In J. C. Mason and M. G. Cox, *Algorithms for Approximation of Functions and Data*, pp. 143–168. 1987.

[36] W. H. Press, S. A. Teukolsky, W. T. Vetterling, and B. P. Flannery. *Numerical Recipes in C: The Art of Scientific Computing*. Cambridge University Press, 2. edition, 1992.

[37] H. Qin and D. Terzopoulos. D-NURBS: A Physics-Based Framework for Geometric Design. *IEEE Transactions on Visualization and Computer Graphics*, 2(1):85–96, March 1996.

[38] R. Schaback. Improved Error Bounds for Scattered Data Interpolation by Radial Basis Functions. *Mathematics of Computation*, 68(225):201–216, January 1999.

[39] F. J. M. Schmitt, B. B. Barsky, and W. Du. An Adaptive Subdivision Method for Surface-Fitting from Sampled Data. In *Computer Graphics (SIGGRAPH '86 Conf. Proc.)*, pp. 179–188, 1986.

[40] L. L. Schumaker. Fitting Surfaces to Scattered Data. In G. G. Lorentz, C. K. Chui, and L. L. Schumaker, *Approximation Theory II*, pp. 203–268. 1976.

[41] D. A. Southard. Piecewise Planar Surface Models from Sampled Data. In *Proc. Computer Graphics International '91)*, pp. 667–680, 1991.

[42] A. J. Stewart. Hierarchical Visibility in Terrains. In *Rendering Techniques '97 (Proc. 8th EG Workshop on Rendering)*, pp. 217–228, 1997.

[43] C. Wiley, A. T. Campbell III, S. Szygenda, D. Fussell, and F. Hudson. Multiresolution BSP Trees Applied to Terrain, Transparency, and General Objects. In *Proc. Graphics Interface '97*, pp. 88–96, 1997.

[44] M. Woo, J. Neider, T. Davis, and D. Shreiner. *OpenGL Programming Guide*. Addison–Wesley, Reading, MA, 1999.

[45] W. Zhang, Z. Tang, and J. Li. Adaptive Hierarchical B-Spline Surface Approximation of Large-Scale Scattered Data. In *Proc. Pacific Graphics '98*, pp. 8–16, 1998.

[46] J. Zhou, N. M. Patrikalakis, S. T. Tuohy, and X. Ye. Scattered Data Fitting with Simplex Splines in Two and Three Dimensional Spaces. *The Visual Computer*, 13(7):295–315, 1997.

Fast Projected Convolution of Piecewise Linear Functions on Non-equidistant Grids

W. Hackbusch

Max-Planck-Institut Mathematik in den Naturwissenschaften, Inselstr. 22, D-04103 Leipzig, Germany, wh@mis.mpg.de

Summary. Usually, the fast evaluation of a convolution integral $\int_{\mathbb{R}} f(y)g(x-y)\mathrm{d}y$ requires that the functions f, g are discretised on an equidistant grid in order to apply FFT. Here we discuss the efficient performance of the convolution in locally refined grids. More precisely, f and g are assumed to be piecewise linear and the convolution result is projected into the space of linear functions in a given locally refined grid. Under certain conditions, the overall costs are still $\mathcal{O}(N \log N)$, where N is the sum of the dimensions of the subspaces containing f, g and the resulting function.

1 Introduction

We consider the convolution integral

$$\omega_{\text{exact}}(x) := (f * g)(x) := \int_{\mathbb{R}} f(y)g(x-y)\mathrm{d}y \tag{1}$$

for functions f, g of bounded support.

The computations are restricted to functions f, g which are piecewise linearly defined on *locally refined meshes* with possible discontinuities at the grid points. A simple example of such a locally refined mesh is depicted below,

$$\tag{2a}$$

showing a typical refinement towards $x = 0$. The depicted mesh can be decomposed into different levels as indicated below:

level 0
level 1
level 2 (2b)

The latter representation uses several levels, but each level ℓ is associated with an equidistant grid of size

$$h_\ell := 2^{-\ell} h \qquad (0 \le \ell \le L). \tag{3}$$

The largest level number appearing in the grid hierarchy is denoted by L.

In principle, one can approximate $f * g$ via Fourier transform: Also for non-equidistant grids there are ways to approximate the Fourier transform $\hat{f}$ and $\hat{g}$ (see [6]) and the back transform of $\hat{f} \cdot \hat{g}$ would yield an approximation of $f * g$. A fast algorithm for a generalised convolution is described in [7]. However, in these approaches the approximation error depends on the interpolation error and we do not guarantee any smoothness of f or g. In contrary, the use of locally refined meshes indicates a nonsmooth situation. In our approach we avoid interpolation errors, since the quantities of interest are computed exactly.

To cover the general case, we will allow that the functions f and g involved in the convolution belong to possibly different locally refined meshes. Also the resulting convolution will be described by a third locally refined grid, which may be different from the grids for f and g.

Since the locally refined meshes have the purpose to approximate some functions f_{exact} and g_{exact} in an adaptive way by f and g, it is only natural to approximate ω_{exact} also by a piecewise linear function ω in a third given locally refined mesh. We use the L^2-orthogonal projection onto this space to obtain the final result ω from ω_{exact}.

Therefore, the goal of the algorithm is to compute ω as the L^2-orthogonal projection of $f * g$. Note that we compute the *exact* L^2-orthogonal projection, i.e., there is no approximation error except the unavoidable projection error.

The computation of $f * g$ for functions from locally refined grids is also discussed in the previous papers [2], [3], [4]. In [2, §§3-6], f, g are assumed to be piecewise constant. [2] describes how to evaluate $\omega_{\text{exact}} := f * g$ at the grid points of the given locally refined grid. [3] explains in general how the convolution of piecewise polynomial functions f, g defined on locally refined grids can be projected L^2-orthogonally into another piecewise polynomial space. However, the concrete algorithm given in [3] is restricted to piecewise constant functions, while for polynomial degree $p > 0$ only a general guideline is given (cf. [3, §8]). The present paper can be considered as the concretisation of the technique for polynomial degree $p = 1$. The paper [4] is also dealing with the piecewise constant case, but describes the necessary modifications in order to obtain local mass conservation for certain population balance models.

The organisation of this paper is as follows.

Section 2 gives a precise definition of locally refined meshes and of the corresponding ansatz spaces $\mathcal{S}^f, \mathcal{S}^g$ which f and g belong to and of the target space $\mathcal{S}^\omega$ for the projection ω of $\omega_{\text{exact}} = f * g$. In particular, the basis functions are introduced in §2.3.

Section 3 introduces some notations and formulates the basic problem.

Section 4 introduces the γ, G, and Γ-coefficients which are essential for the representation of the projected values. The Γ-coefficients will appear in the algorithm.

The main chapter of this paper is Sect. 5 describing the algorithm. Here we have to perform discrete convolutions of sequences by means of the FFT technique. The precise implementation of the discrete convolution is described in [3, §6] and need not repeated here. Also the estimation of the computational cost is contained in [3]. There, under certain conditions, the bound

$$\mathcal{O}\left(N \log N\right)$$

is derived, where N describes the data size of the factors f, g and the projected convolution $\omega = P(f * g)$.

So far we have treated the case of piecewise linear but discontinuous functions. The last section 6 discusses the case of globally continuous piecewise linear functions and proposes an efficient algorithm for the projected convolution. Applications involving convolutions can be found in [1] and [2].

Due to lack of space, some derivations and proofs omitted here can be found in the extended version [5].

2 Spaces

2.1 The Locally Refined Meshes

The grids depicted in (2b) are embedded into infinite grids $\mathcal{M}_\ell$ which are defined below. With h_ℓ from (3) we denote the subintervals of level ℓ by

$$I_\nu^\ell := [\nu h_\ell, (\nu+1) h_\ell) \qquad \text{for } \nu \in \mathbb{Z}, \, \ell \in \mathbb{N}_0. \tag{4}$$

This defines the meshes with the grid points $\{\nu h_\ell : \nu \in \mathbb{Z}\}$:

$$\mathcal{M}_\ell := \{I_\nu^\ell : \nu \in \mathbb{Z}\} \qquad \text{for } \ell \in \mathbb{N}_0. \tag{5}$$

A finite and *locally refined mesh* $\mathcal{M}$ is a set of finitely many disjoint intervals from various levels, i.e.,

$$\mathcal{M} \subset \bigcup_{\ell \in \mathbb{N}_0} \mathcal{M}_\ell, \quad \text{all } I, I' \in \mathcal{M} \text{ with } I \neq I' \text{ are disjoint}, \quad \#\mathcal{M} < \infty. \tag{6}$$

2.2 The Ansatz Spaces

The piecewise linear space $\mathcal{S}$ corresponding to the mesh $\mathcal{M}$ is defined by

$$\mathcal{S} = \mathcal{S}\left(\mathcal{M}\right) = \{\phi \in L^\infty(\mathbb{R}) : \phi|_I \text{ linear if } I \in \mathcal{M}, \, \phi|_{\mathbb{R} \setminus \bigcup_{I \in \mathcal{M}} I} = 0\}.$$

The two factors f, g of the convolution ω_{exact} from (1) as well as the (projected) image ω may be organised by three different locally refined meshes

$$\mathcal{M}^f, \quad \mathcal{M}^g, \quad \mathcal{M}^\omega, \tag{7}$$

which are all of the form (6). These meshes give rise to three spaces

$$\mathcal{S}^f := \mathcal{S}\left(\mathcal{M}^f\right), \quad \mathcal{S}^g := \mathcal{S}\left(\mathcal{M}^g\right), \quad \mathcal{S}^\omega := \mathcal{S}\left(\mathcal{M}^\omega\right).$$

We recall that $f \in \mathcal{S}^f$ and $g \in \mathcal{S}^g$ are the input data, while the result is the exact L^2-orthogonal projection of $f * g$ onto $\mathcal{S}^\omega$.

2.3 Basis Functions

Functions from $\mathcal{S}(\mathcal{M})$ may be discontinuous at the grid points of the mesh. This fact has the advantage that the basis functions spanning $\mathcal{S}(\mathcal{M})$ have minimal support (the support is just one interval of $\mathcal{M}$).

Here, we consider the case of piecewise linear functions. The following basis functions of level ℓ are derived from the Legendre polynomials:

$$\Phi_{i,0}^{\ell}(x) := 1/\sqrt{h_{\ell}} \qquad\qquad \text{if } x \in I_i^{\ell}, \tag{8a}$$

$$\Phi_{i,1}^{\ell}(x) := \sqrt{12}\left(x - x_{i+1/2}^{\ell}\right)/h_{\ell}^{3/2} \quad \text{if } x \in I_i^{\ell}, \tag{8b}$$

and $\Phi_{i,\alpha}^{\ell}(x) = 0$ $(\alpha = 0,1)$ otherwise. $x_{i+1/2}^{\ell} := (i+1/2)h_{\ell}$ is the midpoint of the interval $I_i^{\ell} = \mathrm{supp}(\Phi_{i,\alpha}^{\ell})$. Note that $\Phi_{i,\alpha}^{\ell}$ are orthonormal. Since the intervals $I \in \mathcal{M}$ are non-overlapping, the functions in the right-hand side of

$$\mathcal{S}(\mathcal{M}) = \mathrm{span}\left\{\Phi_{i,\alpha}^{\ell} : \alpha \in \{0,1\}, \; I_i^{\ell} \in \mathcal{M}\right\}$$

form an orthonormal system of $\mathcal{S}(\mathcal{M})$.

$\mathcal{S}_{\ell}$ is the space of piecewise linear functions of level ℓ on the mesh $\mathcal{M}_{\ell}$:

$$\mathcal{S}_{\ell} := \mathrm{span}\left\{\Phi_{i,\alpha}^{\ell} : \alpha \in \{0,1\}, \; i \in \mathbb{Z}\right\} \qquad (\ell \in \mathbb{N}_0). \tag{9}$$

For fixed ℓ, the basis $\left\{\Phi_{i,\alpha}^{\ell} : \alpha \in \{0,1\}, \; i \in \mathbb{Z}\right\}$ is orthonormal due to the chosen scaling.

The spaces $\mathcal{S}_{\ell}$ are nested, i.e., $\mathcal{S}_{\ell} \subset \mathcal{S}_{\ell+1}$. In particular, $\Phi_{i,\alpha}^{\ell}$ can be represented by means of $\Phi_{i,\alpha}^{\ell+1}$:

$$
\begin{aligned}
\Phi_{i,0}^{\ell} &= \tfrac{1}{\sqrt{2}}\left(\Phi_{2i,0}^{\ell+1} + \Phi_{2i+1,0}^{\ell+1}\right), \\
\Phi_{i,1}^{\ell} &= \tfrac{1}{2\sqrt{2}}\left(\Phi_{2i,1}^{\ell+1} + \Phi_{2i+1,1}^{\ell+1} + \sqrt{3}\left(\Phi_{2i+1,0}^{\ell+1} - \Phi_{2i,0}^{\ell+1}\right)\right).
\end{aligned}
\tag{10}
$$

3 Notations and Definition of the Problem

3.1 Representations of $f \in \mathcal{S}^f$ and $g \in \mathcal{S}^g$

Following the definition of $\mathcal{S}^f$, we have $\mathcal{S}^f = \mathrm{span}\left\{\Phi_i^{\ell} : I_i^{\ell} \in \mathcal{M}^f\right\}$. We can decompose the set $\mathcal{M}^f$ into different levels: $\mathcal{M}^f = \bigcup_{\ell=0}^{L^f} \mathcal{M}_{\ell}^f$, where $\mathcal{M}_{\ell}^f := \mathcal{M}^f \cap \mathcal{M}_{\ell}$. This gives rise to the related index set

$$\mathcal{I}_{\ell}^f := \left\{i \in \mathbb{Z} : I_i^{\ell} \in \mathcal{M}_{\ell}^f\right\} \tag{11}$$

and to the corresponding decomposition

$$\mathcal{S}^f = \bigcup_{\ell=0}^{L^f} \mathcal{S}_{\ell}^f \qquad \text{with } \mathcal{S}_{\ell}^f = \mathrm{span}\left\{\Phi_{i,\alpha}^{\ell} : \alpha = 0,1, \; i \in \mathcal{I}_{\ell}^f\right\}.$$

Here, L^f is the largest level ℓ with $\mathcal{M}_\ell^f \neq \emptyset$. Similarly, $\mathcal{I}_\ell^g$, $\mathcal{S}_\ell^g$ and L^g correspond to $\mathcal{M}^g$ and $\mathcal{S}^g$.

We start from the representation

$$f = \sum_{\ell=0}^{L^f} f_\ell, \qquad f_\ell = \sum_{i \in \mathcal{I}_\ell^f} \sum_{\alpha=0}^{1} f_{i,\alpha}^\ell \Phi_{i,\alpha}^\ell \in \mathcal{S}_\ell^f , \qquad (12a)$$

$$g = \sum_{\ell=0}^{L^g} g_\ell, \qquad g_\ell = \sum_{i \in \mathcal{I}_\ell^g} \sum_{\alpha=0}^{1} g_{i,\alpha}^\ell \Phi_{i,\alpha}^\ell \in \mathcal{S}_\ell^g . \qquad (12b)$$

Similarly, the final L^2-orthogonal projection w of $f * g$ will have the form

$$w = \sum_{\ell=0}^{L^w} w_\ell, \qquad w_\ell = \sum_{i \in \mathcal{I}_\ell^w} \sum_{\alpha=0}^{1} w_{i,\alpha}^\ell \Phi_{i,\alpha}^\ell \in \mathcal{S}_\ell^w . \qquad (12c)$$

3.2 Projections P and P_ℓ

The L^2-orthogonal projection P onto $\mathcal{S}_\ell^w$ is defined by

$$P\varphi := \sum_{i \in \mathcal{I}_\ell^w} \sum_{\alpha=0}^{1} \langle \varphi, \Phi_{i,\alpha}^\ell \rangle \, \Phi_{i,\alpha}^\ell \in \mathcal{S}_\ell^w$$

with the L^2-scalar product $\langle \varphi, \psi \rangle = \int_{\mathbb{R}} \varphi\psi \mathrm{d}x$. We will also use the L^2-orthogonal projection P_ℓ onto the space $\mathcal{S}_\ell$ from (9) defined by

$$P_\ell \varphi := \sum_{i \in \mathbb{Z}} \sum_{\alpha=0}^{1} \langle \varphi, \Phi_{i,\alpha}^\ell \rangle \, \Phi_{i,\alpha}^\ell. \qquad (13)$$

3.3 Definition of the Basic Problem

We use the decomposition into scales expressed by $f = \sum_{\ell'=0}^{L^f} f_{\ell'}$ and $g = \sum_{\ell=0}^{L^g} g_\ell$ (see (12a,b)). The convolution $f * g$ can be written as

$$f * g = \sum_{\ell'=0}^{L^f} \sum_{\ell=0}^{L^g} f_{\ell'} * g_\ell .$$

Since the convolution is symmetric, we can rewrite the sum as

$$f * g = \sum_{\ell' \leq \ell} f_{\ell'} * g_\ell + \sum_{\ell < \ell'} g_\ell * f_{\ell'} , \qquad (14)$$

where ℓ', ℓ are restricted to the level intervals $0 \leq \ell' \leq L^f$, $0 \leq \ell \leq L^g$. Hence, the basic task is as follows.

Problem 1. Let $\ell' \le \ell$, $f_{\ell'} \in S_{\ell'}$, $g_\ell \in S_\ell$, and $\ell'' \in \mathbb{N}_0$ a further level. Then, the projection $P_{\ell''}(f_{\ell'} * g_\ell)$ is to be computed. More precisely, only the restriction of $P_{\ell''}(f_{\ell'} * g_\ell)$ to indices related to intervals of $\mathcal{M}_{\ell''}^\omega$ is needed.

Because of the splitting (14), we may assume $\ell' \le \ell$ without loss of generality. In the case of the second sum one has to interchange the rôles of the symbols f and g.

Before we present the solution algorithm in Sect. 5, we introduce some further notations in the next section 4.

4 Auxiliary Coefficients

4.1 γ-Coefficients

For level numbers $\ell'', \ell', \ell \in \mathbb{N}_0$ and integers $i, j, k \in \mathbb{Z}$ we define

$$\gamma_{(i,\alpha),(j,\beta),(k,\varkappa)}^{\ell'',\ell',\ell} := \iint \Phi_{i,\alpha}^{\ell''}(x)\,\Phi_{j,\beta}^{\ell'}(y)\,\Phi_{k,\varkappa}^{\ell}(x-y)\mathrm{d}x\mathrm{d}y \qquad (15)$$

(all integrations over $\mathbb{R}$). The connection to the computation of the projection

$$\omega_{\ell''} = P_{\ell''}(f_{\ell'} * g_\ell) \qquad (16)$$

of the convolution $f_{\ell'} * g_\ell$ from Problem 1 is as follows. $\omega_{\ell''}$ is represented by

$$\omega_{\ell''} = \sum_{i\in\mathbb{Z}}\sum_{\alpha=0}^{1} \omega_{i,\alpha}^{\ell''}\Phi_{i,\alpha}^{\ell''},$$

where the coefficients $\omega_{i,\alpha}^{\ell''}$ result from

$$\omega_{i,\alpha}^{\ell''} = \int (f_{\ell'} * g_\ell)(x)\,\Phi_{i,\alpha}^{\ell''}(x)\,\mathrm{d}x = \sum_{j,k\in\mathbb{Z}}\sum_{\beta,\varkappa=0}^{1} f_{j,\beta}^{\ell'}\,g_{k,\varkappa}^{\ell}\,\gamma_{(i,\alpha),(j,\beta),(k,\varkappa)}^{\ell'',\ell',\ell}. \qquad (17)$$

The recursion formulae (10) can be applied to all three basis functions in the integrand $\Phi_{(i,\alpha)}^{\ell''}(x)\,\Phi_{(j,\beta)}^{\ell'}(y)\,\Phi_{(k,\varkappa)}^{\ell}(x-y)$ of $\gamma_{(i,\alpha),(j,\beta),(k,\varkappa)}^{\ell'',\ell',\ell}$. Some of the resulting recursions for $\gamma_{(i,\alpha),(j,\beta),(k,\varkappa)}^{\ell'',\ell',\ell}$ are given in the next Remark.

Remark 1. For all $\ell'', \ell', \ell \in \mathbb{N}_0$ and all $i, j, k \in \mathbb{Z}$ we have

$$\gamma_{(i,0),(j,\beta),(k,\varkappa)}^{\ell'',\ell',\ell} = \frac{1}{\sqrt{2}}\left(\gamma_{(2i,0),(j,\beta),(k,\varkappa)}^{\ell''+1,\ell',\ell} + \gamma_{(2i+1,0),(j,\beta),(k,\varkappa)}^{\ell''+1,\ell',\ell}\right), \qquad (18\mathrm{a})$$

$$\gamma_{(i,1),(j,\beta),(k,\varkappa)}^{\ell'',\ell',\ell} = \frac{1}{2\sqrt{2}}\left(\begin{array}{l}-\sqrt{3}\,\gamma_{(2i,0),(j,\beta),(k,\varkappa)}^{\ell''+1,\ell',\ell} + \sqrt{3}\,\gamma_{(2i+1,0),(j,\beta),(k,\varkappa)}^{\ell''+1,\ell',\ell} \\ +\,\gamma_{(2i,1),(j,\beta),(k,\varkappa)}^{\ell''+1,\ell',\ell} + \gamma_{(2i+1,1),(j,\beta),(k,\varkappa)}^{\ell''+1,\ell',\ell}\end{array}\right),$$

$$(18\mathrm{b})$$

$$\gamma^{\ell'',\ell',\ell}_{(i,\alpha),(j,0),(k,\varkappa)} = \frac{1}{\sqrt{2}} \left(\gamma^{\ell'',\ell'+1,\ell}_{(i,\alpha),(2j,0),(k,\varkappa)} + \gamma^{\ell'',\ell'+1,\ell}_{(i,\alpha),(2j+1,0),(k,\varkappa)} \right), \tag{18c}$$

$$\gamma^{\ell'',\ell',\ell}_{(i,\alpha),(j,1),(k,\varkappa)} = \frac{1}{2\sqrt{2}} \left(\begin{array}{l} -\sqrt{3}\,\gamma^{\ell'',\ell'+1,\ell}_{(i,\alpha),(2j,0),(k,\varkappa)} + \sqrt{3}\,\gamma^{\ell'',\ell'+1,\ell}_{(i,\alpha),(2j+1,0),(k,\varkappa)} \\ + \gamma^{\ell'',\ell'+1,\ell}_{(i,\alpha),(2j,1),(k,\varkappa)} + \gamma^{\ell'',\ell'+1,\ell}_{(i,\alpha),(2j+1,1),(k,\varkappa)} \end{array} \right). \tag{18d}$$

4.2 Simplified γ-Coefficients

For levels ℓ, ℓ', ℓ'' with $\ell \geq \max\{\ell', \ell''\}$ we set

$$\gamma^{\ell'',\ell',\ell}_{\nu,(\alpha,\beta,\varkappa)} := \iint \Phi^{\ell''}_{0,\alpha}(x)\, \Phi^{\ell'}_{0,\beta}(y)\, \Phi^{\ell}_{\nu,\varkappa}(x-y) \mathrm{d}x \mathrm{d}y \qquad (\nu \in \mathbb{Z}). \tag{19}$$

We call these coefficients simplified γ-coefficients, since only one subindex ν is involved instead of the triple (i, j, k).

Lemma 1. *Let $\ell \geq \max\{\ell', \ell''\}$. Then for any $i, j, k \in \mathbb{Z}$, $\alpha, \beta, \varkappa \in \{0, 1\}$*

$$\gamma^{\ell'',\ell',\ell}_{(i,\alpha),(j,\beta),(k,\varkappa)} = \gamma^{\ell'',\ell',\ell}_{k-i2^{\ell-\ell''}+j2^{\ell-\ell'},(\alpha,\beta,\varkappa)}. \tag{20}$$

Remark 2. The values of $\gamma^{\ell,\ell,\ell}_{\nu,(\alpha,\beta,\varkappa)}$ are zero for $\nu \notin \{0,1\}$ and, otherwise,

$$\gamma^{\ell,\ell,\ell}_{0,(0,0,0)} = \gamma^{\ell,\ell,\ell}_{-1,(0,0,0)} = \sqrt{h_\ell}/2,$$

$$\gamma^{\ell,\ell,\ell}_{0,(1,0,0)} = \sqrt{h_\ell/12}, \quad \gamma^{\ell,\ell,\ell}_{-1,(1,0,0)} = -\sqrt{h_\ell/12},$$

$$\gamma^{\ell,\ell,\ell}_{0,(0,1,0)} = \gamma^{\ell,\ell,\ell}_{0,(0,0,1)} = -\sqrt{h_\ell/12}, \quad \gamma^{\ell,\ell,\ell}_{-1,(0,1,0)} = \gamma^{\ell,\ell,\ell}_{-1,(0,0,1)} = \sqrt{h_\ell/12},$$

$$\gamma^{\ell,\ell,\ell}_{0,(1,1,0)} = \gamma^{\ell,\ell,\ell}_{-1,(1,1,0)} = \gamma^{\ell,\ell,\ell}_{0,(1,0,1)} = \gamma^{\ell,\ell,\ell}_{-1,(1,0,1)} = 0,$$

$$\gamma^{\ell,\ell,\ell}_{0,(0,1,1)} = \gamma^{\ell,\ell,\ell}_{-1,(0,1,1)} = 0,$$

$$\gamma^{\ell,\ell,\ell}_{0,(1,1,1)} = -\sqrt{3h_\ell}/5, \quad \gamma^{\ell,\ell,\ell}_{-1,(1,1,1)} = \sqrt{3h_\ell}/5.$$

4.3 G- and Γ-Coefficients

As stated in (17), we have to compute $\sum_{j,k\in\mathbb{Z}} \sum^1_{\beta,\varkappa=0} f^{\ell'}_{j,\beta}\, g^{\ell}_{k,\varkappa}\, \gamma^{\ell'',\ell',\ell}_{(i,\alpha),(j,\beta),(k,\varkappa)}$. Performing only the summation over k and $\varkappa$, leads us to

$$G^{\ell'',\ell',\ell}_{(i,\alpha),(j,\beta)} = \sum_{k\in\mathbb{Z}} \sum^1_{\varkappa=0} g^{\ell}_{k,\varkappa}\, \gamma^{\ell'',\ell',\ell}_{(i,\alpha),(j,\beta),(k,\varkappa)}. \tag{21}$$

Note that the identity $G^{\ell'',\ell',\ell}_{(i,\alpha),(j,\beta)} = G^{\ell',\ell',\ell}_{(i-j,\alpha),(0,\beta)}$ holds for $\ell'' = \ell'$.
    Using the recursions (18a-d) from Remark 1, one proves

Remark 3. For all $\ell'', \ell', \ell \in \mathbb{N}_0$ and all $i, j \in \mathbb{Z}$ we have

$$
\begin{aligned}
\alpha = 0 : \quad & G_{(i,0),(j,\beta)}^{\ell'',\ell',\ell} = \frac{1}{\sqrt{2}} \left(G_{(2i,0),(j,\beta)}^{\ell''+1,\ell',\ell} + G_{(2i+1,0),(j,\beta)}^{\ell''+1,\ell',\ell} \right), \\
\beta = 0 : \quad & G_{(i,\alpha),(j,0)}^{\ell'',\ell',\ell} = \frac{1}{\sqrt{2}} \left(G_{(i,\alpha),(2j,0)}^{\ell'',\ell'+1,\ell} + G_{(i,\alpha),(2j+1,0)}^{\ell'',\ell'+1,\ell} \right), \\
\alpha = 1 : \quad & G_{(i,1),(j,\beta)}^{\ell'',\ell',\ell} = \frac{1}{2\sqrt{2}} \left(\begin{array}{c} -\sqrt{3}\, G_{(2i,0),(j,\beta)}^{\ell''+1,\ell',\ell} + \sqrt{3}\, G_{(2i+1,0),(j,\beta)}^{\ell''+1,\ell',\ell} \\ + G_{(2i,1),(j,\beta)}^{\ell''+1,\ell',\ell} + G_{(2i+1,1),(j,\beta)}^{\ell''+1,\ell',\ell} \end{array} \right), \\
\beta = 1 : \quad & G_{(i,\alpha),(j,1)}^{\ell'',\ell',\ell} = \frac{1}{2\sqrt{2}} \left(\begin{array}{c} -\sqrt{3}\, G_{(i,\alpha),(2j,0)}^{\ell'',\ell'+1,\ell} + \sqrt{3}\, G_{(i,\alpha),(2j+1,0)}^{\ell'',\ell'+1,\ell} \\ + G_{(i,\alpha),(2j,1)}^{\ell'',\ell'+1,\ell} + G_{(i,\alpha),(2j+1,1)}^{\ell'',\ell'+1,\ell} \end{array} \right).
\end{aligned}
\tag{22}
$$

If the first two levels are equal: $\ell'' = \ell' \le \ell$, the coefficients are denoted by

$$
\Gamma_{i,(\alpha,\beta)}^{\ell',\ell} := G_{(i,\alpha),(0,\beta)}^{\ell',\ell',\ell} = G_{(0,\alpha),(-i,\beta)}^{\ell',\ell',\ell} = \sum_{k \in \mathbb{Z}} \sum_{\varkappa=0}^{1} g_{k,\varkappa}^{\ell}\, \gamma_{k-i2^{\ell-\ell'},(\alpha,\beta,\varkappa)}^{\ell',\ell',\ell}.
\tag{23}
$$

In order to compute $\Gamma_{i,(0,0)}^{\ell',\ell}$ from $\Gamma_{i,(0,0)}^{\ell'+1,\ell}$, one has to combine the first two lines of (22): $\Gamma_{i,(0,0)}^{\ell',\ell} = G_{(i,0),(0,0)}^{\ell',\ell',\ell} = \frac{1}{\sqrt{2}} \left(G_{(2i,0),(0,0)}^{\ell'+1,\ell',\ell} + G_{(2i+1,0),(0,0)}^{\ell'+1,\ell',\ell} \right) = \frac{1}{2}\Gamma_{2i-1,(0,0)}^{\ell'+1,\ell} + \Gamma_{2i,(0,0)}^{\ell'+1,\ell} + \frac{1}{2}\Gamma_{2i+1,(0,0)}^{\ell'+1,\ell}$. Analogously, one obtains for all values of (α,β) the results of Lemma 2b.

Lemma 2. *a) For $\ell' = \ell$, $\Gamma_{i,(\alpha,\beta)}^{\ell,\ell} = \sum_{\varkappa=0}^{1} \left(g_{i,\varkappa}^{\ell}\, \gamma_{0,(\alpha,\beta,\varkappa)}^{\ell,\ell,\ell} + g_{i-1,\varkappa}^{\ell}\, \gamma_{-1,(\alpha,\beta,\varkappa)}^{\ell,\ell,\ell} \right)$ can be computed from the γ-values given in Remark 2.*
b) For $\ell' < \ell$, one can make use of the following recursions:

$$
\Gamma_{i,(0,0)}^{\ell',\ell} = \frac{1}{2}\Gamma_{2i-1,(0,0)}^{\ell'+1,\ell} + \Gamma_{2i,(0,0)}^{\ell'+1,\ell} + \frac{1}{2}\Gamma_{2i+1,(0,0)}^{\ell'+1,\ell},
\tag{24}
$$

$$
\Gamma_{i,(1,0)}^{\ell',\ell} = \frac{\sqrt{3}}{4}\left(\Gamma_{2i+1,(0,0)}^{\ell'+1,\ell} - \Gamma_{2i-1,(0,0)}^{\ell'+1,\ell} \right) + \frac{1}{4}\left(\Gamma_{2i-1,(1,0)}^{\ell'+1,\ell} + \Gamma_{2i+1,(1,0)}^{\ell'+1,\ell} \right) + \frac{1}{2}\Gamma_{2i,(1,0)}^{\ell'+1,\ell},
$$

$$
\Gamma_{i,(0,1)}^{\ell',\ell} = \frac{\sqrt{3}}{4}\left(\Gamma_{2i-1,(0,0)}^{\ell'+1,\ell} - \Gamma_{2i+1,(0,0)}^{\ell'+1,\ell} \right) + \frac{1}{4}\left(\Gamma_{2i-1,(0,1)}^{\ell'+1,\ell} + \Gamma_{2i+1,(0,1)}^{\ell'+1,\ell} \right) + \frac{1}{2}\Gamma_{2i,(0,1)}^{\ell'+1,\ell},
$$

$$
\Gamma_{i,(1,1)}^{\ell',\ell} = -\frac{3}{8}\left(\Gamma_{2i-1,(0,0)}^{\ell'+1,\ell} + \Gamma_{2i+1,(0,0)}^{\ell'+1,\ell} \right) + \frac{3}{4}\Gamma_{2i,(0,0)}^{\ell'+1,\ell} + \frac{1}{8}\left(\Gamma_{2i-1,(1,1)}^{\ell'+1,\ell} + \Gamma_{2i+1,(1,1)}^{\ell'+1,\ell} \right)
$$
$$
+ \frac{1}{4}\Gamma_{2i,(1,1)}^{\ell'+1,\ell} + \frac{\sqrt{3}}{8}\left(-\Gamma_{2i-1,(0,1)}^{\ell'+1,\ell} + \Gamma_{2i+1,(0,1)}^{\ell'+1,\ell} + \Gamma_{2i-1,(1,0)}^{\ell'+1,\ell} - \Gamma_{2i+1,(1,0)}^{\ell'+1,\ell} \right).
$$

4.4 Notation for ℓ^2 Sequences

Let, e.g., $f_{i,\alpha}^{\ell}$ be the coefficients of $f_\ell = \sum_{i \in \mathcal{I}_\ell^f} \sum_{\alpha=0}^{1} f_{i,\alpha}^{\ell} \Phi_{i,\alpha}^{\ell}$. We extend these coefficients by $f_{i,\alpha}^{\ell} := 0$ for $i \notin \mathcal{I}_\ell^f$ and obtain an ℓ^2 sequence

$$
f_{\ell,\alpha} := \left(f_{i,\alpha}^{\ell} \right)_{i \in \mathbb{Z}}.
$$

We use the convention that an upper level index ℓ indicates a coefficient, while the sequence has a lower index ℓ. Here $\alpha \in \{0, 1\}$ is a further parameter.

For general sequences $a, b \in \ell^2$ (i.e, $a = (a_i)_{i \in \mathbb{Z}}$, $b = (b_i)_{i \in \mathbb{Z}}$), the *discrete convolution* $c := a * b$ is defined by

$$c_i = \sum_{j \in \mathbb{Z}} a_j b_{i-j}.$$

For its computation using FFT compare [3, §6].

5 Algorithm

In Problem 1 three level numbers ℓ'', ℓ', ℓ appear. Without loss of generality $\ell' \le \ell$ holds. Below we have to distinguish the following three cases:

$$\textbf{A: } \ell'' \le \ell' \le \ell, \qquad \textbf{B: } \ell' < \ell'' \le \ell, \qquad \textbf{C: } \ell' \le \ell < \ell''. \tag{25}$$

5.1 Case A: $\ell'' \le \ell' \le \ell$

The convolution $f_{\ell'} * g_\ell$ is a piecewise linear function, where the pieces correspond to the smaller step size h_ℓ. The projection $P_{\ell''}$ of $f_{\ell'} * g_\ell$ is required in two intervals. Because of $\ell'' \le \ell' \le \ell$ the step size $h_{\ell''}$ is equal or larger than the other ones.

The following algorithm has to compute the projection of $\omega_{\ell''} = P_{\ell''}\omega_{\text{exact}}$ of $\omega_{\text{exact}} := f_{\ell'} * g_\ell$. A straightforward but naive approach would be to compute ω_{exact} first and then its projection. The problem is that in the case $\ell' \ll \ell$, the product $f_{\ell'} * g_\ell$ requires too many data (corresponding to the fine grid in the third line of the figure above). The projection $P_{\ell''}$ would map the many data into few ones. The essence of the following algorithm is to incorporate the projection before a discrete convolution is performed.

Step 1 computes the Γ-Coefficients. We start with the sequences $\Gamma_{\ell,\ell,(\alpha,\beta)} = (\Gamma^{\ell,\ell}_{i,(\alpha,\beta)})_{i \in \mathbb{Z}}$. Following Lemma 2a, we obtain

$$\Gamma^{\ell,\ell}_{i,(0,0)} = \frac{\sqrt{h_\ell}}{2}\left(g^\ell_{i,0} + g^\ell_{i-1,0}\right) + \frac{\sqrt{h_\ell}}{12}\left(g^\ell_{i-1,1} - g^\ell_{i,1}\right), \tag{26a}$$

$$\Gamma^{\ell,\ell}_{i,(1,0)} = -\Gamma^{\ell,\ell}_{i,(0,1)} = \frac{\sqrt{h_\ell}}{12}\left(g^\ell_{i,0} - g^\ell_{i-1,0}\right), \tag{26b}$$

$$\Gamma^{\ell,\ell}_{i,(1,1)} = \frac{\sqrt{3h_\ell}}{5}\left(g^\ell_{i-1,1} - g^\ell_{i,1}\right) \tag{26c}$$

for all $i \in \mathbb{Z}$. Then we compute the sequences $\Gamma_{\ell',\ell,(\alpha,\beta)} := (\Gamma^{\ell',\ell}_{i,(\alpha,\beta)})_{i \in \mathbb{Z}}$ for $\ell' = \ell - 1, \ell - 2, \ldots, 0$ using the recursions from Lemma 2b.

Step 2a: Let ℓ' be any level in $[0, \ell]$. For each $\ell'' = \ell', \ell' - 1, \ldots, 0$ the projection $P_{\ell''}(f_{\ell'} * g_\ell)$ is to be computed (see Problem 1). Following (16) and (17), the coefficients $\omega_{i,\alpha}^{\ell''} = \sum_{j,k,\beta,\varkappa} f_{j,\beta}^{\ell'} g_{k,\varkappa}^{\ell} \gamma_{(i,\alpha),(j,\beta),(k,\varkappa)}^{\ell'',\ell',\ell}$ are needed. The sequence is denoted by $\omega_{\ell'',\alpha} = (\omega_{i,\alpha}^{\ell''})_{i\in\mathbb{Z}}$.

For the starting value $\ell'' = \ell'$ we have

$$
\begin{aligned}
\omega_{i,\alpha}^{\ell'} &= \sum_{j,k\in\mathbb{Z}}\sum_{\beta,\varkappa=0}^{1} f_{j,\beta}^{\ell'} g_{k,\varkappa}^{\ell} \gamma_{(i,\alpha),(j,\beta),(k,\varkappa)}^{\ell',\ell',\ell} \\
&\underset{(20)}{=} \sum_{j,k\in\mathbb{Z}}\sum_{\beta,\varkappa=0}^{1} f_{j,\beta}^{\ell'} g_{k,\varkappa}^{\ell} \gamma_{k-(i-j)2^{\ell-\ell'},(\alpha,\beta,\varkappa)}^{\ell',\ell',\ell} \\
&\underset{(23)}{=} \sum_{\beta=0}^{1}\sum_{j\in\mathbb{Z}} f_{j,\beta}^{\ell'} \Gamma_{i-j,(\alpha,\beta)}^{\ell',\ell} \qquad \text{for all } i \in \mathbb{Z}.
\end{aligned}
$$

The four sums $\sum_{j\in\mathbb{Z}} f_{j,\beta}^{\ell'} \Gamma_{i-j,(\alpha,\beta)}^{\ell',\ell}$ for all combinations of $\alpha,\beta \in \{0,1\}$ describe the discrete convolution of the two sequences $f_{\ell',\beta} := (f_{j,\beta}^{\ell'})_{j\in\mathbb{Z}}$ and $\Gamma_{\ell',\ell,(\alpha,\beta)} := (\Gamma_{k,(\alpha,\beta)}^{\ell',\ell})_{k\in\mathbb{Z}}$. Concerning the performance of the following discrete convolutions we refer to [3, §6]:

$$
\omega_{\ell',\alpha} = \sum_{\beta=0}^{1} f_{\ell',\beta} * \Gamma_{\ell',\ell,(\alpha,\beta)} \qquad (\alpha = 0,1;\ 0 \le \ell' \le \ell). \tag{27}
$$

Step 2b: Given $\omega_{\ell',\alpha}$ from (27), we compute $\omega_{\ell'',\alpha}$ for $\ell'' = \ell' - 1, \ldots, 0$ by the following recursions.

Lemma 3. *The recursions*

$$
\begin{aligned}
\omega_{i,0}^{\ell''} &= \frac{1}{\sqrt{2}} \left(\omega_{2i,0}^{\ell''+1} + \omega_{2i+1,0}^{\ell''+1} \right), \\
\omega_{i,1}^{\ell''} &= \frac{\sqrt{3}}{2\sqrt{2}} \left(\omega_{2i+1,0}^{\ell''+1} - \omega_{2i,0}^{\ell''+1} \right) + \frac{1}{2\sqrt{2}} \left(\omega_{2i,1}^{\ell''+1} + \omega_{2i+1,1}^{\ell''+1} \right)
\end{aligned} \tag{28}
$$

holds for all $i \in \mathbb{Z}$ and for all $0 \le \ell'' \le \ell'$.

Proof. Use $\omega_{i,0}^{\ell''} = \sum_{j,k,\beta,\varkappa} f_{j,\beta}^{\ell'} g_{k,\varkappa}^{\ell} \gamma_{(i,0),(j,\beta),(k,\varkappa)}^{\ell'',\ell',\ell}$ and apply (18a). Similarly, $\omega_{i,1}^{\ell''} = \sum_{j,k,\beta,\varkappa} f_{j,\beta}^{\ell'} g_{k,\varkappa}^{\ell} \gamma_{(i,1),(j,\beta),(k,\varkappa)}^{\ell'',\ell',\ell}$ and (18b) yields the result for $\omega_{i,1}^{\ell''}$. $\square$

Intertwining the Computations for all $\ell'' \le \ell' \le \ell$

The superindex ℓ in $\Gamma_{i,(\alpha,\beta)}^{\ell',\ell}$ indicates that this sequence at level ℓ' is originating from the data $g_{\ell,\varkappa}$. Since the further treatment of $\Gamma_{i,(\alpha,\beta)}^{\ell',\ell}$ does not depend on ℓ, we can gather all $\Gamma_{i,(\alpha,\beta)}^{\ell',\ell}$ into

$$
\Gamma_{i,(\alpha,\beta)}^{\ell'} := \sum_{\ell=\ell'}^{L^g} \Gamma_{i,(\alpha,\beta)}^{\ell',\ell} \qquad (0 \le \ell' \le L^g). \tag{29}
$$

Hence, their computation is performed by the loop

$$
\boxed{
\begin{array}{l}
\textbf{for } \ell' := L^g \textbf{ downto } 0 \textbf{ do} \\
\textbf{begin if } \ell' = L^g \textbf{ then } \Gamma^{L^g}_{i,(\alpha,\beta)} := 0 \\
\qquad \textbf{else compute } \Gamma^{\ell'}_{i,(\alpha,\beta)} \textbf{ from } \Gamma^{\ell'+1}_{i,(\alpha,\beta)} \textbf{ using (24);} \\
\quad \Gamma^{\ell'}_{i,(\alpha,\beta)} := \Gamma^{\ell'}_{i,(\alpha,\beta)} + \Gamma^{\ell',\ell'}_{i,(\alpha,\beta)} \\
\textbf{end;}
\end{array}
}
\qquad
\begin{array}{l}
\textit{explanations:} \\
\text{starting values,} \\
\text{see Lemma 2b,} \\
\Gamma^{\ell',\ell'}_{i,(\alpha,\beta)} \text{ defined} \\
\text{in (26a-c).}
\end{array}
\tag{30}
$$

Having available $\Gamma^{\ell'}_{i,(\alpha,\beta)}$ for all $0 \le \ell' \le L^g$, we can compute $\omega_{\ell',\alpha}$ for any ℓ' (cf. (27)). For a moment, we use the symbols $\omega_{\ell',\ell',\alpha}$, $\omega_{\ell'-1,\ell',\alpha}, \ldots, \omega_{\ell'',\ell',\alpha}$ for the quantities computed in Step 2a,b. Here, the additional second index ℓ' expresses the fact that the data stem from $f_{\ell'}$ at level ℓ' (see (27)).

The coarsening $\omega_{\ell',\ell',\alpha} \mapsto \omega_{\ell'-1,\ell',\alpha} \mapsto \ldots \mapsto \omega_{\ell'',\ell',\alpha}$ can again be done jointly for the different ℓ', i.e., we form

$$
\omega_{\ell'',\alpha} := \sum_{\ell'=\ell''}^{L^g} \omega_{\ell'',\ell',\alpha} \qquad \left(0 \le \ell'' \le \min\{L^\omega, L^f, L^g\}\right).
$$

The algorithmic form is

$$
\boxed{
\begin{array}{l}
\textbf{for } \ell'' := \min\{L^f, L^g\} \textbf{ downto } 0 \textbf{ do} \\
\textbf{begin if } \ell'' = \min\{L^f, L^g\} \textbf{ then } \omega_{\ell'',\alpha} := 0 \\
\qquad \textbf{else compute } \omega^{\ell''}_{i,\alpha} \textbf{ from } \omega^{\ell''+1}_{i,\alpha} \textbf{ via (28);} \\
\quad \omega_{\ell'',\alpha} := \omega_{\ell'',\alpha} + \omega_{\ell'',\ell',\alpha} \\
\textbf{end;}
\end{array}
}
\qquad
\begin{array}{l}
\textit{explanations:} \\
\text{starting values,} \\
\text{see Lemma 3,} \\
\omega_{\ell'',\ell',\alpha} \text{ defined} \\
\text{in (27).}
\end{array}
\tag{31}
$$

This algorithm yields $\omega_{\ell''} = P_{\ell''} \sum_{\ell',\ell \text{ with } \ell'' \le \ell' \le \ell} f_{\ell'} * g_{\ell}$ involving all combinations of indices with $\ell'' \le \ell' \le \ell$.

5.2 Case B: $\ell' < \ell'' \le \ell$

In Case B the step size $h_{\ell''}$ used by the projection $P_{\ell''}$ is smaller than the step size $h_{\ell'}$ but larger than h_ℓ.

We use a loop of ℓ'' from $\ell'+1$ to ℓ. First, we discuss the first value $\ell'' = \ell'+1$ and assume $\ell'+1 \le \ell$.

We recall that $S_{\ell'} \subset S_{\ell'+1}$. The function $f_{\ell'} = \sum_{j,\beta} f^{\ell'}_{j,\beta} \Phi^{\ell'}_{j,\beta} \in S_{\ell'}$ can be rewritten as a function of level $\ell'+1$ by using (10):

$$
f_{\ell'} = \sum_{\beta=0}^{1} \sum_j \hat{f}^{\ell'+1}_{j,\beta} \Phi^{\ell'+1}_{j,\beta} \qquad \text{with} \tag{32}
$$

$$
\hat{f}^{\ell'+1}_{2j,0} := \tfrac{1}{\sqrt{2}} f^{\ell'}_{j,0} - \tfrac{\sqrt{3}}{2\sqrt{2}} f^{\ell'}_{j,1}, \qquad \hat{f}^{\ell'+1}_{2j+1,0} := \tfrac{1}{\sqrt{2}} f^{\ell'}_{j,0} + \tfrac{\sqrt{3}}{2\sqrt{2}} f^{\ell'}_{j,1},
$$

$$
\hat{f}^{\ell'+1}_{2j,1} := \tfrac{1}{2\sqrt{2}} f^{\ell'}_{j,1}, \qquad\qquad \hat{f}^{\ell'+1}_{2j+1,1} := \tfrac{1}{2\sqrt{2}} f^{\ell'}_{j,1}.
$$

Let $\hat{f}_{\ell'+1,\beta} := (\hat{f}_{j,\beta}^{\ell'+1})_{j \in \mathbb{Z}}$ be the sequences of the newly defined coefficients. Since $\ell'' = \ell'+1 \le \ell$, the three level numbers $\ell'', \ell'+1, \ell$ satisfy the inequalities of Case A. As in Step 2a of Case A (see §5.1) the desired coefficients of the projection at level $\ell'' = \ell'+1$ are $\omega_{i,\alpha}^{\ell'+1} = \sum_{j,\beta} \hat{f}_{j,\beta}^{\ell'+1} \Gamma_{i-j,(\alpha,\beta)}^{\ell'+1,\ell}$, i.e., discrete convolutions $\hat{f}_{\ell'+1,\beta} * \Gamma_{\ell'+1,\ell,\beta}$ are to be performed.

Now we consider the complete recursion. Step 1 in Case A has already produced the coefficients $\Gamma_{j,(\alpha,\beta)}^{\ell'}$ gathering all $\Gamma_{j,(\alpha,\beta)}^{\ell',\ell}$ ($\ell \ge \ell'$, cf. (29)). For $\ell'' = \ell'+1, \ell'+2, \ldots, \ell$ we represent the function $f_{\ell'}$ at these levels ℓ'' by computing the coefficients $\hat{f}_{j,\beta}^{\ell''}$ as in (32):

$$\hat{f}_{j,\beta}^{\ell'} := f_{j,\beta}^{\ell'} \qquad \text{(starting value)}, \tag{33a}$$

$$\text{compute } \hat{f}_{j,\beta}^{\ell''} \text{ from } \hat{f}_{j,\beta}^{\ell''-1} \text{ via (32)} \qquad (\ell'+1 \le \ell'' \le \ell). \tag{33b}$$

Note, however, that only those coefficients are to be determined which are really needed in the next step, which are four discrete convolutions

$$\omega_{\ell'',\alpha} = \sum_{\beta=0}^{1} \hat{f}_{\ell'',\beta} * \Gamma_{\ell'',(\alpha,\beta)} \qquad (\ell'+1 \le \ell'' \le \ell) \tag{33c}$$

of the sequences $\hat{f}_{\ell'',\beta} := (\hat{f}_{j,\beta}^{\ell''})_{j \in \mathbb{Z}}$ with $\Gamma_{\ell'',(\alpha,\beta)}$.

The combined computations for all $\ell' < \ell'' \le \ell$ is performed by

$$\begin{array}{l|l}
\hat{f}_j^0 := 0; & \textit{explanations:} \\
\textbf{for } \ell'' := 1 \textbf{ to } \min\{L^\omega, L^g\} \textbf{ do} & \\
\textbf{begin } \hat{f}_{j,\beta}^{\ell''-1} := \hat{f}_{j,\beta}^{\ell''-1} + f_{j,\beta}^{\ell''-1}; & \text{starting value (33a),} \\
\qquad \text{compute } \hat{f}_{j,\beta}^{\ell''} \text{ from } \hat{f}_{j,\beta}^{\ell''-1} \text{ via (32);} & \text{see (33b),} \\
\qquad \omega_{\ell'',\alpha} := \sum_{\beta=0}^{1} \hat{f}_{\ell'',\beta} * \Gamma_{\ell'',(\alpha,\beta)} & \text{see (33c).} \\
\textbf{end;} &
\end{array} \tag{34}$$

The sum $\hat{f}_{j,\beta}^{\ell''-1} + f_{j,\beta}^{\ell''-1}$ in the third line defines $\hat{f}_{j,\beta}^{\ell''-1}$ as coefficients of $\sum_{\ell'=0}^{\ell''-1} f_{\ell'} = \sum_{j,\beta} \hat{f}_{j,\beta}^{\ell''-1} \Phi_{j,\beta}^{\ell''-1}$. Therefore the next two lines consider all combinations of $\ell' < \ell''$. Since $\Gamma_{\ell''}$ contains all contributions from $\ell \ge \ell''$, $\omega_{\ell''}$ is the projection $P_{\ell''} \sum_{\ell',\ell \text{ with } \ell' < \ell'' \le \ell} f_{\ell'} * g_\ell$.

5.3 Case C: $\ell' \le \ell < \ell''$

Now the step size $h_{\ell''}$ used in the projection is smaller than $h_{\ell'}$ and h_ℓ.

Explanations

The exact convolution $\omega_{\text{exact}}(x) := \int f_{\ell'}(y) g_\ell(x-y) dy$ ($x \in \mathbb{R}$) is a piecewise cubic and globally continuous function with possible jumps of the derivative at the grid points νh_ℓ ($\nu \in \mathbb{Z}$) of the grid at level ℓ. The projection $P_{\ell''} \omega_{\text{exact}} = \sum_{i,\alpha} \omega_{i,\alpha}^{\ell''} \Phi_{i,\alpha}^{\ell''}$ involves all scalar products

$$\omega_{i,\alpha}^{\ell''} = \int \Phi_{i,\alpha}^{\ell''}(x)\omega_{\text{exact}}(x)\mathrm{d}x.$$

The whole support of $\Phi_{i,\alpha}^{\ell''}$ belongs to one of the intervals $[\nu h_\ell, (\nu+1)h_\ell]$, where $\omega_{\text{exact}}(x)$ is a cubic function.

We define the point values and one-sided derivatives

$$\delta_\nu^\ell := \omega_{\text{exact}}(\nu h_\ell), \quad \delta_{\nu,+}^\ell := \omega_{\text{exact}}'(\nu h_\ell + 0), \quad \delta_{\nu,-}^\ell := \omega_{\text{exact}}'(\nu h_\ell - 0).$$

Then ω_{exact} can be represented in the interval $I_i^\ell \in \mathcal{M}$ by the cubic polynomial

$$\delta_i^\ell + (x - ih_\ell)\frac{\delta_{i+1}^\ell - \delta_i^\ell}{h_\ell}$$
$$-\frac{(x-ih_\ell)(x-(i+1)h_\ell)}{h_\ell}\left(\delta_{i,+}^\ell - (x-ih_\ell)\frac{\delta_{i,+}^\ell + \delta_{i+1,-}^\ell}{h_\ell} + \frac{2x-(2i+1)h_\ell}{h_\ell}\frac{\delta_{i+1}^\ell - \delta_i^\ell}{h_\ell}\right). \qquad (35)$$

Its values at the midpoint $x_{i+1/2}^\ell := (i+1/2)h_\ell$ are

$$\begin{aligned}
\omega_{\text{exact}}(x_{i+1/2}^\ell) &= \tfrac{1}{2}\left(\delta_i^\ell + \delta_{i+1}^\ell\right) + \tfrac{h_\ell}{8}\left(\delta_{i,+}^\ell - \delta_{i+1,-}^\ell\right), \\
\omega_{\text{exact}}'(x_{i+1/2}^\ell \pm 0) &= \tfrac{3}{2h_\ell}\left(\delta_{i+1}^\ell - \delta_i^\ell\right) - \tfrac{1}{4}\left(\delta_{i,+}^\ell + \delta_{i+1,-}^\ell\right).
\end{aligned} \qquad (36)$$

Pointwise Evaluations

Let ω_{exact} be described in I_ν^ℓ by the data $\delta_\nu^\ell, \delta_{\nu+1}^\ell, \delta_{\nu,+}^\ell, \delta_{\nu+1,-}^\ell$ (cf. (35)). Then

$$\omega_{i,\alpha}^\ell = \int \Phi_{i,\alpha}^\ell(x)\omega_{\text{exact}}(x)\mathrm{d}x \qquad (37)$$

$$= \begin{cases}
\frac{\sqrt{h_\ell}}{2}\left(\delta_\nu^\ell + \delta_{\nu+1}^\ell\right) + \frac{h_\ell^{3/2}}{12}\left(\delta_{\nu,+}^\ell - \delta_{\nu+1,-}^\ell\right) & \text{for } \alpha = 0, \\
\frac{\sqrt{3h_\ell}}{5}\left(\delta_{\nu+1}^\ell - \delta_\nu^\ell\right) - \frac{\sqrt{3}h_\ell^{3/2}}{60}\left(\delta_{i,+}^\ell + \delta_{i+1,-}^\ell\right) & \text{for } \alpha = 1,
\end{cases}$$

yields the coefficients of the projection. It remains to determine $\delta_\nu^\ell, \delta_{\nu,\pm}^\ell$.

Computation of $\delta, \delta_\pm$

We define new γ-coefficients

$${}^0\gamma_{i,(j,\beta),(k,\varkappa)}^\ell := \int \Phi_{j,\beta}^\ell(y)\Phi_{k,\varkappa}^\ell(ih_\ell - y)\mathrm{d}y \qquad (i,j,k \in \mathbb{Z}, \beta, \varkappa \in \{0,1\})$$

involving only one level ℓ. Simple substitutions yield

$${}^0\gamma_{i,(j,\beta),(k,\varkappa)}^\ell = {}^0\gamma_{k-i+j,(\beta,\varkappa)}^\ell$$

for the "simplified" γ-coefficient ${}^0\gamma_{\nu,(\beta,\varkappa)}^\ell := {}^0\gamma_{\nu,(0,\beta),(0,\varkappa)}^\ell$.

The δ-values of $f_\ell * g_\ell$ are

$$\delta_i^\ell = (f_\ell * g_\ell)(ih_\ell) = \sum_{j,k\in\mathbb{Z}} \sum_{\beta,\varkappa=0}^1 f_{j,\beta}^\ell g_{k,\varkappa}^\ell \int \Phi_{j,\beta}^\ell(y)\Phi_{k,\varkappa}^\ell(ih_\ell - y)dy \qquad (38a)$$

$$= \sum_{j,k\in\mathbb{Z}} \sum_{\beta,\varkappa=0}^1 f_{j,\beta}^\ell g_{k,\varkappa}^\ell \, {}^0\gamma_{i,(j,\beta),(k,\varkappa)}^\ell = \sum_{j,k\in\mathbb{Z}} \sum_{\beta,\varkappa=0}^1 f_{j,\beta}^\ell g_{k,\varkappa}^\ell \, {}^0\gamma_{k-i+j,(\beta,\varkappa)}^\ell$$

$$= \sum_{\beta=0}^1 \sum_{j\in\mathbb{Z}} f_{j,\beta}^\ell \, {}^0\Gamma_{i-j,\beta}^\ell \,, \qquad \text{where} \quad {}^0\Gamma_{i,\beta}^\ell := \sum_{\varkappa=0}^1 \sum_{k\in\mathbb{Z}} g_{k,\varkappa}^\ell \, {}^0\gamma_{k-i,(\beta,\varkappa)}^\ell \,.$$

Analogously, we set $\pm\gamma_{i,(j,\beta),(k,\varkappa)}^\ell = \lim_{\varepsilon\searrow 0} \frac{d}{dx}\int \Phi_{j,\beta}^\ell(y)\Phi_{k,\varkappa}^\ell(ih_\ell \pm \varepsilon - y)dy$ and obtain

$$\delta_{i,\pm}^\ell = \sum_{\beta=0}^1 \sum_{j\in\mathbb{Z}} f_{j,\beta}^\ell \, {}^\pm\Gamma_{i-j,\beta}^\ell, \qquad \text{where} \quad {}^\pm\Gamma_{i-j,\beta}^\ell := \sum_{\varkappa=0}^1 \sum_{k\in\mathbb{Z}} g_{k,\varkappa}^\ell \, {}^\pm\gamma_{k-i+j,(\beta,\varkappa)}^\ell.$$

$$\qquad\qquad (38b)$$

Remark 4. Coefficients not indicated below are zero:

$$\begin{array}{ll}
{}^0\gamma_{-1,(0,0)}^\ell = 1, & {}^0\gamma_{-1,(1,1)}^\ell = -1, \\
{}^+\gamma_{0,(0,0)}^\ell = 1/h_\ell, & {}^+\gamma_{-1,(0,0)}^\ell = -1/h_\ell, \\
{}^+\gamma_{0,(1,0)}^\ell = {}^+\gamma_{0,(0,1)}^\ell = -\sqrt{3}/h_\ell, & {}^+\gamma_{-1,(1,0)}^\ell = {}^+\gamma_{-1,(0,1)}^\ell = \sqrt{3}/h_\ell, \\
{}^+\gamma_{0,(1,1)}^\ell = {}^+\gamma_{-1,(1,1)}^\ell = 3/h_\ell, & \\
{}^-\gamma_{-1,(0,0)}^\ell = 1/h_\ell, & {}^-\gamma_{-2,(0,0)}^\ell = -1/h_\ell, \\
{}^-\gamma_{-1,(1,0)}^\ell = {}^-\gamma_{-1,(0,1)}^\ell = \sqrt{3}/h_\ell, & {}^-\gamma_{-2,(1,0)}^\ell = {}^-\gamma_{-2,(0,1)}^\ell = -\sqrt{3}/h_\ell, \\
{}^-\gamma_{-1,(1,1)}^\ell = {}^-\gamma_{-2,(1,1)}^\ell = -3/h_\ell. &
\end{array}$$

Remark 5. Remark 4 implies

$$\begin{aligned}
&{}^0\Gamma_{i,0}^\ell = g_{i-1,0}^\ell, \qquad {}^0\Gamma_{i,1}^\ell = -g_{i-1,0}^\ell\,, \\
&{}^+\Gamma_{i,0}^\ell = \frac{1}{h_\ell}\left(g_{i,0}^\ell - g_{i-1,0}^\ell + \sqrt{3}\left(g_{i-1,1}^\ell - g_{i,1}^\ell\right)\right), \\
&{}^+\Gamma_{i,1}^\ell = \frac{1}{h_\ell}\left(\sqrt{3}\left(g_{i-1,0}^\ell - g_{i,0}^\ell\right) + 3\left(g_{i-1,1}^\ell + g_{i,1}^\ell\right)\right), \\
&{}^-\Gamma_{i,0}^\ell = \frac{1}{h_\ell}\left(g_{i-1,0}^\ell - g_{i-2,0}^\ell + \sqrt{3}\left(g_{i-1,1}^\ell - g_{i-2,1}^\ell\right)\right), \\
&{}^-\Gamma_{i,1}^\ell = \frac{1}{h_\ell}\left(\sqrt{3}\left(g_{i-1,0}^\ell - g_{i-2,0}^\ell\right) - 3\left(g_{i-1,1}^\ell + g_{i-2,1}^\ell\right)\right).
\end{aligned}$$

Since $f_\ell * g_\ell$ is cubic in I_i^ℓ, the recursions derived from (36) are

$$\begin{aligned}
\delta_{2i}^{\ell''} &= \delta_i^{\ell''-1}, & \delta_{2i+1}^{\ell''} &= \frac{1}{2}\left(\delta_i^{\ell''-1} + \delta_{i+1}^{\ell''-1}\right) + \frac{h_{\ell''-1}}{8}\left(\delta_{i,+}^{\ell''-1} - \delta_{i+1,-}^{\ell''-1}\right), \\
\delta_{2i,\mp}^{\ell''} &= \delta_{i,\pm}^{\ell''-1}, & \delta_{2i+1,\mp}^{\ell''} &= \frac{3}{2h_{\ell''-1}}\left(\delta_{i+1}^{\ell''-1} - \delta_i^{\ell''-1}\right) - \frac{1}{4}\left(\delta_{i,+}^{\ell''-1} + \delta_{i+1,-}^{\ell''-1}\right)
\end{aligned}$$

$$\qquad\qquad (39)$$

for $\ell'' > \ell$.

Combined Computations for all $\ell' \leq \ell < \ell''$

The data $\hat{f}_{i,\alpha}^{\ell}$ have the same meaning as in Case B. Similarly, $\hat{\delta}_{i}^{\ell}, \hat{\delta}_{i,\pm}^{\ell}$ collect all δ-data from $\ell' < \ell$.

$$
\begin{aligned}
&\hat{f}_{i,\alpha}^{0} := 0; \; \hat{\delta}_{i}^{0} := \hat{\delta}_{i,\pm}^{0} := 0; \\
&\text{for } \ell := 0 \text{ to } L^{\omega} \text{ do} \\
&\text{begin if } \ell > 0 \text{ then} \\
&\qquad \text{begin compute } \hat{f}_{i,\alpha}^{\ell} \text{ from } \hat{f}_{i,\alpha}^{\ell-1} \text{ by (32)}; \\
&\qquad\qquad \text{compute } \hat{\delta}_{i}^{\ell}, \hat{\delta}_{i,\pm}^{\ell} \text{ from } \hat{\delta}_{i}^{\ell-1}, \hat{\delta}_{i,\pm}^{\ell-1} \text{ by (39)}; \\
&\qquad\qquad \text{compute } \omega_{i,\alpha}^{\ell} \text{ from } \hat{\delta}_{i}^{\ell} \text{ by (37)} \\
&\qquad \text{end}; \\
&\qquad \text{if } \ell \leq \min\{L^{f}, L^{\omega} - 1\} \text{ then } \hat{f}_{i,\alpha}^{\ell} := \hat{f}_{i,\alpha}^{\ell} + f_{i,\alpha}^{\ell}; \\
&\qquad \text{if } \ell \leq \min\{L^{g}, L^{\omega} - 1\} \text{ then} \\
&\qquad \text{begin compute } \delta_{i}^{\ell}, \delta_{i,\pm}^{\ell} \text{ by the convolutions (38a,b)}; \\
&\qquad\qquad \hat{\delta}_{i}^{\ell} := \hat{\delta}_{i}^{\ell} + \delta_{i}^{\ell}; \; \hat{\delta}_{i,\pm}^{\ell} := \hat{\delta}_{i,\pm}^{\ell} + \delta_{i,\pm}^{\ell} \\
&\text{end end};
\end{aligned}
\tag{40}
$$

The quantities $\hat{f}_{i,\alpha}^{\ell}$ used in the lines 4-6 are the coefficients of $\sum_{\ell'=0}^{\ell-1} f_{\ell'} = \sum_{i,\alpha} \hat{f}_{i,\alpha}^{\ell} \Phi_{i,\alpha}^{\ell}$. The convolutions in (39) called at line 5 involve the Γ-sequences defined in Remark 5. The coefficients $\omega_{i,\alpha}^{\ell}$ in line 6 belong to the projection $P_{\ell''} \sum_{\ell',\lambda \text{ with } 0 \leq \ell' \leq \lambda \leq \ell} f_{\ell'} * g_{\lambda}$ at level ℓ'', where ℓ'' is the actual value ℓ of the loop index.

5.4 Range of Products

In the previous subsections we have reduced the problem to a number of specific discrete convolutions (the first example is (27)). The resulting products are infinite sequences $(c_{\nu})_{\nu \in \mathbb{Z}}$. The first reasonable reduction would be to determine $(c_{\nu})_{\nu=\nu_1}^{\nu_2}$ only in the support $[\nu_1, \nu_2] \cap \mathbb{Z}$ of the sequence. But it is essential to go a step further. Even if we need the function $f_{\ell'} * g_{\ell}$ (see (14)) in the whole support $S := \mathrm{supp}(f_{\ell'} * g_{\ell})$, the projections $P_{\ell''}(f_{\ell'} * g_{\ell})$ are required in disjoint subsets $S_{\ell''} \subset S$. In terms of the sequences $(c_{\nu})_{\nu \in \mathbb{Z}}$ this means that we are interested in the components c_{ν} in an index interval $[\nu_1', \nu_2'] \cap \mathbb{Z}$ which is possibly much smaller than the support $[\nu_1, \nu_2]$. The restriction to the minimal range of the discrete convolution is an essential part of the algorithm. The appropriate treatment of the fast discrete convolution is explained in [3].

6 Globally Continuous and Piecewise Linear Case

The space $\mathcal{S} = \mathcal{S}(\mathcal{M})$ consists of *discontinuous* piecewise linear functions. An alternative is the subspace

$$
\mathcal{S}^{1} := \mathcal{S}(\mathcal{M}) \cap C^{0}(\mathbb{R})
$$

of *globally continuous* and piecewise linear functions. Next we consider the projection $\omega^{\mathcal{S}^1}$ of the convolution $\omega_{\text{exact}} := f * g$ into the space $\mathcal{S}^1$. The direct computation of $\omega^{\mathcal{S}^1} \in \mathcal{S}^1$ cannot follow the same lines as before, since the standard basis functions of $\mathcal{S}^1$ (the usual hat functions) are not orthogonal and any orthonormal basis has a larger support.

Nevertheless, there is a simple indirect way of computing $\omega^{\mathcal{S}^1}$. The inclusions

$$\mathcal{S}^1 \subset \mathcal{S} \subset L^2(\mathbb{R})$$

allow the following statement: Let $P_{\mathcal{S}} : L^2(\mathbb{R}) \to \mathcal{S}$ be the L^2-orthogonal projection onto $\mathcal{S}$ and $P_{\mathcal{S}^1} : \mathcal{S} \to \mathcal{S}^1$ the L^2-orthogonal projection onto $\mathcal{S}^1$. Then the product

$$P := P_{\mathcal{S}^1} \circ P_{\mathcal{S}} : L^2(\mathbb{R}) \to \mathcal{S}^1$$

is the L^2-orthogonal projection onto $\mathcal{S}^1$. This leads to the following algorithm.

Step 1: Let $f, g \in \mathcal{S}^1$. Because of $\mathcal{S}^1 \subset \mathcal{S}$, the data f, g can be used as input of the algorithm described in the previous part. The result is the projection $\omega = P_{\mathcal{S}} (f * g) \in \mathcal{S}$.

Step 2: The projection $\omega \mapsto \omega^{\mathcal{S}^1} = P_{\mathcal{S}^1} \omega \in \mathcal{S}^1$ can be computed by solving a tridiagonal system (the system matrix is the Gram matrix $\left(\int b_i b_j \mathrm{d}x \right)_{i,j=1,\dots,n}$ generated by the piecewise linear hat functions b_i).

References

[1] M.V. Fedorov, H.-J. Flad, L. Grasedyck, and B.N. Khoromskij: *Low-rank wavelet solver for the Ornstein-Zernike integral equation.* Max-Planck-Institut für Mathematik in den Naturwissenschaften. Preprint 59/2005, Leipzig 2005.

[2] W. Hackbusch: *On the efficient evaluation of coalescence integrals in population balance models.* Computing **72** (to appear).

[3] W. Hackbusch: *Fast and exact projected convolution for non-equidistant grids.* Computing (submitted) - *Extended version.* Max-Planck-Institut für Mathematik in den Naturwissenschaften. Preprint 102/2006, Leipzig 2006.

[4] W. Hackbusch: *Approximation of coalescence integrals in population balance models with local mass conservation.* Numer. Math (submitted).

[5] W. Hackbusch: *Fast and exact projected convolution of piecewise linear functions on non-equidistant grids. Extended version.* Max-Planck-Institut für Mathematik in den Naturwissenschaften. Preprint 110/2006, Leipzig 2006.

[6] D. Potts: *Schnelle Fourier-Transformationen für nichtäquidistante Daten und Anwendungen.* Habilitation thesis, Universität zu Lübeck, 2003.

[7] D. Potts, G. Steidl, and A. Nieslony: *Fast convolution with radial kernels at nonequispaced knots.* Numer. Math. **98** (2004) 329-351.

Intrusive versus Non-Intrusive Methods for Stochastic Finite Elements

M. Herzog[1], A. Gilg[2], M. Paffrath[3], P. Rentrop[4], and U. Wever[5]

[1] Technische Universität München, Centre for Mathematical Sciences, Numerical
   Analysis, Boltzmannstrasse 3, D-85748 Garching, Germany
   `martin.herzog@mytum.de`
[2] Siemens AG, Corporate Technology, Otto-Hahn-Ring 6, D-81730 Munich,
   Germany `albert.gilg@siemens.com`
[3] Siemens AG, `meinhard.paffrath@siemens.com`
[4] Technische Universität München, `toth-pinter@ma.tum.de`
[5] Siemens AG, `utz.wever@siemens.com`

Keywords: MSC 41A10 60Gxx 65Cxx

Summary. In this paper we compare an intrusive with an non-intrusive method
for computing Polynomial Chaos expansions. The main disadvantage of the non-
intrusive method, the high number of function evaluations, is eliminated by a special
Adaptive Gauss-Quadrature method. A detailed efficiency and accuracy analysis is
performed for the new algorithm. The Polynomial Chaos expansion is applied to a
practical problem in the field of stochastic Finite Elements.

1 Introduction

In this paper we deal with solving mechanical systems with multiple stochastic inputs. We restrict ourselves to stationary problems solved by the Finite
Element method. The described techniques may be applied to other fields as
well, e.g. in fluid dynamics or chemistry. The uncertainties may occur e.g.
for material parameters, system parameters, boundary conditions, layout and
outer forces. They can be modeled by using stochastic fields or in simple cases
stochastic variables. Therefore we have to solve a stochastic Finite Element
system. There are two approaches to treat such kind of problems. The first
is a statistic one using sampling methods. The simplest is the Monte-Carlo
method requiring several ten-thousands samples which is by far too time consuming for solving a Finite Element problem. There are improved sampling
methods such as Latin Hypercube Sampling [1], Quasi-Monte-Carlo Method
[2], Markov Chain Monte Carlo Method [3] and Response Surface Method
[4] but they have restrictions on the stochastic modeling. We use instead the

Polynomial Chaos (PC) expansion which discretizes the stochastic problem. Ghanem first applied this to Finite Elements based on Wiener-Hermite Chaos [5]. Later this work was generalized to the Wiener-Askey PC in [6], [7] and [8]. The principle of the PC expansion is to approximate stochastic fields by series of orthogonal polynomials. The PC expansion allows to cheaply compute interesting stochastic quantities such as mean value, deviation or skewness. Computing PC expansions is much cheaper than sampling methods in terms of function evaluations. Therefore it can be used for solving stochastic Finite Element Systems.

The PC expansion was also successfully applied to other fields: e.g. for stochastic advection equation see [9] or for stochastic perturbations in fluid flows see [10] and [11].

In this paper first we give a short explanation of the PC expansion. Next we describe the intrusive and non-intrusive method which are used to set up the expansion series. Subsequently we compare the both methods. In the following section we present a new algorithm for the non-intrusive method called Adaptive Gauss-Quadrature. We perform a detailed efficiency and accuracy analysis for this algorithm applied to nonlinear stochastic functions. Finally we demonstrate our method by a stochastic Finite Element application.

2 Polynomial Chaos Expansion

Wiener [12] introduced the Homogeneous PC, an infinite series expansion of orthogonal Hermite Polynomials, to express stochastic processes. Due to the theorem of Cameron-Martin [13] such series expansions are convergent for any L^2 functional. Therefore the Homogeneous PC converges for every stochastic process with finite second moments. In [6] and [7] this is generalized to orthogonal polynomials of the Askey scheme.

Let $X(\omega)$ be a stochastic field mapping $X : \Omega \to V$ from a probability space $(\Omega, \mathcal{A}, \mathcal{P})$ to a function space V. Let V be a Hilbert space with inner product $\langle .,. \rangle : V \times V \to \mathbb{R}$. Then $X : \Omega \to V$ is a stochastic field with a finite second moment if

$$\langle X(\omega)X(\omega) \rangle = \int_{\omega \in \Omega} X(\omega)^2 \, dP(\omega) < \infty. \tag{1}$$

With a PC expansion it is possible to represent such a random field parametrically through a set of independent stochastic variables $\{\xi_i(\omega)\}_{i=1}^N, N \in \mathbb{N}$

$$X(\boldsymbol{\xi}(\omega)) = \sum_{j=0}^{\infty} a_j \Phi_j(\boldsymbol{\xi}(\omega)). \tag{2}$$

The polynomials $\{\Phi_j(\boldsymbol{\xi}(\omega))\}$ are orthogonal satisfying the relation

$$\langle \Phi_i \Phi_j \rangle = \langle \Phi_i^2 \rangle \delta_{ij} \tag{3}$$

The inner product in (3) respective to the probability measure of the random vector $\boldsymbol{\xi}$ is

$$\langle f(\boldsymbol{\xi})g(\boldsymbol{\xi})\rangle = \int_{\omega \in \Omega} f(\boldsymbol{\xi})g(\boldsymbol{\xi})\, dP(\omega) = \int_{\omega \in \Omega} f(\boldsymbol{\xi})g(\boldsymbol{\xi})w(\boldsymbol{\xi})\, d\boldsymbol{\xi}, \qquad (4)$$

where $P(\omega)$ is the probability measure with respect to the density function $w(\boldsymbol{\xi}(\omega))$.

For computational use the expansion (2) is truncated. The highest order P of the polynomials Φ_j is set according to the accuracy requirements and the dimension N of the random vector $\boldsymbol{\xi}$ is the number of stochastic inputs. The truncated expansion has the form

$$X(\boldsymbol{\xi}(\omega)) \approx \sum_{j=0}^{M} a_j \Phi_j(\boldsymbol{\xi}(\omega)) \qquad (5)$$

where $(M+1)$ is the number of the N-dimensional orthogonal basis polynomials with maximum degree P

$$(M+1) = \sum_{i=0}^{P} \frac{(N+i-1)!}{i!(N-1)!} = \frac{(N+P)!}{N!P!}. \qquad (6)$$

The orthogonal polynomials Φ_j can be constructed as a tensor product of one-dimensional orthogonal polynomials $\phi_k(\xi_i)$

$$\Phi_j(\boldsymbol{\xi}) = \prod_{i=1}^{N} \phi_{\eta_{ji}}(\xi_i), \quad 0 \le j \le M \qquad (7)$$

where η is a $(M+1) \times N$ dimensional matrix which includes all multi-indices satisfying

$$\sum_{i=1}^{N} \eta_{ji} \le P, \quad 0 \le j \le M. \qquad (8)$$

For more details setting up the multidimensional orthogonal polynomials, see [6], [7] and [8].

In the following two sections we present the intrusive and non-intrusive method based on the truncated PC expansion (5).

2.1 Intrusive Method

We consider an intrusive method where the PC expansion is applied to the stochastic input and output of Finite Elements. The resulting system is projected onto the basis given by the PC expansion. The general application to stochastic differential equations is described in [7]. In case of elastic solid bodies we have to solve the d-dimensional linear equation system

$$KU = F \tag{9}$$

where K is the stiffness matrix, U is the solution vector of the displacement and F is the load vector. The stiffness matrix K depends e.g. in the isotropic case on the Young's modulus E and on the Poisson ratio ν (material parameters). We assume the material parameters to be stochastic. Therefore also the stiffness matrix K and the solution vector U (displacements) become stochastic. The stochastic quantities K and U may be approximated by a truncated PC expansion

$$K(\boldsymbol{\xi}) \approx \sum_{j=0}^{M} K_j \Phi_j(\boldsymbol{\xi}) \tag{10}$$

$$U(\boldsymbol{\xi}) \approx \sum_{i=0}^{M} U_i \Phi_i(\boldsymbol{\xi}) \tag{11}$$

Inserting the expansions (10) and (11) into the linear system (9) leads to

$$\sum_{i=0}^{M} \sum_{j=0}^{M} K_j U_i \Phi_i \Phi_j \approx F.$$

Next we perform a Galerkin projection onto the orthogonal basis polynomials of the Chaos space

$$\sum_{i=0}^{M} \sum_{j=0}^{M} K_j U_i \langle \Phi_i \Phi_j \Phi_k \rangle = \langle F \Phi_k \rangle = F \delta_{0k} \quad k = 0, 1, \ldots, M.$$

This system of equations can be written as linear system

$$\underbrace{\begin{bmatrix} \sum_{j=0}^{M} K_j \langle \Phi_0 \Phi_j \Phi_0 \rangle & \cdots & \sum_{j=0}^{M} K_j \langle \Phi_M \Phi_j \Phi_0 \rangle \\ \vdots & & \vdots \\ \sum_{j=0}^{M} K_j \langle \Phi_0 \Phi_j \Phi_M \rangle & \cdots & \sum_{j=0}^{M} K_j \langle \Phi_M \Phi_j \Phi_M \rangle \end{bmatrix}}_{K_{pc}} \underbrace{\begin{bmatrix} U_0 \\ \vdots \\ U_M \end{bmatrix}}_{U_{pc}} = \underbrace{\begin{bmatrix} F \\ 0 \\ \vdots \\ 0 \end{bmatrix}}_{F_{pc}}. \tag{12}$$

We can easily see that the dimension of the linear system increases linear with the number of basis polynomials of the PC space. The matrix K_{pc} is block-symmetric. For the solution of system (12) we developed an efficient Block-Gauss-Seidel algorithm in [14] which utilizes the properties of the matrix K_{pc}, see appendix B. A similar algorithm was developed on [15] for steady state diffusion problems. In our algorithm only the right upper blocks need to be saved in the original FE sparse structure. In addition only the diagonal blocks need to be LU factorized. Due to the fact, that the Block-Gauss-Seidel converges in a few iterations the complexity increases almost linear with the number of basis polynomials.

Another advantage of the intrusive method is that we do not make use of an expensive multi-dimensional quadrature formula. Appendix A shows that the multi-dimensional inner products $\langle \Phi_i \Phi_j \Phi_k \rangle$ can be calculated as a product of one-dimensional inner products.

A general application of the method to other physical quantities (than the displacements) e.g. for the von Mises stress can only be realized by large effort. Even for simple cases a separate algorithm must be developed, thus the computations can not be performed by the existing Finite Element tools.

Advantages and disadvantages of intrusive method:

+ size of linear system of equations grows linear with number of basis polynomials
+ no multi-dimensional Gauss-Quadrature necessary
+ better handling for transient problems
− changes in the source code of deterministic solver
− high memory usage
− difficult to implement for some interesting values (e.g von Mises stress)

2.2 Non-Intrusive Method

An alternative is the non-intrusive method, which is the common Fourier projection in a Hilbert space. This method can be applied for arbitrary output quantities. Again we are interested in the random displacements in terms of a a PC expansion

$$U(\boldsymbol{\xi}) \approx \sum_{i=0}^{M} U_i \Phi_i(\boldsymbol{\xi}). \tag{13}$$

Equation (13) is a Fourier type series and we obtain the coefficients by projection of $U(\boldsymbol{\xi})$ to the PC space. This space is spanned by the multi-dimensional orthogonal basis polynomials:

$$U_i = \frac{\langle U(\boldsymbol{\xi}) \Phi_i(\boldsymbol{\xi}) \rangle}{\langle \Phi_i(\boldsymbol{\xi}) \Phi_i(\boldsymbol{\xi}) \rangle} \tag{14}$$

$$= \frac{\int_{\omega \in \Omega} U(\boldsymbol{\xi}) \Phi_i(\boldsymbol{\xi}) w(\boldsymbol{\xi}) \, d\boldsymbol{\xi}(\omega)}{\int_{\omega \in \Omega} \Phi_i(\boldsymbol{\xi}) \Phi_i(\boldsymbol{\xi}) w(\boldsymbol{\xi}) \, d\boldsymbol{\xi}(\omega)} \tag{15}$$

Due to the theorem of Cameron-Martin [13] the above series expansion converges to every L^2-functional for $M \to \infty$. This means that the second central moment of the stochastic process approximated by the PC expansion needs to be finite

$$\langle U(\boldsymbol{\xi})^2 \rangle < \infty. \tag{16}$$

The big advantage of this method is that we can use a deterministic Finite Element tool as a black box in order to obtain realizations of $U(\boldsymbol{\xi})$. We use these realizations to calculate the numerator of the coefficients U_i (14) with a multi-dimensional Gauss-Quadrature formula. The problem is that the number of function evaluations for the quadrature grows exponentially with the number of stochastic input variables which is also reported in [16]. We have developed an algorithm (see Sect. 3) which drastically reduces the number of function evaluations.

Advantages and disadvantages of non-intrusive method:

+ usage of deterministic solver as black box
+ easily applicable to different quantities
+ independent coefficients of PC expansion
+ little additional memory usage
− number of function evaluations increases exponentially with independent stochastic inputs
− number of abscissae values must be chosen in advance
− discretization of time for transient problems necessary before PC expansion

3 Adaptive Gauss-Quadrature

The Adaptive Gauss-Quadrature algorithm, we present in this section, is suitable for solving multi-dimensional integrals as they occur in the non-intrusive method. The goal is to reduce the number of function evaluations of the multi-dimensional Gauss-Quadrature. The quadrature formula with S abscissae values in each of the N dimensions approximates the integral with the weighted sum

$$\mathbb{E}(f) = \int_\Omega f(\boldsymbol{\xi})w(\boldsymbol{\xi})\,d\boldsymbol{\xi} \approx \sum_{i=1}^{S^N} W_i f(\boldsymbol{\xi}_i). \tag{17}$$

The formula is exact if $f(\boldsymbol{\xi})$ is a polynomial of degree $2S-1$. The idea is that we evaluate the function f only for a few abscissae values and approximate it with $\tilde{f}$ for the remaining evaluation points. We use a PC expansion for the approximation $\tilde{f} = \sum_{i=0}^M a_i \Phi_i(\boldsymbol{\xi})$. Therefore we have to solve the minimization problem

$$\min_{a_0,\ldots,a_M} \int_\Omega \left(f(\boldsymbol{\xi}) - \sum_{i=0}^M a_i \Phi_i(\boldsymbol{\xi}) \right)^2 w(\boldsymbol{\xi})\,d\boldsymbol{\xi}. \tag{18}$$

The corresponding discrete Weighted-Least-Square problem is

$$\min_a \left\| \sqrt{W}Ba - \sqrt{W}y \right\|_2 \tag{19}$$

with

$$B = \begin{bmatrix} \Phi_0(\boldsymbol{\xi}_1) & \cdots & \Phi_M(\boldsymbol{\xi}_1) \\ \vdots & & \vdots \\ \Phi_0(\boldsymbol{\xi}_{S^N}) & \cdots & \Phi_M(\boldsymbol{\xi}_{S^N}) \end{bmatrix}, \qquad \sqrt{W} = \begin{bmatrix} \sqrt{w(\boldsymbol{\xi}_1)} & & 0 \\ & \ddots & \\ 0 & & \sqrt{w(\boldsymbol{\xi}_{S^N})} \end{bmatrix},$$

$$a = [a_0, \ldots, a_M]^T, \qquad\qquad y = [f(\boldsymbol{\xi}_1), \ldots, f(\boldsymbol{\xi}_{S^N})]^T.$$

Splitting up the quadrature formula into k exact function evaluations and $S^N - k$ approximations of the function f, the following approximation of $\mathbb{E}(f)$ is obtained:

$$\mathbb{E}_k(f) = \sum_{i=1}^{k} W_i f(\boldsymbol{\xi}_i) + \sum_{i=k+1}^{S^N} W_i \tilde{f}(\boldsymbol{\xi}_i) \tag{20}$$

On the other hand PC expansion leads to a different approximation of $\mathbb{E}(f)$:

$$\mathbb{E}(\tilde{f}) = \left\langle \sum_{i=0}^{M} a_i \Phi_i(\boldsymbol{\xi}) \right\rangle = a_0 \langle \Phi_0 \rangle \tag{21}$$

Also for the variance of f we get two approximations:

$$\mathrm{Var}_k(f) = \mathbb{E}_k(f^2) - (\mathbb{E}_k(f))^2 \tag{22}$$

$$\mathrm{Var}(\tilde{f}) = \sum_{i=1}^{M} a_i^2 \langle \Phi_i(\boldsymbol{\xi}) \Phi_i(\boldsymbol{\xi}) \rangle. \tag{23}$$

The two embedded methods (20) and (21) are utilized to obtain information for the accuracy of the approximation.

The Adaptive Gauss-Quadrature solves the following minimization problem:

$$\min k$$

$$\left| \mathbb{E}_k(f) - \mathbb{E}(\tilde{f}) \right| \leq \varepsilon_1 \tag{24}$$

$$\left| \mathrm{Var}_k(f) - \mathrm{Var}(\tilde{f}) \right| \leq \varepsilon_2$$

In the next sections, we show, that for multi-dimensional problems many function evaluations may be saved by this procedure.

Efficiency and Accuracy Analysis

First we examine a 2-dimensional modified Gauss curve

$$f(x, y) = \frac{1}{2\pi \cdot 0.9^2} e^{-\frac{(x-0.6)^2}{2 \cdot 0.9^2} - \frac{(y-0.4)^2}{2 \cdot 0.9^2}} + \frac{5}{2\pi \cdot 1.5^2} e^{-\frac{(x+0.2)^2}{2 \cdot 1.5^2} - \frac{(y+0.8)^2}{2 \cdot 1.5^2}}$$

$$+ \frac{2}{2\pi} e^{-\frac{(x-0.07)^2}{2} - \frac{(y-0.09)^2}{2}} \tag{25}$$

with standard normal distributed inputs x and y. The convergence of the Monte-Carlo analysis is presented in Fig. 1. The comparison with different degrees for the Adaptive Gauss-Quadrature may be observed in Table 1.

More interesting for analyzing the efficiency of the Adaptive Gauss-Quadrature is the higher dimensional problem

$$d(p, I, E, b, l) = -\frac{pbl^4}{8EI}. \tag{26}$$

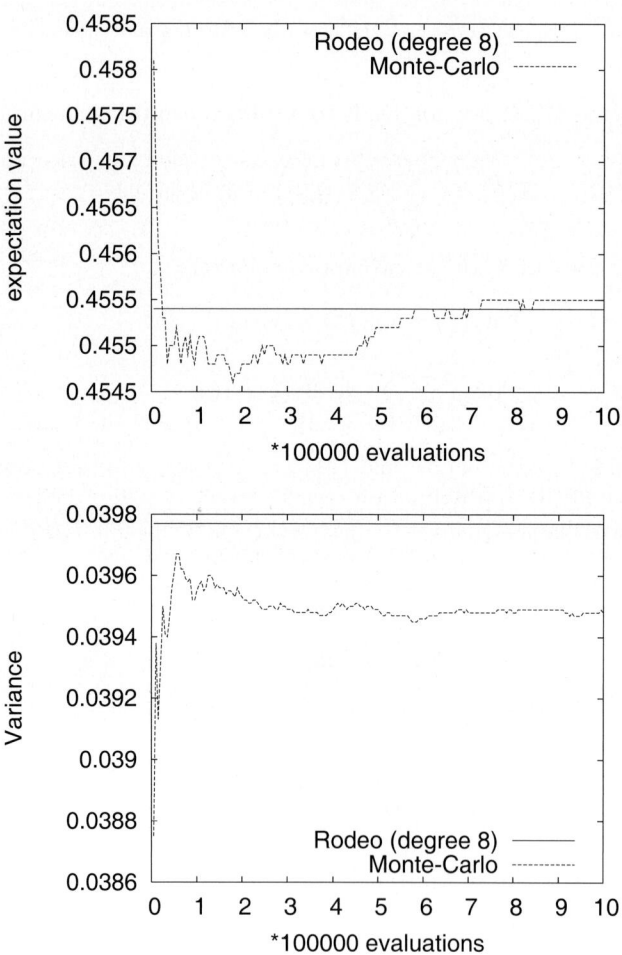

Fig. 1. Number of function calls for Monte-Carlo

This formula describes the displacement of a cantilever beam at its tip. All 5 input values are uniformly distributed with the following expectation values $\mathbb{E}$ and deviations σ: $\mathbb{E}(p) = 1000$, $\sigma(p) = 28.88$; $\mathbb{E}(I) = 15.625$, $\sigma(I) = 0.3608$; $\mathbb{E}(E) = 26000$, $\sigma(E) = 57.735$; $\mathbb{E}(b) = 12$, $\sigma(b) = 0.2888$; $\mathbb{E}(l) = 6$, $\sigma(l) = 5.7735 \cdot 10^{-3}$.

We used RODEO [17] where we implemented the Adaptive Gauss-Quadrature for the non-intrusive method and compared the calculated expectation value and variance with Monte-Carlo results. In Fig. 2 we plot the expectation value, the variance and the density of the output for different approximation degrees of the Adaptive Gauss-Quadrature. The results are compared with the Monte-Carlo analysis.

In Table 1 we show the relative error of the Adaptive Gauss-Quadrature for different approximation degrees with respect to a converged Monte-Carlo analysis with 10^6 samples. In addition the table presents the used function evaluations and the saved ones in comparison to an ordinary multi-dimensional Gauss-Quadrature. We see that the accuracy increases for higher orders and that the efficiency gain respective to multidimensional quadrature is huge for high orders and dimensions. For our 5-dimensional example with approximation degree 6 we save 92% evaluations and still get similar results for expectation and deviation. By this way we improved the non-intrusive method by substantially reducing the exponential increase of function evaluations with dimension.

Table 1. Efficiency and accuracy of the two test examples

Problem	Adaptive Gauss-Quad.	evaluations	saved evaluations	rel. error mean	rel. error deviation
Gauss curve (Dim. 2)	degree 2	9	0	0.04830812	0.32042313
	degree 3	16	0	0.01446631	0.19581190
	degree 4	25	0	0.00396731	0.08667579
	degree 5	36	0	0.00154345	0.04201181
	degree 6	49	0	5.2692e-05	0.01840407
	degree 7	57	7	0.00039080	0.00803699
	degree 8	66	15	0.00026785	0.00362344
Beam (Dim. 5)	degree 2	161	82	7.3102e-05	0.08381494
	degree 3	654	370	4.8039e-05	0.03109469
	degree 4	361	2764	2.0886e-06	0.01391660
	degree 5	809	6967	4.4279e-04	0.00204725
	degree 6	1351	15456	1.2531e-05	0.00296498

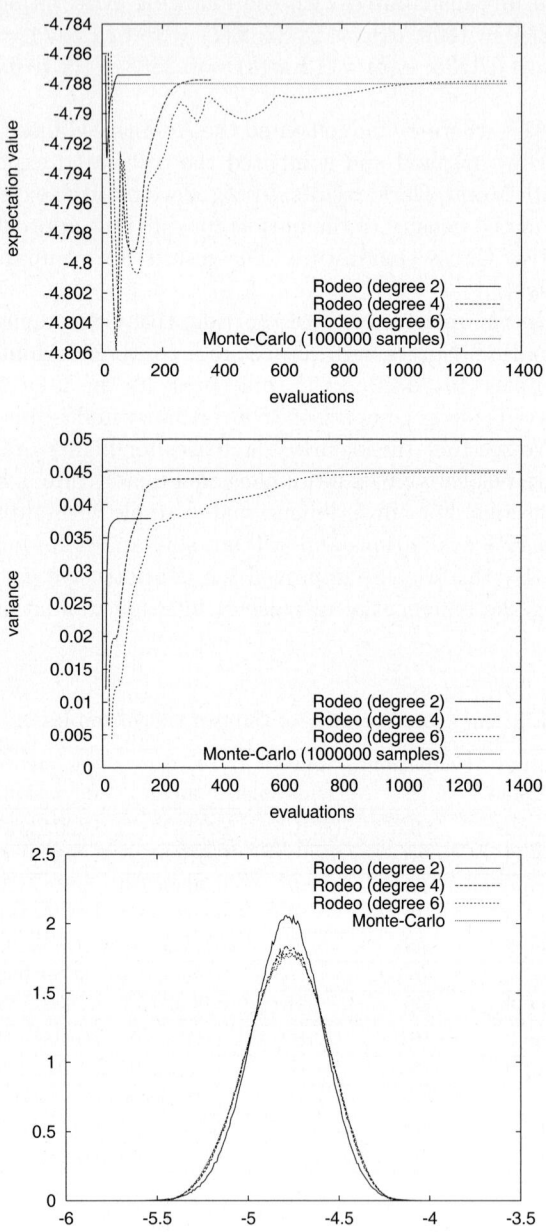

Fig. 2. Adaptive Gauss-Quadrature versus Monte-Carlo: Upper picture: Expectation value, Middle picture: Deviation, Lower picture: Density

4 Stochastic Finite Elements

Now we want to show that the described methods are applicable to static Finite Element problems. Therefore we examine an elastic isotropic elbow consisting of two different materials with uniformly distributed Young's modulus E. It is fixed at its bottom and a force is applied to the upper edge. The Young's modulus of the elbow is uniformly distributed on the interval $[10000; 12000]$ and the one of the cross beam on the interval $[50000; 60000]$. The Poisson ratio of the elbow material is set to 0.25 and the one of the cross beam to 0.3.

With the described methods we can compute the random behavior of different quantities of the elbow. Figure A.20 on page 336 shows the expectation value of the von Mises stress and Fig. A.21 on page 336 its deviation. We used an approximation degree of 6 in each dimension for the Adaptive Gauss-Quadrature. In comparison to multi-dimensional Gaussian quadrature requiring 49 function evaluations, the proposed Adaptive Gauss-Quadrature only needs 32 function evaluations for a relative error tolerance of 10^{-4}.

Accuracy Analysis of Intrusive versus Non-intrusive Method

We analyzed the accuracy of the intrusive and the non-intrusive method for the above described elbow. Therefore the relative error $\varepsilon(\boldsymbol{\xi})$ is defined as

$$\varepsilon(\boldsymbol{\xi}) = \frac{\left\|\left(\sum_{i=0}^{M} \mathbf{U}_i \Phi_i(\boldsymbol{\xi})\right) - \mathbf{U}(\boldsymbol{\xi})\right\|_\infty}{\|\mathbf{U}(\boldsymbol{\xi})\|_\infty} \tag{27}$$

giving the possibility to measure the error of the Polynomial Chaos approximation of the displacements $\mathbf{U}(\boldsymbol{\xi}) \approx \sum_{i=0}^{M} \mathbf{U}_i \Phi_i(\boldsymbol{\xi})$. In Table 2 the relative error of the approximation $\varepsilon(\boldsymbol{\xi})$ is shown for the intrusive and non-intrusive method each with different orders of the PC expansion and different realizations $\boldsymbol{\xi}$. ξ_1 and ξ_2 may be varied in the interval $[0; 1]$ because the random inputs of the regarded example are uniformly distributed and therefore we used modified Legendre Polynomials for the basis of the PC.

In summary, both methods provide similar accuracy for fixed orders of the PC expansion. Thus, accuracy reasons do not give a preference to one of the methods.

5 Conclusion

In this paper we have shown that the Polynomial Chaos expansion is superior to the classical sampling approach for our applications because the number of saved function evaluations is immense. The two described methods (intrusive and non-intrusive) for setting up the PC expansion provide similar results for

Table 2. Accuracy comparison of the intrusive and non-intrusive method with the relative error $\varepsilon(\boldsymbol{\xi})$ of the best approximation

		intrusive		non-intrusive		
ξ_1	ξ_2	order 2	order 4	order 2	order 4	order 6
0.00	0.00	0.312E-03	0.833E-06	0.309E-03	0.836E-06	0.245E-08
0.00	0.50	0.234E-03	0.656E-06	0.232E-03	0.675E-06	0.148E-08
0.00	1.00	0.223E-03	0.604E-06	0.222E-03	0.603E-06	0.152E-08
0.25	0.25	0.127E-03	0.602E-07	0.128E-03	0.593E-07	0.324E-09
0.50	0.00	0.473E-04	0.937E-07	0.468E-04	0.757E-07	0.121E-09
0.50	0.50	0.294E-10	0.478E-08	0.118E-06	0.423E-08	0.153E-10
0.50	1.00	0.413E-04	0.919E-07	0.409E-04	0.131E-06	0.317E-09
0.75	0.75	0.127E-03	0.523E-07	0.128E-03	0.570E-07	0.348E-09
1.00	0.00	0.199E-03	0.574E-06	0.197E-03	0.743E-06	0.232E-08
1.00	0.50	0.225E-03	0.661E-06	0.223E-03	0.611E-06	0.158E-08
1.00	1.00	0.312E-03	0.861E-06	0.310E-03	0.122E-05	0.413E-08

all examples we treated. Therefore accuracy criteria do not influence the selection of method. The advantages and disadvantages of both methods however, listed in Sect. 2, have a great impact on the practical usability. We prefer the non-intrusive method for our static Finite Element problems because it is possible to use existing deterministic solvers (in our case FEM tools) as black box. On the other hand for the intrusive method changes in the source code of the deterministic solver are necessary. In addition we have reduced the exponential increase of function evaluations with the dimension of the random input by developing the *Adaptive Gauss-Quadrature*. Thus, the worst disadvantage of the non-intrusive method is eliminated.

For transient problems the relevance of intrusive method increases, if not only the final state is computed but the whole transient trajectories. In this case a time discretization has to be performed before applying the Polynomial Chaos expansion.

References

[1]   W. Loh: On Latin hypercube sampling, Ann. Stat. **24** 2058–2080 (1996).
[2]   B. L. Fox: Strategies for Quasi-Monte Carlo, Springer, (1999).
[3]   N. N. Madras: Lectures on Monte Carlo Methods, American Mathematical Society, (2002).
[4]   M. Rajaschekhar, B. Ellingwood: A new look at the response surface approach for reliability analysis, Struc. Safety **123** 205–220 (1993).
[5]   R. G. Ghanem, P. D. Spanos: Stochastic Finite Elements. A Spectral Approach., Springer, Heidelberg, (1991).
[6]   D. Xiu, G. E. Karniadakis: The Wiener–Askey Polynomial Chaos for Stochastic Differential Equations, SIAM Journal on Scientific Computing **24** (2) 619–644 (2003).

[7]   G. Karniadakis, C.-H. Su, D. Xiu, D. Lucor, C. Schwab, R. Todor: Generalized Polynomial Chaos Solution for Differential Equations with Random Inputs, ETH Zürich, Seminar für Angewandte Mathematik Research Report No. **01** (2005).

[8]   M. Schevenels, G. Lombaert, G. Degrande: Application of the stochastic finite element method for Gaussian and non-Gaussian systems, in: ISMA, Katholieke Universiteit Leuven, Belgium, (2004).

[9]   M. Jardak, C.-H. Su, G. E. Karniadakis: Spectral Polynomial Chaos Solutions of the Stochastic Advection Equation, Journal of Scientific Computing **17** 319–338 (2002).

[10] O. P. L. Maître, O. M. Knio, H. N. Najm, R. G. Ghanem: A Stochastic Projection Method for Fluid Flow I. Basic Formulation, Journal of Computational Physics **173** 481–511 (2001).

[11] D. Lucor, G. E. Karniadakis: Noisy Inflows Cause a Shedding-Mode Switching in Flow Past an Oscillating Cylinder, Physical Review Letters **92** 154501 (2004).

[12] N. Wiener: The Homogeneous Chaos, Amer. J. Math. **60** 897–936 (1938).

[13] R. Cameron, W. Martin: The orthogonal development of nonlinear functionals in series of Fourier-Hermite functionals, Ann. Math. **48** 385 (1947).

[14] M. Herzog: Zur Numerik von stochastischen Störungen in der Finiten Elemente Methode, Master's thesis, TU München (2005).

[15] D. Xiu, G. E. Karniadakis: Modeling uncertainty in steady state diffusion problems via generalized polynomial chaos, Comput. Methods Appl. Mech. Engrg. **191** 4927–4948 (2002).

[16] R. Ghanem, S. Masri, M. Pellissetti, R. Wolfe: Identification and prediction of stochastic dynamical systems in a polynomial chaos basis, Comput. Methods Appl. Mech. Engrg. **194** 1641–1654 (2005) .

[17] M. Paffrath, U. Wever: The Probabilistic Optimizer RODEO, Internal Report Siemens AG (2005).

A Multi-Dimensional Inner Products $\langle \Phi_i \Phi_j \Phi_k \rangle$

We can split the multi-dimensional inner products $\langle \Phi_i \Phi_j \Phi_k \rangle$ into a product of one-dimensional inner products if its density function $w(\boldsymbol{\xi})$ meets

$$w(\boldsymbol{\xi}) = \prod_{i=1}^{N} w_i(\xi_i). \tag{28}$$

Then we get with Eq. (7)

$$\langle \Phi_i(\boldsymbol{\xi}) \Phi_j(\boldsymbol{\xi}) \Phi_k(\boldsymbol{\xi}) \rangle = \prod_{l=1}^{N} \langle \phi_{\eta_{il}}(\xi_l) \phi_{\eta_{jl}}(\xi_l) \phi_{\eta_{kl}}(\xi_l) \rangle \tag{29}$$

for the multi-dimensional inner products. Therefore we have not to perform an expensive multi-dimensional quadrature.

B Block-Gauss-Seidel Algorithm

We developed an Block-Gauss-Seidel algorithm to solve Eq. (12) which has the structure

$$\underbrace{\begin{bmatrix} A_{11} & \cdots & A_{1n} \\ \vdots & & \vdots \\ A_{n1} & \cdots & A_{nn} \end{bmatrix}}_{=:A} \begin{bmatrix} x_1 \\ \vdots \\ x_n \end{bmatrix} = \begin{bmatrix} b_1 \\ \vdots \\ b_n \end{bmatrix} \tag{30}$$

with the block-symmetric matrix A

$$A_{ij} = A_{ji}, \quad 1 \leq i, j \leq n. \tag{31}$$

Algorithm

1. LU factorization of the diagonal matrices $A_{ii}, i = 1, \ldots, n$.
2. Get a good starting value for $x^{(0)} = [x_1^{(0)}, \ldots, x_n^{(0)}]^T$. Set $x_1^{(0)}$ to the deterministic solution. This is a good approximation because x_1 is the expectation value. Set the vectors $x_3^{(0)}, \ldots, x_n^{(0)}$ to zero. Approximate $x_2^{(0)}$ with the solution of $A_{22}x_2^{(0)} = b_2 - A_{21} - x_1^{(0)}$.
3. Iterate $x^{(k)}, k = 1, 2, 3, \ldots$ until convergence:

$$A_{nn}x_n^{(k)} = b_n - \sum_{j=1}^{n-1} A_{nj}x_j^{(k-1)}$$

$$\vdots$$

$$A_{ii}x_i^{(k)} = b_i - \sum_{j=1}^{i-1} A_{ij}x_j^{(k-1)} - \sum_{j=i+1}^{n} A_{ij}x_j^{(k)}$$

$$\vdots$$

$$A_{11}x_1^{(k)} = b_1 - \sum_{j=2}^{n} A_{1j}x_j^{(k)}$$

Walking, Running and Kicking of Humanoid Robots and Humans

M. Stelzer and O. von Stryk

Simulation, Systems Optimization and Robotics Group, Department of Computer
Science, Technische Universität Darmstadt, Hochschulstr.10, D-64289 Darmstadt
[stelzer|stryk]@sim.tu-darmstadt.de

Summary. In this paper key aspects and several methods for modeling, simulation, optimization and control of the locomotion of humanoid robots and humans are discussed. Similarities and differences between walking and running of humanoid robots and humans are outlined. They represent several, different steps towards the ultimate goals of understanding and predicting human motion by validated simulation models and of developing humanoid robots with human like performance in walking *and* running. Numerical and experimental results are presented for model-based optimal control as well as for hardware-in-the-loop optimization of humanoid robot walking and for forward dynamics simulation and optimization of a human kicking motion.

1 Introduction

A large variety of bipedal motions are known from humans whereas today's humanoid robots can only realize a small fraction of them. All motions on two legs have in common that maintaining stability and balance is a critical issue and that there are redundancies in the actuation of the respective system. For typical humanoid robots actuation redundancies lie in the level of joint angles: One overall locomotion goal (i.e., a certain walking trajectory and contact situation history of the feet during walking) usually may be achieved by an infinite number of joint angle trajectories. For humans, an additional level of redundancy must be considered in comparison with today's humanoid robots which usually have one actuator per rotational joint in the leg: Even for given joint angle trajectories, the involvement of the various muscles which actuate the respective human joints is not uniquely defined.

A widely accepted hypothesis in biomechanics is that for trained leg or whole body motions among all possible muscle actuation strategies the one is selected which minimizes or maximizes a certain objective [23]. Selecting the best possible walking trajectories is also mandatory for *autonomous* humanoid robots which must carry not only all of their actuators but also onboard computing, additional sensors and energy supplies.

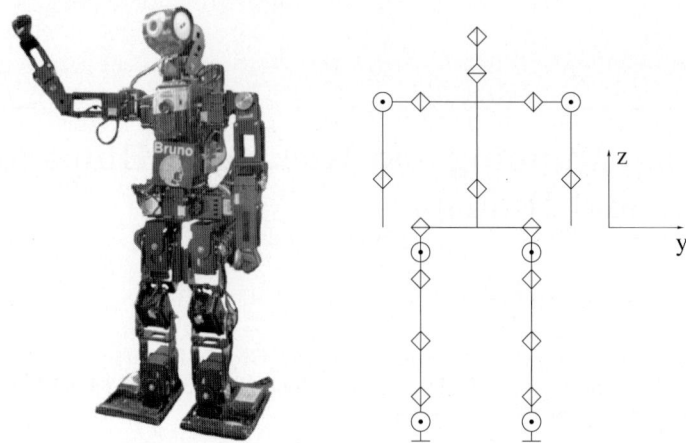

Fig. 1. The 55 cm tall, autonomous humanoid robot Bruno developed at TU Darmstadt (left) and its kinematic structure with 21 servo motor driven joints (right).

The research presented in this paper has been inspired by Roland Z. Bulirsch in manifold ways. For example, mathematical models of the locomotion dynamics of humanoid robots and humans result in medium to large systems of nonlinear ordinary or differential-algebraic equations. The determination of optimal control variable trajectories lead to large-scale nonlinear optimal control problems which can only be solved numerically. Bulirsch had already in the 1960s during his affiliation with the University of California, San Diego, pioneered one of the very first numerical methods for solving such problems, the so-called indirect multiple shooting method (in the terminology of [39]). This method enabled to solve trajectory optimization problems in aeronautics and astronautics numerically in a satisfactory manner which had not been possible before. A variety of efficient numerical methods for solving optimal control problems which nowadays are used all over the world has directly evolved from his pioneering work, e.g., [4]. Furthermore, Bulirsch has always emphasized strongly that numerical methods must be mature enough to enable the treatment of real-world problems and not only simplified academical problem statements. However, this requires significant efforts for deriving a sophisticated and validated mathematical model of the problem in question.

2 Kinematical Leg Design

During the last decade significant advances in humanoid robotics concerning autonomous walking and hardware and software design have been achieved. The humanoid robot *H7* (137 cm, 55 kg, 35 degrees of freedom (DoF)) [24] is able to execute reaching motions based on the implemented whole body motion control. Footstep planning and balancing compensation is used for

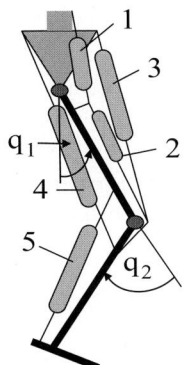

1. Ilio Psoas group
2. Vastus group
3. Rectus Femoris
4. Hamstring group
5. Gastrocnemius group

Fig. 2. Kinematic structure of a planar human leg model with five muscle groups actuating the hip and knee joints.

adaptive walking. The German humanoid robot *Johnnie* (180 cm, 40 kg, 17 DoF) [22] can walk with a maximum speed of 2 km/h. The control and computational power is onboard, whereas the power supply is outside of the robot. In the Japanese Humanoid Robot Project the robot *HRP-3* (160 cm, 65 kg, 36 DoF) has been developed with special skills for water and dust resistivity. It can walk with a maximum speed of 2.5 km/h [20]. The Korean robot *KHR-2* (120 cm, 54 kg, 41 DoF) [21] walks with a maximum speed of 1 km/h. The robots *QRIO* (50 cm, 5 kg, 24 DoF) by Sony and *ASIMO* (120 cm, 52 kg, 26 DoF) by Honda are two commercial humanoid robot platforms. *QRIO* [25] can walk stable, jump and "jog" (i.e., fast walking with small flight phases) including the transitions between them. It can also execute many special motions, among them coordinated dancing, squatting and getting up. *ASIMO* [14] is the humanoid robot with the currently highest walking speed of 6 km/h and the most costly development. The autonomous humanoid robot *Bruno* (55 cm, 3.3 kg, 21 DoF, Fig. 1 left) can play soccer and walks with more than 40 cm/s, almost 1.5 km/h, in permanent operation [8, 15].

All of these humanoid robots which today can walk flexible, stable and reliably in repeatable experiments share the same basic kinematic leg structure: Six (sometimes seven) rigid, rotational joints per leg are connected with rigid links. Usually three joints are used in the hip, one in the knee and two in the ankle (cf. Fig. 1 right) to reach a general position and orientation with the foot within its range. A joint typically consists of an electric actuator with gear which are designed to be as rigid, powerful and lightweight as possible.

The human leg, however, does not have one rigid rotary actuator in each joint, e.g., the knee joint. Redundant, elastic linear actuators, i.e., the contracting muscles, antagonistically arranged around the leg joints result in a very compliant leg design. In general, in biomechanical motion systems each joint is often driven by more than two muscles. Also there are muscles that have effect on more than one joint, e.g., the rectus femoris, the gastrocnemius and the hamstring group (cf. Fig. 2).

3 Modeling and Simulation of Locomotion Dynamics

Current humanoid robots can be modeled as a kinematical tree structure consisting of rigid links and joints, e.g., Fig. 1 right, and changing contact situation of the feet with the ground during walking. The locomotion dynamics describes the relationship between the joint angles $q^T = (q_1(t), \ldots, q_n(t))$ and the joint torques $\tau^T = (\tau_1, \ldots, \tau_n)$. It is represented by a multibody system (MBS) dynamics model with contact constraints

$$\mathcal{M}(q)\,\ddot{q} = \tau - \mathcal{C}(q, \dot{q}) - \mathcal{G}(q) + J_c^T f_c \tag{1}$$

$$0 = g_c(q), \tag{2}$$

where $\mathcal{M}$ denotes the positive definite mass matrix, $\mathcal{C}$ the Coriolis and centrifugal forces, $\mathcal{G}$ the gravitational forces, and $J_c^T f_c$ the contact forces. The ground contact constraints $g_c \in \mathbb{R}^{n_c}$ represent holonomic constraints on the system from which the constraint Jacobian $J_c = \partial g_c / \partial q \in \mathbb{R}^{n_c \times n}$ may be obtained, while $f_c \in \mathbb{R}^{n_c}$ is the ground constraint force.

For formulating the second order differential equations (1) different methods exist ranging from recursive methods based on force-moment relations as Newton-Euler to energy based, analytic methods as Euler-Lagrange [5]. For efficiently formulating these equations in case of a large number of joints n the recursive $\mathcal{O}(n)$ articulated body algorithm (ABA) [7] has been shown to be an accurate, numerically stable and efficient algorithm which computes $\mathcal{M}, \mathcal{C}, \mathcal{G}, J_c$ in three, resp. five in case of contact forces, forward or backward iterations (sweeps).

An alternative formulation of the ABA has been derived which results in an $\mathcal{O}(n)$ closed-form expression for the inverse mass matrix [29] (see also [18] for details). This recursive approach is modular and flexible as it facilitates the exchange of submodels and the reuse of other model parts without having to reformulate the complete model as it is the case, e.g., with the Euler-Lagrange method. An object oriented implementation has been developed based on this formulation of the ABA (cf. Sect. 5.3 and [17]) which enables a flexible and efficient computation in different applications of the dynamics model (1), e.g., simulation, optimization or control. The method can be extended to an efficient recursive, computation of partial derivatives of the dynamics model which are required for optimal control and trajectory optimization (cf. Sect. 5.3) or optimal parameter estimation.

In contrast to humanoid robots the torques τ_i acting in human joints do not stem from a single actuator but from contracting muscle groups whose dynamic behavior must be considered in addition to the dynamics model (1) of the skeleton and the wobbling masses. For modeling of the dynamic motion and force behavior of muscles as contracting actuators with serial and parallel elasticities and active contractile elements a number of well investigated approaches have been developed. They describe the muscle forces in relation to muscle length, muscle velocity and muscle activation as the many models based on the fundamental approaches of Hill and Huxley, cf. [23, 26]. Almost all models from literature assume that the muscle forces act at a point. For non-punctual areas of force application the muscles are divided into several muscle models with single points of actuation. Several approaches exist for modeling the muscle paths as the straight line method (modeling the muscle path to connect the points of application in a straight line), the centroid line method (modeling the muscle path to connect the centers of mass of the muscle cross sectional areas) or the obstacle set method (modeling the muscle path to move freely sliding along the bones). A survey of these approaches may be found, e.g., in [26].

4 Control and Stability of Bipedal Gaits

Several criteria have been established to ensure postural stability of a bipedal robot walking trajectory either during offline computation or online using feedback control. Two basic groups are distinguished: criteria for static and for dynamic stability. *Static stability* is present if the center of gravity (CoG) of the robot projected along the direction of gravity lies in the convex hull of the support area consisting of all foot-ground contact points. *Dynamic stability* is defined as a bipedal walking motion which is not statically stable but the robot does not fall over. This is the case for running, jogging and even medium fast walking of humans. Several constructive criteria are used to realize dynamic walking in humanoid robots, among them the nowadays widely used zero moment point (ZMP) [38] and the foot rotation indicator (FRI) [12], an extension of the ZMP definition. Both indices for dynamic stability basically induce that no moments around the two possible axes that might lead to a falling of the robot occur while taking into account not only the mass distribution of the robot (as the static projected CoG criterion) but also dynamic effects, i.e., acting forces and moments.

For online evaluation of the stability indices for feedback control, the humanoid robot needs additional sensors to the standard joint angle position encoders. Usually accelerometers and gyroscopes are used in the upper body close to the CoG to determine the robot's pose and and force/moment sensors are used in the feet to measure foot-ground contact for ZMP-based stability control. In small to medium sized humanoid robots ZMP-based control can be implemented successfully also without ground contact force sensing, e.g., [8]. On the other hand for human-sized, high-grade humanoid robots in addition force/moment sensing in the joints is provided.

Commonly trajectory tracking control is applied in a hierarchical, decentralized scheme where the setpoints for the joint angles of the humanoid robot are updated in a constant frequency between 1 and 10 ms [18]. The feedback control schemes of the joints, e.g., PD or PID, are usually operated independently of each other. Ideally, a nonlinear feedback control scheme would be applied based on a full nonlinear MBS dynamics model (1) of the humanoid

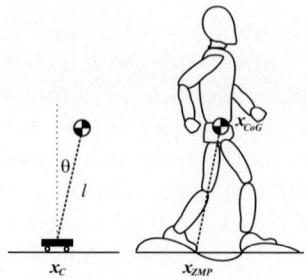

Fig. 3. The inverted pendulum model is used as a simplified model to describe walking properties of humanoid robots and humans.

robot. However, this would require to evaluate such a model for at least 18 bodies and joints on an embedded processor and at least faster than the update frequency of the joint angle setpoints which is difficult to achieve.

Therefore, simplified motion dynamic models are used for stabilizing control schemes based on the ZMP or its precursor, the inverted pendulum (IP) model (Fig. 3), which was used in balancing control of the first bipedal walking robots. The IP model approximates the basic behavior of the whole walking robot and is easy to handle with low computational efforts [32].

The IP model has also been used to describe slow human walking as balancing on stiff legs. Measured ground reaction forces are reproduced well. But this rigid model can not be extended to human running. A spring-mass system describes much better rebounding on compliant legs during running or jogging. Therefore, elasticities must be included in a robot leg design to describe dynamic walking and running as in Raibert's hopping robots [27]. However, the latter can not be used for slow bipedal walking or even standing. Therefore, as the grand challenge in research of humanoid robot locomotion the question remains to be answered how humanoid walking and running can be realized well using *one* robot leg design.

An important difference between today's humanoid robots and humans is that any motion of a humanoid robot is under full feedback control at any time. To enable bipedal running using today's robot technology, the motors in the joints must be more powerful to enable higher joint velocities (like the wheels in a fast car). But the motion can only be as fast as the joint sensors are able to measure joint position and velocity to enable feedback control. This is not the case for many types of fast human motions. By suitable training, human motions can be generated feed-forward which are much faster than the internal sensing of an arm or leg but nevertheless are of high quality, e.g., a fast serve in tennis or fast piano playing. These fast motions make effective use of the elastic, compliant design of legs, arms and fingers like shooting an arrow with an elastic bow. From a robot control point of view, however, elasticity in a kinematical leg or arm design is avoided as much as possible because control becomes severely difficult. Compliant joint behavior is instead simulated using joint force/moment sensor and compliant control schemes.

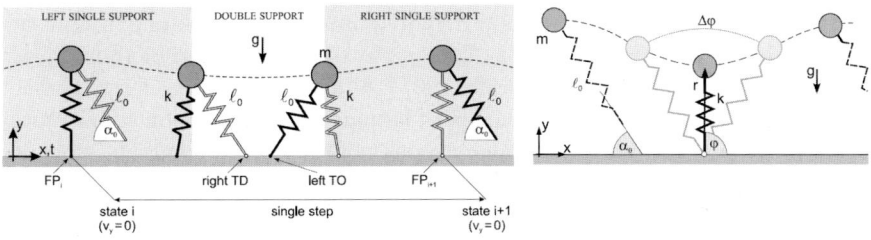

Fig. 4. A basic compliant leg model exhibits properties of both human-like walking (left) and running (right) in the same mechanical design (due to [9]).

Recent results in biomechanics showed that by adding suitable elasticities to the most simple bipedal leg model properties of both fast, elastic running and slow, stiff walking can be represented quite well (Fig. 4) [9, 13]. In numerical simulations bipedal locomotion using such basic compliant leg models has demonstrated to be quite robust against external disturbances like uneven terrain. This motivates to investigate bipedal walking machines with compliant, three-segmented legs like the ones of Sect. 5.2 as a step towards solving the grand challenge problem of humanoid robot locomotion. However, it remains yet unsolved how an elastic, possibly underactuated, bipedal leg design can be stabilized under different walking and running conditions.

In this context, passive dynamic bipedal walking should also be mentioned which has recently gained much interest. Low powered bipedal walking machines have demonstrated stable bipedal locomotion on flat terrain with low or even no actuation and therefore claimed a highly energy-efficient and natural way of walking [6]. A closer inspection reveals that the kinematical leg design consists of two-segmented, rigid legs with thigh and shank only. There is no need for an articulated foot with ankle joints. A foot point contact or knob-like foot and a rolling motion along the foot knob's surface during stance phase of the leg is sufficient when the bipedal walkers swing over a straightened knee joint. Moreover, such leg designs only enable a certain constant walking speed. A larger variation in walking speed is not possible, not even mentioning running. Passive dynamic walkers share these properties with current passive above-knee prostheses. It is very difficult using them to walk at very different speeds or to use them in uneven or steep terrain.

5 Methods and Case Studies

5.1 Forward Dynamics Simulation of Human Kicking Motion

There are approximately 650 skeletal muscles in the human body which are anchored by tendons to bone and affect skeletal movement such as locomotion. Considering the many individual muscles involved in locomotion and a mathematical model of locomotion dynamics and the redundancy in joint actuation by muscles involved (Sect. 3) then the secret of human biodynamics is explained in N.A. Bernstein's words (1935): "As in orchestra, each instrument plays its individual score, so in the act of human walking each joint reproduces its own curve of movements and each center of gravity performs its sequence of accelerations, each muscle produces its melody of efforts, full with regularly changing but stable details. And in like manner, the whole of this ensemble acts in unison with a single and complete rhythm, fusing the whole enormous complexity into clear and harmonic simplicity. The consolidator and manager of this complex entity, the conductor and at the same time the composer of the analyzed score, is of course the central nervous system".

A widely accepted hypothesis in biomechanics is that for trained motions among all possible actuation strategies of the many muscles involved the one is selected which minimizes or maximizes a suitable objective [23]. Such an actuation strategy must then essentially coincide with the much more simpler compliant leg model depicted in Fig. 4 which describes well the observed overall leg behavior in human locomotion but cannot explain the behavior of the many individual muscles involved. The modeling and numerical solution of optimization problems for the system dynamics to reliably predict system behavior and motion is nowadays well established for vehicle and robot dynamics. It is a grand challenge in human biodynamics research to develop such methodologies for the human musculoskeletal system and requires the development and validation of assumptions, models and methods.

The central problem statement addressed in this section is to find the activations $\boldsymbol{u}(t) = (u_1(t), ..., u_{n_m}(t))^T$ of each of the n_m muscles involved so that the resulting calcium ion concentration γ_i caused by the activation u_i of each muscle i leads to forces F_i, $i = 1, \ldots, n_m$, which cause a motion of all n joints (i.e., joint angle trajectories $\boldsymbol{q}(t) = (q_1(t), ..., q_n(t))^T$, $0 \leq t \leq t_f$) which

1. is equal or as "close" as possible to the kinematic and/or kinetic data of a human body motion measured in experiments (inverse problem), or
2. best fulfills some motion goal like maximum jump height or width or fastest possible walking or running (forward problem).

While in the first case only the redundancy of the muscles must be considered, the second case incorporates also the additional level of redundancy with respect to the overall motion. "Close" in the first case may be measured by an objective function, e.g., the integral over the difference of measured and calculated joint angle trajectories. The goal achievement in the second case can be measured by a suitable objective function as time or energy required.

Generally, the two different approaches of forward and of inverse dynamics simulation exist [37]. The *forward dynamics simulation* of a human motion leads to high dimensional, nonlinear optimal control problems. Current approaches in this field are usually based on direct shooting techniques [4, 39] with finite difference gradient approximations. They require even for problems with reduced models of the whole human body computational times of days

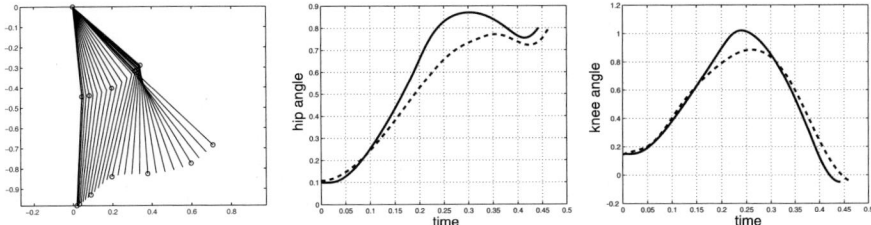

Fig. 5. Numerical results for the minimum time kicking motion: Visualization in phase space (left), measured (dashed line) and optimized (solid line) joint angle trajectories of hip (middle) and knee (right).

or weeks on workstations, cf. [2, 31]. Forward dynamics simulation based on a validated dynamics model and model parameters has the important potential of *predicting* certain motions.

On the other hand, *inverse dynamics simulation* investigates given kinematic position and velocity trajectories of a human motion, e.g., obtained from measurements. Together with special approximate modeling approaches for the dynamics model and the objective it allows a comparatively fast numerical computation of the controls of each muscle group if restrictive assumptions on the underlying model like special objective functions for control of the muscles involved are made, e.g., [28]. Inverse dynamics simulation for a measured human motion gives an *interpretation* of the acting forces and torques on the level of the single muscles involved.

To overcome the drawback of high computational burden an efficient forward dynamics simulation and optimization approach for human body dynamics has been suggested [37]. It is based on an efficient $\mathcal{O}(n)$ modeling of the dynamics of the musculoskeletal system consisting of the MBS (1) and suitable models of the activation dynamics $\hat{\gamma}_i$, the force-velocity and tension-length and further muscle properties. Instead of using one of the direct shooting approaches which require feasibility with respect to the MBS dynamics constraints in each iteration of the optimization method a simultaneous approach for solving the MBS dynamics integration and optimization problems inherent in the optimal control problem is selected. In direct collocation [39] the implicit integration for a sequence of discretization steps from initial to final time is included as a set of explicit nonlinear equality constraints in the optimization problem and the optimal control problem is transformed into a sparse and large-scale nonlinearly constrained optimization problem (NLP). Without the restriction to feasibility to the ODE constraints in each iteration as in direct shooting only the final solution of direct collocation iterations must satisfy them. Without the restriction to feasible iterates and with much easier computable gradients the solution may be obtained much faster.

As one first example, a time optimal kicking motion has been investigated, i.e., $t_f \rightarrow$ min!. Kinematic and kinetic data of the musculoskeletal system

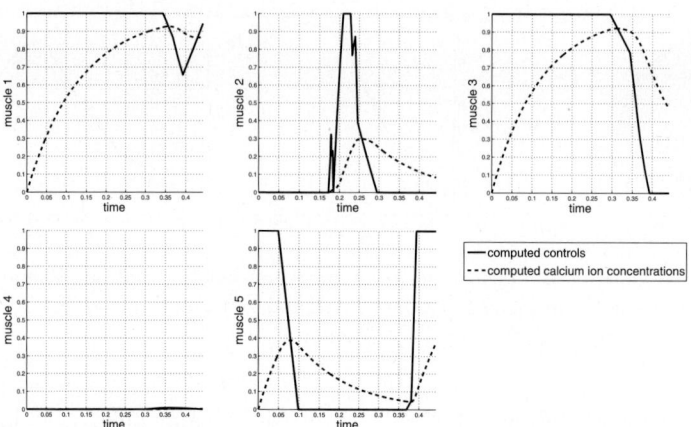

Fig. 6. Numerical results for the minimum time kicking motion: The control variables are the muscle activations (which correspond to EMG, solid lines) and the resulting calcium ion concentrations of the muscles (dashed lines).

as well as muscle model parameters and measured reference data have been taken from [33, 31]. The model consists of two joints, two rigid links and five muscle groups (Fig. 2). The problem is formulated as an optimal control problem with 9 first order state variables (hip angle q_1, knee angle q_2, the corresponding joint velocities and 5 calcium ion concentrations) and 5 control variables, i.e., the activations of the muscles. The muscle lengths and velocities that are needed for the force-velocity and tension-length relationship of the Hill-type muscle model are calculated according to [33]. The resulting lever arms used for transforming the linear muscle force into joint-torques depend on the joint angle and are also taken from [33], as well as the passive moments. Suitable boundary conditions at t_0 and t_f must be meet by q_i, $\dot{q}_i$, γ_j and box constraints are imposed on $q_i(t)$, $\gamma_j(t)$, $u_j(t)$ during $[t_0, t_f]$ [37].

Compared to the data of the measured human kick (and the results of [33, 31] which matched the measured data very well), our results show a shorter time t_f and larger maximum angles (Fig. 5). This is because in [33] the maximum muscle forces were modified successively in a way that the computed minimum time for the motion matches the measured time closely. Now solving the same problem with our approach a better (local) minimum can be computed. But, the controls (Fig. 6) show the same characteristics. Time discretizations with, e.g., 10 resp. 60 grid points lead to NLPs with 129 resp. 531 nonlinear variables and 81 resp. 829 nonlinear constraints. The resulting computing time of SNOPT [11] was 1.2 s resp. 6.3 s on an Athlon XP1700+ PC. The direct shooting approach of [33, 31] for 11 grid points required hours on a workstation to compute the solution [34]. Compared with our approach and considering how processor performance has progressed since 1996, we obtain a speed up of two orders of magnitude.

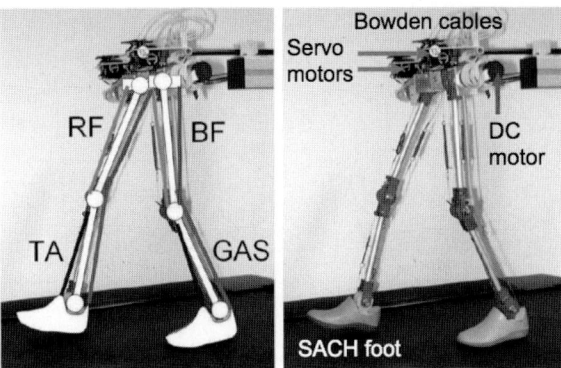

Fig. 7. The arrangement of adjustable elastic structures (springs) spanning the ankle, knee and hip joints in the JenaWalker II as a demonstrator for bipedal walking with three-segmented passively compliant robot legs [19, 36].

5.2 Passively Compliant Three-Segmented Bipedal Robot Legs

For validating hypotheses in human locomotion [13] and to overcome the limitations of current robot legs three-segmented bipedal robot legs have been developed by the Locomotion Laboratory of the University of Jena in cooperation with TETRA GmbH, Ilmenau [19]. Four major leg muscle groups (cf. Fig. 2) are represented in the leg by passive elastic structures: tibialis anterior (TA), gastrocnemius (GAS), rectus femoris (RF) and biceps femoris (BF) (Fig. 7). Only the hip joints are directly actuated by a central pattern generator through sinusoidal oscillations. The motions of the knee and the ankle is only due to gravity, foot-ground contact forces and the interaction by the elastic leg structure. The bipedal walker locomotes on a treadmill while its upper body is attached to a holder to avoid sidewards falling.

A computational model has been developed to optimize nine different parameters including parameters of the hip central pattern generator like frequency, amplitude and offset angle [36]. As derivatives of the solution of the computational model implemented in Matlab are not available, different methods for robust parameter optimization are investigated. In the first optimization study implicit filtering [10] was used as a robust, projected Newton-type method to maximize the walking speed. A final speed of 1.6 m/s was achieved which was about 60 % faster than the one achieved with a manually adjusted parameter set. Both speeds were below the estimated maximum walking speed of about 3 m/s which results from the product of the angular velocity of the hip joint multiplied by the leg length. The drawback is that the hip joint torque increases dramatically to 700 Nm for a robot model of assumed total mass 80 kg. Thus, in a second study the torques were bounded to be below 500 Nm. Nomad [1] was used to optimize for walking speed w.r.t. to the torque constraint. A walking speed of 3.6 m/s was found which is remarkable for two reasons: the speed was increased after constraining the problem (as a conse-

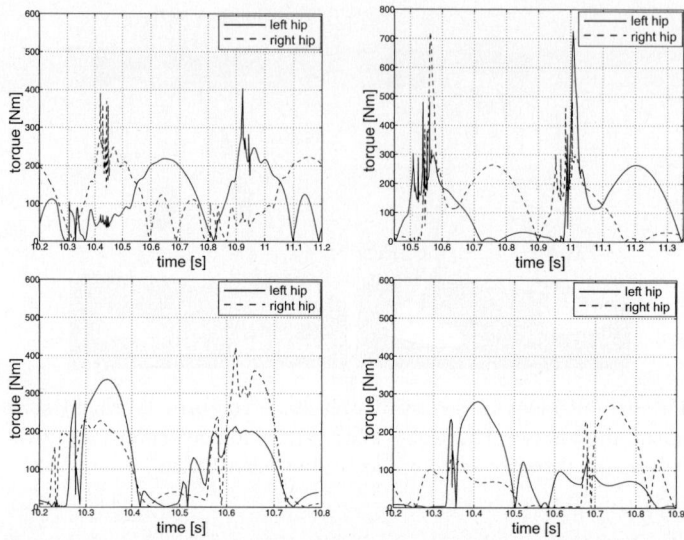

Fig. 8. Bipedal walking with three-segmented passively compliant robot legs. Resulting hip torque trajectories for different problem statements (from top left to bottom right): (0) initial guess, (1) maximum velocity, (2) maximum velocity with limited torque, (3) minimum torque with constrained minimum velocity.

quence, implicit filtering must earlier have gotten stuck in a local minimum) and the speed is even higher than the predicted maximum walking speed. Indeed, flight phases have been observed in the model that lead to a speed higher than achievable by walking. In a third study, the hip torques were minimized and the minimum walking speed was bounded to be $\geq 2\,\mathrm{m/s}$. This problem formulation results in a reduction of the hip torques to less than $300\,\mathrm{Nm}$. The resulting hip torques are displayed in Fig. 8. Experimental results have shown similar behavior like the simulation results which are visualized in Fig. A.22 on page 337. Also the flight phases can be observed.

5.3 Model-Based Trajectory Optimization and Control for Humanoid Robots

MBS dynamics models for humanoid robots are typically characterized by a high number of degrees of freedom, relatively few contact constraints or collision events, and a variety of potential ground contact models, actuator models, and mass-inertial parameter settings due to changing load conditions (cf. Sect. 3). Better suited for practical use in simulation, optimization and control than closed-form dynamical expressions of Eqs. (1) - (2) is an recursive modeling approach. The latter can be similarly efficient but permits in addition the easy interchangeability of joint types, parameters, and the introduction of external forces without repeated extensive preprocessing. Among the various, recursive, numerical algorithms which satisfy these criteria the ABA [7] has been selected in the alternative formulation of [29] (cf. [18]). It can be implemented highly modular as all calculations and parameters can occur as an exchange of information between links.

To avoid the problem in industrial robotics of having to re-code the MBS dynamics model in different applications for different purposes like simulation during robot design, model-based trajectory optimization and model-based control an object-oriented modeling architecture has been proposed [16, 17]. A central aim was to generate modular and reconfigurable MBS dynamics code for legged robots which can be used efficiently and consistently in different applications. In [16] it was demonstrated that using this approach a full, 3D MBS dynamics model of a humanoid robot consisting of 12 mechanical DoF for the torso and two legs and 1D drive-train models can be computed on a typical onboard computer in less than 1 ms. Thus, it is well suited for developing novel, nonlinear model-based humanoid robot control schemes.

Computing optimal locomotion trajectories is equally important for the design and operation of humanoid robots. The question of finding optimal trajectories of the n joint motor torques $\boldsymbol{\tau}(t)$, $0 \leq t \leq t_f$, w.r.t. to time, energy and/or stability leads to optimal control problems [18]. A forward dynamics simulation and optimization approach analogous to the biomechanical optimization in Sect. 5.1 is applied. To improve the efficiency and robustness of the numerical optimization the contact dynamics of the DAE system (1)-(2) can be computed using a reduced dynamics approach. This is based on coordinate partitioning and results in an equivalent ODE system of minimal size by projecting the dynamics onto a reduced set of independent states. The approach requires solving the inverse kinematics problem for the dependent states which, in the case of legged systems, are generally the contact leg states. For most leg configurations, this problem is easily solved using knowledge of the relative hip and foot contact locations [18].

Besides the system dynamics and the objective, further details must be considered in the optimal control problem formulation for humanoid robot locomotion. To determine optimal walking motions, only one half-stride has to be considered due to the desired symmetry and periodicity of the walking motion. Further constraints to be considered include a) box constraints on joint angles and control, b) symmetry resp. anti-symmetry of the joint angles at the boundaries of the half-stride, c) lift-off force equals zero at the end of a half-stride, d) stability (one of the criteria of Sect. 4), e) foot orientation and position, and f) avoidance of slipping (cf. [3, 18] for details). The object oriented modeling approach allows easy exchange of parts of the model as well as reuse of code, in this case for the differential equations as well as for different constraints.

Using the outlined approach a design study for optimal motor and gear selection for an 80 cm tall humanoid robot has been conducted [18]. For the manufactured prototype optimal joint trajectories for walking have been computed which enabled the robot to perform stable walking experiments on level ground using only PID-control of the joint angle without using inertial sensors in the upper body and contact sensors in the feet for stabilization [3]. More flexible and stable walking motions however can be realized using the approaches based on additional sensors as outlined in Sect. 4.

5.4 Hardware-in-the-Loop Optimization of the Walking Speed of an Autonomous Humanoid Robot

Optimization based on simulation models for humanoid robot dynamics has many advantages. For example, the optimization process can be performed unsupervised, it can run overnight and there is no hardware deterioration. Nevertheless, it has a major drawback. The relevance of the optimization result for the application on a real robot depends critically on the quality and accuracy of the simulation model. For legged robots obtaining accurate kinetical data is difficult. This data may even vary from one robot to another of the same production. Moreover, effects like gear backlash, joint elasticity and temperature dependent joint friction as well as different ground contact properties are difficult and cumbersome to model accurately. All these effects may accumulate to a significant simulation model error.

The solution of an optimal control problem for maximizing walking speed or stability will utilize all "resources" available in the model. The numerical solution is then likely to be found in a region where the above mentioned modeling errors significantly affect the applicability of the numerically computed trajectories to the real robot. For obtaining best robot performance it is therefore advisable to use the robot itself as its best model and to perform optimization based on experiments which replace the evaluation of the simulation model. But then the optimization method must be able to cope with a noisy function evaluation as no walking experiment will give exactly the same results if repeated even in the same setting. Moreover, the method should use as few as possible function evaluations as every experiment may cause not only time for human operators, too many experiments will wear out the robot's hardware and make the results useless. A disadvantage which is often observed with evolutionary type optimization methods.

For optimization of the walking speed for the autonomous humanoid robot Bruno (Fig. 1), the distance the robot covers during a walking experiment for a certain walking parameter set is used as the objective function. The robot starts walking with a small step length and increases it linearly during the experiment until the robot falls or reaches a final step length. The distance obtained by a large, constant number of steps (e.g., 52) is then measured. The walking motion is generated by prescribing trajectories for the hip and the feet and solving the inverse kinematics for the joint angles. Thus, the walking motion is parameterized by a large number of parameters for the trajectories of hip and feet. By experimental investigation the most relevant parameters affecting walking performance have been identified: the relation of the distances of the front and of the rear leg to the robot's CoG, the lateral position, the roll angle and the height above ground of the foot during swing phase, and the pitch of the upper body. Starting from an initial, stable but slow walking motion these five parameters are varied in each iteration of the optimization method and a walking experiment is carried out. It should be noted that only lower and upper bound constraints on the parameters are applied. It is not needed to incorporate explicit constraints for maintaining walking stability as stability is implicitly included in the objective function.

To solve the arising non deterministic black-box optimization problem, where besides of a noise function value no further information, especially no objective gradient, is provided, a surrogate modeling approach has been selected. An initial set of experiments is generated around the initial motion by varying each parameter on its own. This set builds the basis points for the use of design and analysis of computer experiments, [35], which is applied to approximate the original objective function on the whole feasible parameter domain. The sequential quadratic programming method [11] is applied to rapidly compute the maximizer of the smooth surrogate function resulting in the current iteration. For this parameter set the corresponding walking experiment is performed. If the distance of a found maximizer to a point already evaluated by experiments falls below a defined limit, not the actual maximizer, but the maximizer of the expected mean square error of the surrogate function is searched, evaluated, and added to the set of basis points for approximation. This procedure improves the approximation quality of the surrogate function in unexplored regions of the parameter domain and avoids to get stuck in a local maximum. After a new point is added, a new surrogate function is approximated, and the optimization starts again. From our experience this approach for online optimization of walking speed is much more efficient than genetic or evolutionary algorithms which are usually applied to cope with the robust minimization of noisy functions.

After about 100 walking experiments in less than four hours a very stable and fast walking motion with a speed of 30 cm/s has been obtained for the first version of the humanoid robot. The distance the robot covered before falling down or reaching the end of the velocity slope is plotted in Fig. 9. A sequence of a resulting walking motion is depicted in Fig. A.23 on page 337. Later the robot has been modified to reduce weight in the upper body and the optimization procedure has been repeated resulting in an even further improved speed of 40 cm/s [15]. This is so far the fastest walking motion of any humanoid robot of any size in the humanoid robot league of the RoboCup (www.robocup.de).

6 Conclusions and Outlook

In this paper, several methods and case studies on modeling, simulation, optimization and control of motion dynamics of humanoid robot and humans have been presented. They constitute steps towards the grand challenges of

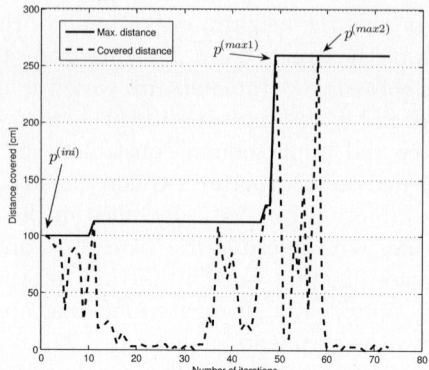

Fig. 9. Hardware-in-the-loop optimization of the walking speed of an autonomous humanoid robot: The distance the robot covers in each iteration.

understanding and predicting human biodynamics by numerical simulation and of developing humanoid robots being able to walk and run with human-like efficiency.

Acknowledgement. The successful investigation of the methods and case studies presented in this paper would not have been possible without the major contributions of the following coauthors whose most valuable contributions are deeply acknowledged: A. Seyfarth, R. Tausch, F. Iida, A. Karguth (Sect. 5.2 [36]), M. Hardt, R. Höpler (Sect. 5.3 [18, 16, 17]), Th. Hemker, H. Sakamoto (Sect. 5.4 [15]).

References

[1]   M.A. Abramson. *Pattern Search Filter Algorithms for Mixed Variable General Constrained Optimization Problems*. PhD thesis, Rice University, 2002.

[2]   F.C. Anderson and M.G. Pandy. A dynamic optimization solution for vertical jumping in three dimensions. *Comp. Meth. Biomech. Biomed. Eng.*, 2:201–231, 1999.

[3]   M. Buss, M. Hardt, J. Kiener, M. Sobotka, M. Stelzer, O. von Stryk, and D. Wollherr. Towards an autonomous, humanoid, and dynamically walking robot: modeling, optimal trajectory planning, hardware architecture, and experiments. In *Proc. IEEE Intl. Conf. on Humanoid Robots*, page to appear, Karlsruhe, München, Sept. 30 - Oct. 3 2003. Springer Verlag.

[4]   R. Bulirsch and D. Kraft, editors. *Computational Optimal Control*, volume 115 of *Intl. Series in Numer. Math.* Birkhäuser, 1994.

[5]   J.J. Craig. *Introduction to Robotics*. Pearson Education, 3rd edition, 2005.

[6]   S. Collins, A. Ruina, R. Tedrake, and M. Wisse. Efficient bipedal robots based on passive dynamic walkers. *Science Magazine*, 307:1082–1085, 2005.

[7]   R. Featherstone. *Robot Dynamics Algorithms*. Kluwer Academic Publishers, 1987.

[8]   M. Friedmann, J. Kiener, S. Petters, H. Sakamoto, D. Thomas, and O. von
      Stryk. Versatile, high-quality motions and behavior control of humanoid soccer
      robots. In *Proc. Workshop on Humanoid Soccer Robots of the 2006 IEEE-RAS
      Intl. Conf. on Humanoid Robots*, pages 9–16, Genoa, Italy, Dec. 4-6 2006.

[9]   H. Geyer. *Simple Models of Legged Locomotion based on Compliant Limb
      Behavior*. PhD thesis, Friedrich-Schiller-University Jena, 2005.

[10] P. Gilmore and C.T. Kelley. An implicit filtering algorithm for optimization
      of functions with many local minima. *SIAM J. Optim.*, 5:269–285, 1995.

[11] P.E. Gill, W. Murray, and M.A. Saunders. SNOPT: An SQP algorithm for
      large-scale constrained optimization. *SIAM J. Optim.*, 12:979–1006, 2002.

[12] A. Goswami. Foot rotation indicator (FRI) point: A new gait planning tool
      to evaluate postural stability of biped robots. In *IEEE Intl. Conf. on Robotics
      and Automation (ICRA)*, pages 47–52, 1999.

[13] H. Geyer, A. Seyfarth, and R. Blickhan. Compliant leg behaviour explains basic
      dynamics of walking and running. *Proc. Roy. Soc. Lond. B*, 273(1603):2861–
      2867, 2006.

[14] M. Hirose, Y. Haikawa, T. Takenaka, and K. Hirai. Development of humanoid
      robot ASIMO. In *IEEE/RSJ Intl. Conf. Intelligent Robots and Systems, Work-
      shop 2*, Maui, HI, USA, 29 Oct 2001.

[15] Th. Hemker, H. Sakamoto, M. Stelzer, and O. von Stryk. Hardware-in-the-loop
      optimization of the walking speed of a humanoid robot. In *9th Intl. Conf. on
      Climbing and Walking Robots (CLAWAR)*, pages 614–623, Brussels, Belgium,
      September 11-14 2006.

[16] R. Höpler, M. Stelzer, and O. von Stryk. Object-oriented dynamics modeling
      for legged robot trajectory optimization and control. In *Proc. IEEE Intl. Conf.
      on Mechatronics and Robotics (MechRob)*, pages 972–977, Aachen, Sept. 13-15
      2004. Sascha Eysoldt Verlag.

[17] R. Höpler, M. Stelzer, and O. von Stryk. Object-oriented dynamics modeling of
      walking robots for model-based trajectory optimization and control. In I. Troch
      and F. Breitenecker, editors, *Proc. 5th MATHMOD Vienna*, number 30 in
      ARGESIM Reports, Feb. 8-10 2006.

[18] M. Hardt and O. von Stryk. Dynamic modeling in the simulation, optimization
      and control of legged robots. *ZAMM: Z. Angew. Math. Mech. (Appl. Math.
      Mech.)*, 83(10):648–662, 2003.

[19] F. Iida, Y. Minekawa, J. Rummel, and A. Seyfarth. Toward a human-like biped
      robot with compliant legs. *Intelligent Autonomous Systems*, 9:820–827, 2006.

[20] K. Kaneko, F Kanehiro, S. Kajita, H. Hirukawa, T. Kawasaki, M. Hirata,
      K. Akachi, and T. Isozumi. Humanoid robot HRP-2. In *IEEE Intl. Conf. on
      Robotics and Automation (ICRA)*, pages 1083–1090, New Orleans, LA, USA,
      Apr. 26 Apr - May 1 2004.

[21] J. Y. Kim, I. W. Park, J. Lee, M. S. Kim, B. K. Cho, and J. H. Oh. System
      design and dynamic walking of humanoid robot KHR-2. In *IEEE Intl. Conf.
      on Robotics and Automation (ICRA)*, pages 1443–1448, Barcelona, Spain, 18
      - 22 Apr 2005.

[22] K. Löffler, M. Gienger, and F. Pfeiffer. Sensor and control design of a dynam-
      ically stable biped robot. In *IEEE Intl. Conf. on Robotics and Automation
      (ICRA)*, pages 484–490, Taipei, Taiwan, Sep. 14-19 2003.

[23] B.M. Nigg and W. Herzog. *Biomechanics of the Musculo-skeletal System*. Wi-
      ley, 1999.

[24] K. Nishiwaki, M. Kuga, S. Kagami, M. Inaba, and Hirochika Inoue. Whole-body cooperative balanced motion generation for reaching. In *IEEE/RAS Intl. Conf. on Humanoid Robots*, pages 672–689, Los Angeles, USA, Nov. 10-12 2004.

[25] K. Nagasaka, Y. Kuroki, S. Suzuki, Y. Itoh, and J. Yamaguchi. Integrated motion control for walking, jumping and running on a small bipedal entertainment robot. In *IEEE Intl. Conf. on Robotics and Automation (ICRA)*, volume 4, pages 3189–3194, New Orleans, LA, USA, Apr. 26 - May 1 2004.

[26] M.G. Pandy. Computer modeling and simulation of human movement. *Annu. Rev. Biomed. Eng.*, 3:245–273, 2001.

[27] M.H. Raibert. *Legged Robots That Balance*. MIT Press, 1986.

[28] J. Rasmussen, M. Damsgaard, and M. Voigt. Muscle recruitment by the min/max criterion? A comparative numerical study. *J. Biomechanics*, 34(3):409–415, 2001.

[29] G. Rodriguez, K. Kreutz-Delgado, and A. Jain. A spatial operator algebra for manipulator modeling and control. *Intl. J. Robotics Research*, 40:21–50, 1991.

[30] RoboCup. The RoboCup federation. www.robocup.org.

[31] T. Spägele, A. Kistner, and A. Gollhofer. Modelling, simulation and optimisation of a human vertical jump. *J. Biomechanics*, 32(5):521–530, 1999.

[32] T. Sugihara, Y. Nakamura, and H. Inoue. Realtime humanoid motion generation through ZMP manipulation based on inverted pendulum control. In *IEEE Intl. Conf. on Robotics and Automation (ICRA)*, pages 1404–1409, May 2002.

[33] T. Spägele. *Modellierung, Simulation und Optimierung menschlicher Bewegungen*. PhD thesis, Universität Stuttgart, 1998.

[34] T. Spägele. personal communication. 2005.

[35] J. Sacks, S. B. Schiller, and W.J. Welch. Design for computer experiments. *Technometrics*, 31:41–47, 1989.

[36] A. Seyfarth, R. Tausch, M. Stelzer, F. Iida, A. Karguth, and O. von Stryk. Towards bipedal running as a natural result of optimizing walking speed for passively compliant three-segmented legs. In *9th Intl. Conf. on Climbing and Walking Robots (CLAWAR)*, pages 396–401, Brussels, Belgium, September 12-14 2006.

[37] M. Stelzer and O. von Stryk. Efficient forward dynamics simulation and optimization of human body dynamics. *ZAMM: Z. Angew. Math. Mech. (Appl. Math. Mech.)*, 86(10):828–840, 2006.

[38] M. Vukobratovic and B. Borovac. Zero-moment point - thirty five years of its life. *Intl. J. Humanoid Robotics*, 1(1):157–173, 2004.

[39] O. von Stryk and R. Bulirsch. Direct and indirect methods for trajectory optimization. *Annals of Operations Research*, 37:357–373, 1992.

Numerical Simulation of Shape Memory Actuators in Mechatronics

G. Teichelmann and B. Simeon

M2 – Zentrum Mathematik, Technische Universität München, Boltzmannstr.3, 85748 Garching, Germany, {teichelmann|simeon} @ma.tum.de

Summary. This paper deals with shape memory alloys (SMA) as temperature controlled actuators in mechatronic applications. The mathematical model consists of a coupled system of partial differential and differential-algebraic equations where continuum equations describe the evolution of deformation and temperature while rigid body equations and control laws define the interaction with the mechatronic device. Both modeling and numerical issues will be addressed. In particular, by applying the method of lines with finite elements, the overall problem is reduced to a differential-algebraic system in time that is shown to be of index two. Simulation results for a robotics application illustrate the approach.

1 Introduction

The use of shape memory alloys (SMA) in industrial applications demands for new simulation methods that are able to handle the typically highly heterogeneous and multi-scale problem structure. A variety of mathematical models and appropriate numerical simulation schemes are currently being developed world-wide, see, e.g., Arndt/Griebel [1], Carstensen/Plechac [2], Jung/Papadopoulos/Ritchie [6], and Mielke/Roubiček [7].

In mechatronic applications, the microstructural features of SMA are not relevant. Instead, a macroscopic approach based on homogenization is usually sufficient, and the main challenge are fast numerical methods that can be used to test controllers or to optimize the device. In this paper, we employ the model introduced by Helm [4]. It consists of a heterogeneous coupled system of partial differential and differential-algebraic equations (PDAEs) where continuum equations describe the evolution of deformation and temperature. We extend the model by interfaces with a mechatronic device such as rigid body equations or control laws. Moreover, the temperature is explicitly taken into account by means of a modified heat equation that depends on material stress, strain rate, and internal variables.

The development of numerical simulation methods for SMA is strongly influenced by the related fields of plasticity and inelastic materials. State of the art methods mostly make use of the return mapping algorithm [13], which is comparable to low order implicit time integration with fixed stepsize. However, it should be remarked that the overall time-dependent process evolves locally and is solely loosely coupled via the balance of momentum. Thus, adaptive time integration or even dynamic iteration may become attractive in the future.

In this paper, we apply finite elements in space to transform the coupled PDAE system into a system of differential-algebraic equations (DAEs) of index two. Particular attention is paid to the discretization of the heat equation and to an implicitly given approximation of the strain rate by an additional DAE that acts as a differentiator. While the finite element grid is fixed, adaptive time integration by means of the standard Matlab code `ode15s` [12] is chosen to resolve the transient behavior of the SMA device.

An example of an SMA actuator in mechatronics is given by the artificial finger depicted in Fig. 1. It is driven by NiTi wires that are heated by electric current and act as flexor and extensor, similar to biological muscles. The device was developed at the Institute of Applied Mechanics of TU Munich and analyzed in extensive test runs by Pfeifer, Ulbrich and Schleich [11, 10].

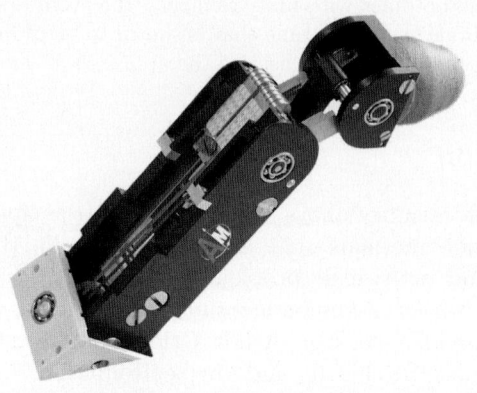

Fig. 1. Artificial finger actuated by SMA wires. Courtesy of H. Ulbrich.

Along with theory and experiment, simulation represents an indispensable partner in the advance of such mechatronic applications. In case of the finger example, one of the key goals is to maximize the speed of the finger motion while minimizing the risk of failure and heat damage. For this purpose, the SMA model needs to be coupled with the equations of rigid body motion for the finger and the control law for the temperature.

The paper is organized as follows. In Sect. 2 we discuss the mathematical model of Helm that can be viewed as generalized viscoplastic constitutive law. Furthermore, as mentioned above, a differential-algebraic approach to obtain rate information on quasistationary modeled variables is presented. Section 3 deals with the Galerkin projection for semidiscretization in space. In view of the mechatronic application field with mostly NiTi wires as SMA structure, one dimension in space suffices usually, which leads to small scale systems and meets the efficiency requirements. Finally, in Sect. 4 we combine the SMA model as substructure with rigid body equations and simulate the behavior of a device based on the above finger example.

2 Mathematical Model

In this section, we sketch the SMA model recently proposed by Helm [4, 5], analyze its structure, and extend it by a modified heat equation that depends on material stress, strain rate, and internal variables. Though not resolving the crystal grid, this macroscale model keeps track of the phase changes between austenite and the twinned and oriented martensite stages and is able to reproduce, depending on the temperature, the main effects of SMA structures. Since our main interest lies in the simulation of the longitudinal contraction and prolongation of an SMA wire, it is sufficient to restrict the model to the one-dimensional case.

2.1 Balance and Evolution Equations

Let for simplicity $\Omega = [0, 1] \subset \mathbb{R}$ denote the domain that is occupied by the SMA material in the undeformed referential configuration and let $[0, T]$ be the time interval in which we study its behavior. The deformation of the material maps each point $x \in \Omega$ at time $t \in [0, T]$ to its current position $x + u(t, x)$ with displacement $u : [0, T] \times \Omega \to \mathbb{R}$. We consider small strains only, and consequently the (total) strain ϵ is given by

$$\epsilon(t, x) = \frac{1}{2} \left(\nabla u(t, x) + \nabla u(t, x)^T \right) = \frac{\partial u(t, x)}{\partial x} \tag{1}$$

due to the one-dimensional model.

After the kinematics has been specified, we proceed with the balance of linear momentum

$$\rho \ddot{u}(t, x) = \frac{\partial \sigma(t, x)}{\partial x} + \beta(t, x) \tag{2}$$

with stress $\sigma : [0, T] \times \Omega \to \mathbb{R}$, density of body forces $\beta : [0, T] \times \Omega \to \mathbb{R}$, and mass density $\rho \in \mathbb{R}$. While the function β is given, constitutive equations are required to specify the relation between stress σ and strain ϵ. Using a model structure common in elasto-plasticity and visco-plasticity applications, the stress is assumed to depend on the displacement u, on a so-called inelastic strain $\epsilon_p : [0, T] \times \Omega \to \mathbb{R}$, and also on the temperature $\theta : [0, T] \times \Omega \to \mathbb{R}$. For the moment, we assume the temperature θ to be given. A standard additive split [15] defines the total strain as sum of inelastic and elastic strains,

$$\epsilon(t,x) = \epsilon_p(t,x) + \epsilon_e(t,x), \quad \epsilon_p, \epsilon_e : [0,T] \times \Omega \rightarrow \mathbb{R}. \tag{3}$$

The relation between stress σ and elastic strain $\epsilon_e = \epsilon - \epsilon_p$ is then given by the model of linear thermoelasticity

$$\sigma(t,x) = 2\mu(\theta)\epsilon_e(t,x) + \kappa(\theta)\,\epsilon_e(t,x) - 3\alpha(\theta)\kappa(\theta)(\theta - \theta_0) \tag{4}$$

with material parameters $\mu(\theta), \alpha(\theta), \kappa(\theta) \in \mathbb{R}$ and the reference temperature $\theta_0 \in \mathbb{R}$. Thus, given total strain ϵ and inelastic strain ϵ_p, the stress can be evaluated by (4), which reduces to Hooke's law for constant temperature $\theta = \theta_0$.

The total strain has already been defined in (1) as the spatial derivative of the displacement. What remains to be specified is the inelastic strain. Similar to plasticity, this quantity satisfies an evolution equation

$$\dot{\epsilon}_p(t,x) = \psi(u(t,x), \epsilon_p(t,x), \eta(t,x), \theta(t,x)) \tag{5}$$

with additional unknowns η, the internal variables. In the model of Helm one has two internal variables $\eta = (X, z_{\text{TIM}})$ where $X : [0,T] \times \Omega \rightarrow \mathbb{R}$ stands for the internal stress and $z_{\text{TIM}} : [0,T] \times \Omega \rightarrow \mathbb{R}$ for temperature-induced martensite. In particular the latter variable z_{TIM} expresses the different phases in each point of the SMA material. In combination with $z_{\text{SIM}} = z_{\text{SIM}}(\epsilon_p)$, the stress induced martensite, the total martensite fraction is given by $z = z_{\text{TIM}} + z_{\text{SIM}}$. For $z = 0$ the martensite phase vanishes while for $z = 1$ one has a purely martensitic material. The internal variables η are also governed by an evolution equation, which can be summarized as

$$\dot{\eta}(t,x) = \chi(u(t,x), \epsilon_p(t,x), \eta(t,x), \theta(t,x)). \tag{6}$$

Both right hand sides ψ and χ of the evolution equations involve several case statements that determine the phase transitions. The interested reader is referred to the Appendix for more details on this part of the model. We point out that the case statements lead to discontinuities in the right hand side. Moreover, the function ψ stems from a regularized yield condition with a yield function of the form $\phi(\sigma, X, \theta) = |\sigma - X| - \sqrt{2/3}\,k(\theta)$ and the yield value $k(\theta)$.

The model outlined so far is, assuming the temperature θ to be known, completely specified except for the boundary conditions. With respect to the displacement, we formulate mixed Dirichlet and Neumann boundary conditions as

$$u(t,0) = u_L(t) \quad \text{and} \quad \begin{cases} \sigma(t,1) = \tau(t) & \text{if Neumann boundary at } x = 1, \\ u(t,1) = u_R(t) & \text{if Dirichlet boundary at } x = 1. \end{cases}$$

Note that the balance equation (2) will below be used in quasi-stationary form $0 = \partial\sigma/\partial x + \beta$. Hence there is no need for an initial value of the displacement. Initial and boundary data of the other unknowns such as ϵ_p are implicitly provided in the computational method and thus can be omitted here.

2.2 Coupling the Heat Equation with the Evolution Equations

As described in the previous paragraph, the temperature θ determines strongly the material behavior. Most approaches in SMA simulation assume θ to be a given quantity that may vary as $\theta = \theta(t, x)$. For the application we have in mind, however, it is mandatory to include the heat equation in the model in order to better understand the behavior of temperature-controlled SMA actuators.

We consider here a formulation of the heat equation that is derived from convective heat fluxes and takes the deformation of the material into account. Omitting the arguments (t, x) and writing $u_x = \partial u / \partial x$ for ease of notation, the equation reads

$$\rho c_0 \frac{\partial}{\partial t} \theta = \frac{1}{1 + u_x} \frac{\partial}{\partial x} \left(\frac{\lambda}{1 + u_x} \cdot \frac{\partial}{\partial x} \theta \right) + \frac{1}{1 + u_x} f(\dot{\epsilon}, \sigma, \dot{\epsilon}_{\mathrm{p}}, \eta, \theta) \qquad (7)$$

with parameters $\rho, c_0, \lambda \in \mathbb{R}$ and the initial- and boundary conditions

$$\theta(0, x) = \theta_0(x), \quad x \in \Omega;$$
$$\theta(t, x) = g(t, x), \quad x \in \{0, 1\}, \ t \in [0, T].$$

Despite the nonlinear dependence on the displacement derivatives, the modified heat equation (7) is still of parabolic type following the definition of [3]. More details on its derivation can be found in the Appendix.

Besides the nonlinear coupling with the displacement, the most interesting aspect of the heat equation (7) is the source term f. More specifically, it can be written as

$$f = r - \frac{3\alpha\kappa}{\rho} \theta (\dot{\epsilon} - \dot{\epsilon}_{\mathrm{p}}) + \frac{1}{\rho} \sigma \dot{\epsilon}_{\mathrm{p}} - \Delta e_0 \dot{z} + \frac{1}{\rho} \varsigma(\theta, \dot{\epsilon}_{\mathrm{p}}, z_{\mathrm{SIM}}) X^2 , \qquad (8)$$

where $r \in \mathbb{R}$ stands for the pre-defined volume source, $\Delta e_0 \in \mathbb{R}$ is a specific energy gap of the material, and ς a given function. Moreover, z is the total martensite fraction. In contrast to the derivative $\dot{\epsilon}_{\mathrm{p}}$, the rate $\dot{\epsilon}$ of the strain is not available in a quasi-stationary computation of the displacements. To overcome this problem without making use of finite differences, we introduce a differential-algebraic equation that acts as a differentiator. Consider the system

$$\dot{\epsilon} = \delta , \quad 0 = \epsilon - \partial u / \partial x . \qquad (9)$$

Obviously, the exact solution is $\epsilon = \partial u / \partial x$ and $\delta = \dot{\epsilon}$. The Lagrange multiplier δ in (9), in other words, is the desired derivative of the total strain. Coupling this DAE to our model, the quantity $\dot{\epsilon} = \delta$ is now available on the right hand side. Note that this additional DAE is of differentiation and perturbation index two and therefore the choice of an appropriate integration scheme is crucial.

For convenience, the coupled system of model equations for a one-dimensional SMA structure is finally summarized as

$$0 = \partial\sigma/\partial x + \beta \qquad \text{(balance eq.)}, \quad (10a)$$

$$\epsilon = \epsilon_e + \epsilon_p \qquad \text{(additive split)}, \quad (10b)$$

$$\dot{\epsilon} = \delta \qquad \text{(strain rate)}, \quad (10c)$$

$$0 = \epsilon - \partial u/\partial x \qquad \text{(kinematic equation)}, \quad (10d)$$

$$\sigma = 2\mu(\theta)\epsilon_e + \kappa(\theta)\,\epsilon_e - 3\alpha(\theta)\kappa(\theta)(\theta - \theta_0) \qquad \text{(material law)}, \quad (10e)$$

$$\dot{\epsilon}_p = \psi(u, \epsilon_p, \eta, \theta) \qquad \text{(evolution inelastic strain)}, \quad (10f)$$

$$\dot{\eta} = \chi(u, \epsilon_p, \eta, \theta) \qquad \text{(evolution internal variables)}, \quad (10g)$$

$$\rho c_0 \cdot \dot{\theta} = -\frac{\lambda\, u_{xx}}{(1 + u_x)^3} \cdot \theta_x + \frac{\lambda}{(1 + u_x)^2} \cdot \theta_{xx} + \frac{1}{1 + u_x}\, f(\delta, \sigma, \dot{\epsilon}_p, \eta, \theta) \qquad (10h)$$

$$\text{(heat equation)}$$

with the unknown variables

u	displacement,	η	internal variables X and z_{TIM},
σ	stress,	X	internal stress,
ϵ_p	inelastic strain,	z_{TIM}	temperature induced martensite,
ϵ	total strain,	δ	total strain rate,
$\epsilon_e = \epsilon - \epsilon_p$,	elastic strain,	θ	temperature.

Remarks

- In the more general 3D case, the model equations (10) still have the same structure though, obviously, the differential operators are more involved and stress as well as strain variables need to be formulated as tensors.
- It can be shown that this model is consistent with the laws of thermodynamics. In particular, the Clausius-Duhem inequality is satisfied [4]. However, there are no results about existence and uniqueness of solutions available so far.
- The coupled system of equations (10) can be viewed as a partial differential-algebraic equation (PDAE). Further details of the mathematical model can be found in [4].

3 Method of Lines

For the numerical treatment of (10), we employ the method of lines based on linear finite elements for the spatial variable x, which eventually leads to a system of DAEs in time t. Particular attention is paid to the discretization of the evolution equations by numerical quadrature.

In order to set up a weak formulation we will use the Sobolev space $H^1(\Omega) = \{\, w \in L_2(\Omega),\ \mathcal{D}^1_w w \in L_2(\Omega)\}$, which is the set of all functions $w \in L_2(\Omega)$ that possess first order generalized derivatives $\mathcal{D}^1_w w \in L_2(\Omega)$.

3.1 Discretization of the Displacement by Finite Elements

The weak formulation of the balance equation is formulated using test functions $v \in V := \{ w \in H^1(\Omega) : w(t,0) = 0 \}$, where we assume a Dirichlet boundary condition at the left boundary $x = 0$ of Ω and, for simplicity, a Neumann boundary condition at the right $x = 1$. In the usual way, we obtain $\forall v \in V$

$$0 = \int_\Omega v(x) \frac{\partial}{\partial x} \sigma(t,x) \, dx - \int_\Omega v(x) \beta(t,x) \, dx ,$$

$$\Leftrightarrow \quad 0 = v(1)\sigma(t,1) - \int_\Omega \frac{d}{dx} v(x) \, \sigma(t,x) \, dx - \int_\Omega v(x)\beta(t,x) \, dx \qquad (11)$$

using integration by parts. This weak form characterizes the displacement $u(t,\cdot) \in H^1(\Omega)$.

A finite dimensional problem results from the Galerkin projection both for test functions v and the unknown displacement u. For this purpose, we define the subspace $S \subset H^1(\Omega)$ as span of a finite set of basis functions v_i.

Restricting the discussion without loss of generality to linear finite elements, the basis functions have the representation $v_i(x_j) = \delta_{ij}$ with the Kronecker symbol δ_{ij} and the nodes x_j of the finite element mesh, see Fig. 2.

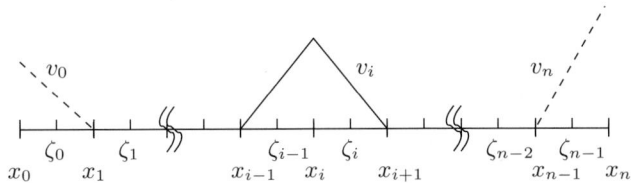

Fig. 2. FE grid of onedimensional wire with hat functions v_i

The approximation u_S of the displacement field by the Galerkin projection can then be written as

$$u_S(t,x) = v_D(t,x) + \sum_{i=1}^n v_i(x) q_i(t) = v_D(t,x) + \Phi(x)^T q(t) ,$$

where $\Phi = (v_1, \ldots, v_n)^T$ is the vector of basis functions. Observe that $\Phi(x)^T q(t)$ vanishes at $x = 0$ and that v_D satisfies the prescribed Dirichlet boundary condition $v_D(t,0) = u_L(t)$. Correspondingly, the kinematic equation results in

$$\epsilon(u_S) = \frac{\partial}{\partial x} u_S(t,x) = \epsilon(v_D) + \frac{d}{dx} \Phi(x)^T q(t) = \epsilon(v_D) + B(x)q(t) ,$$

the linearized material (or Lagrangian) strain. Notice that $\epsilon(v_i) = B \cdot e_i$ with the unit vector e_i. Inserting $v = v_i$ for $i = 1, \ldots, n$ in the weak form (11) and exploiting the material law (10e), the discretized momentum balance equation is found to read

$$0 = K(\theta) \cdot q(t) - b(t, \theta) - \int_{\Omega} B(x)^T (2\mu(\theta) + \kappa(\theta))\epsilon_p(t, x) \, dx \qquad (12)$$

with $n \times n$ stiffness matrix $K(\theta) := \int_{\Omega} B(x)^T (2\mu(\theta) + \kappa(\theta))B(x) \, dx$ and force vector

$$b(t, \theta) := \int_{\Omega} \Phi(x)\beta(t, x) \, dx - \int_{\Omega} B(x)^T (2\mu(\theta) + \kappa(\theta))\epsilon(v_D) \, dx$$
$$+ \int_{\Omega} B(x)^T 3\alpha(\theta)\kappa(\theta)(\theta - \theta_0) \, dx + (0, \ldots, 0, \tau(t))^T .$$

The extra integral term on the right of (12) represents the coupling term for inelastic deformation, which is in analogy to the treatment of elasto- and viscoplastic problems [8].

3.2 Discretization of the Heat Conduction

Choosing test functions $\nu \in \{w \in H^1(\Omega), \ w(t, x) = 0 \text{ for } x \in \{0, 1\}\}$ that vanish at the boundary, we transform the non-linear heat equation (7) by partial integration into

$$\rho c_0 \int_{\Omega} \nu \dot{\theta} \, dx = \int_{\Omega} \frac{\nu}{1 + u_x} \frac{\partial}{\partial x} \left(\frac{\lambda}{1 + u_x} \frac{\partial}{\partial x} \theta \right) dx + \int_{\Omega} \frac{1}{1 + u_x} \nu f \, dx$$
$$= -\lambda \int_{\Omega} \frac{\nu_x \theta_x}{(1 + u_x)^2} - \frac{u_{xx}}{(1 + u_x)^3} \nu \theta_x \, dx + \int_{\Omega} \frac{\nu f}{1 + u_x} \, dx . \quad (13)$$

For Galerkin projection we again need a finite dimensional subspace of $H^1(\Omega)$. Letting $\nu_1, \ldots, \nu_{n-1}$ denote the corresponding basis functions, the numerical approximation of the temperature can be written as

$$\theta_S(t, x) = \nu_D(t, x) + \sum_{i=1}^{n-1} \nu_i(x)\xi_i(t) = \nu_D(t, x) + \Theta(x)^T \xi(t)$$

where the function $\nu_D(t, x)$ fulfills the Dirichlet boundary conditions and where $\Theta(x) := (\nu_1(x), \ldots, \nu_{n-1}(x))^T$. Inserting this representation and applying the test functions in (13), we arrive in the standard way at the semi-discrete system

$$M \cdot \dot{\xi}(t) = -A(q) \cdot \xi(t) + d(t, q, \xi). \qquad (14)$$

More precisely, we replace the displacement u by its Galerkin approximation and get thus $1 + u_x(t, x) \doteq 1 + \epsilon(v_D)|_x + B(x)q(t)$. Then the deformation dependent stiffness matrix A, the source vector d and the constant mass matrix M read

$$A(q) := \lambda \int_\Omega \Theta_x \Theta_x^T \frac{1}{(1 + \epsilon(v_D) + Bq)^2} \, dx, \qquad M := \rho c_0 \int_\Omega \Theta \Theta^T \, dx \ ,$$

$$d(t, q, \xi) := \int_\Omega \frac{1}{1 + \epsilon(v_D) + Bq} \Theta f(\delta, q, \dot\epsilon_p, \eta, \xi) \, dx$$

$$- \int_\Omega \left(\Theta \dot v_D + \frac{\lambda}{(1 + \epsilon(v_D) + Bq)^2} \Theta_x \frac{\partial}{\partial x} v_D \right) dx \ .$$

Due to the dependence on the deformation, the stiffness matrix has to be recomputed in each timestep. In case of linear finite elements, the derivative u_x is constant in each element and the computation of $A(q)$ is an easy task as all terms involving the second derivative u_{xx} vanish. It is convenient to choose the same finite element mesh $\{x_0, \ldots, x_n\}$ and the same basis functions as introduced in Sect. 3.1 for both temperature and structural mechanics.

Concerning the existence and uniqueness of solutions of (13) and the convergence of its Galerkin discretization (14) we refer to results shown in [14, § 23].

3.3 Discretization of the Evolution Equations

The space discretization of the evolution equations follows from the treatment of the remaining integral term in the momentum balance (12). We apply a quadrature rule

$$\int_\Omega B(x)^T (2\mu(\theta) + \kappa(\theta))\epsilon_p(t, x) \, dx \doteq G(\theta_S) \cdot (\epsilon_p(t, \zeta_1), \ldots, \epsilon_p(t, \zeta_m))^T$$

and approximate the integral by m evaluations of the plastic strain ϵ_p in specific quadrature nodes ζ_i, which can be written as matrix-vector product with the $n \times m$ matrix $G(\theta_S)$ for the weights. Correspondingly, the remaining internal variables z_{TIM} and X are solely required at the same quadrature nodes, and thus the infinite-dimensional system of evolution equations has been discretized.

We do not introduce a new notation for the discretized internal variables and use ϵ_p and $\eta = (X, z_{TIM})$ both for the continuous representation as well as the discretized analogue. The different meanings should become apparent from the context. Summing up, in each node ζ_i it holds

$$\dot\epsilon_p(t, \zeta_i) = \psi(u_S(t, \zeta_i), \epsilon_p(t, \zeta_i), \eta(t, \zeta_i), \theta_S(t, \zeta_i)) \ ,$$
$$\dot\eta(t, \zeta_i) = \chi(u_S(t, \zeta_i), \epsilon_p(t, \zeta_i), \eta(t, \zeta_i), \theta_S(t, \zeta_i)) \ .$$

Here, the displacement u and the temperature θ in the right hand sides have been replaced by evaluations of the corresponding approximations u_S and θ_S.

Which quadrature order should be used? If we disregard the temperature and its space dependence for the moment, the evolution of the internal variables depends only on the strain. Since the displacements are modeled by linear finite elements, the strain is a constant function between each two finite element nodes. Therefore the midpoint quadrature rule with one node per element is sufficient $(m = n)$, see Fig. 2.

In view of the outlined quadrature technique, we stress that the computation of the nonlinear source term $d(t, q, \xi)$ in the discretized heat equation (14) deserves special attention. In fact, there is no way to express the integral in terms of a matrix-vector multiplication with a quadrature matrix. However, by employing a tensor notation it is possible to circumvent this difficulty.

3.4 Discretization of the Differentiator DAE

As discussed in Sect. 2.2, the time derivatives $\dot{\epsilon}$ of the total strain are required in the source term of the heat equation, which in turn is subject to numerical quadrature. Thus it is sufficient to compute $\dot{\epsilon}$ in the same quadrature nodes ζ_i. The appropriate DAE system reads

$$\dot{\epsilon}(t, \zeta_i) = \delta_i,$$
$$0 = \epsilon(\zeta_i) - \epsilon(v_D)|_{\zeta_i} - B(\zeta_i)\, q(t) \ ,$$

where the index runs over all nodes $i = 1, \ldots, m$.

Finally, the overall system of semidiscretized partial differential and differential algebraic equations comprising the balance of momentum (15a), the heat conduction (15b), the evolution of internal variables (15c,d) and the differentiator DAE (15e,f) can be written as

$$0 = -K(\xi) \cdot q(t) + b(t, \xi) + G(\xi) \cdot \epsilon_{\mathrm{p}}(t) \ , \tag{15a}$$
$$M \cdot \dot{\xi}(t) = A(q) \cdot \xi(t) - d(t, q, \delta, \dot{\epsilon}_{\mathrm{p}}, \eta, \xi) \ , \tag{15b}$$
$$\dot{\epsilon}_{\mathrm{p}}(t, \zeta_i) = \psi(q(t), \epsilon_{\mathrm{p}}(t, \zeta_i), \eta(t, \zeta_i), \xi(t)) \quad i = 1..m \ , \tag{15c}$$
$$\dot{\eta}(t, \zeta_i) = \chi(q(t), \epsilon_{\mathrm{p}}(t, \zeta_i), \eta(t, \zeta_i), \xi(t)) \quad i = 1..m \ , \tag{15d}$$
$$\dot{\epsilon}(t, \zeta_i) = \delta_i \quad i = 1..m \ , \tag{15e}$$
$$0 = \epsilon(\zeta_i) - \epsilon(v_D)|_{\zeta_i} - B(\zeta_i)\, q(t) \quad i = 1..m \ . \tag{15f}$$

In the right hand sides of the discretized evolution equations, the evaluation of the different phase changes requires the knowledge of internal variables and temperature both at the FEM nodes and the quadrature nodes. Since the temperature is described by linear finite elements it is straightforward to compute its values at the quadrature nodes by linear interpolation. On the other hand the values of the internal variables at the FEM nodes can be determined by extrapolation.

4 Time Integration and Simulation Results

The differential-algebraic system (15) is now subject to numerical integration. We discuss first the application of BDF methods and continue thereafter with simulation results for a model of the artificial finger mentioned in the Introduction.

4.1 Solving the Semi-discrete Equations

Overall, the semi-discrete equations (15) are of index two, as a short inspection shows. The discretized balance equation (15a) forms an index one constraint if the stiffness matrix $K(\xi)$ is invertible while the last two equations (15e) and (15f) are a semi-implicit DAE of index two. Thus, a direct application of DAE solvers is still feasible. However, the discontinuous right hand sides ψ, χ of the evolution equations are a severe drawback for the time integration, in particular if implicit methods are applied. As a remedy, we use a smoothing of these discontinuities at the cost of small stepsizes due to strong non-linearities in these regions.

In Sect. 2.2 we have introduced a differentiator DAE for the determination of the derivative information $\dot{\epsilon}_i(t) = \dot{\epsilon}(t, \zeta_i)$. If we use a BDF scheme, the question arises which finite difference scheme the results can be compared to. In the simplest case of the implicit Euler we can write the DAE subsystem

$$\epsilon_i(t_{n+1}) - \Delta t \cdot \delta_i(t_{n+1}) = \epsilon_i(t_n)$$
$$0 = \epsilon_i(t_{n+1}) - \epsilon(v_D(t_{n+1}))|_{\zeta_i} - B(\zeta_i) q(t_{n+1})$$

and get

$$\delta_i(t_{n+1}) = (\epsilon_i(t_{n+1}) - \epsilon_i(t_n))/\Delta t \ ,$$

which is just a backward difference scheme of first order. Accordingly the associated difference scheme for the BDF2 integration reads

$$\delta_i(t_{n+1}) = (3\epsilon_i(t_{n+1}) - 4\epsilon_i(t_n) + \epsilon_i(t_{n-1}))/2\Delta t \ ,$$

which is the backward difference formula of second order. Note that these two schemes depend only on the values of ϵ_i, which are exact due to the algebraic condition. Furthermore, if variable stepsizes are used, the finite difference approximation will be adjusted automatically.

The following example was computed by means of the Matlab DAE solver ode15s, a multistep scheme based on the backward differentiation formulas up to order 5. Although designed for ODEs and systems of index one, the code copes well with the DAE (15) if the error estimation for the algebraic variables δ is scaled appropriately.

4.2 Manipulator Example

The colored Fig. A.24 on page 338 shows a manipulator that is actuated by a shape memory wire (green line). The initial configuration at time $t = 0$ is a closed manipulator with the SMA wire consisting of pure temperature-induced martensite. In the beginning, the wire is pulled by an external threshold force of 300N and the martensite transforms completely into stress-induced

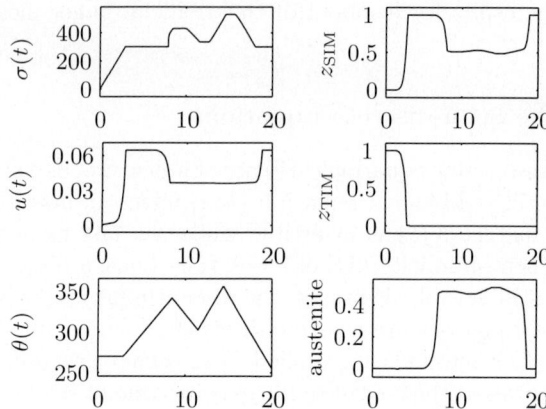

Fig. 3. Simulation example: manipulator

martensite. This is achieved by engaging the spring (red) to the manipulator mechanics.

Figure 3 displays the evolution of the main variables over time. The temperature rises next due to an external heat source and the wire shortens by building up austenite. Thus the manipulator closes until it hits an object at $t \approx 8$. Now the stress boundary condition is replaced by a displacement boundary at the present displacement. Applying some heat cycles shows the ability to control the grip by changing the temperature. Finally, cooling the wire leads to a decrease of the exerted force below the threshold level. From $t \approx 18$, the displacement boundary condition is replaced by the stress boundary condition again. The wire then lengthens and the manipulator releases the object.

As the simulation results demonstrate, the mathematical model is able to reproduce the phenomenological effects of shape memory alloys. Thus the behavior of NiTi materials in special applications, like the actuation of robot manipulators, can be simulated. Thereby all relevant temperature regions are covered.

The technical parameters for this simulation are listed in the Appendix. It turned out that $n = 10$ finite elements yield already adequate accuracy, which makes this approach also attractive for real-time applications. The integration statistics of `ode15s` at tolerance 10^{-6} and time intervall $[0, T] = [0, 20]$ are 767 steps, 233 rejections, and 51 Jacobian evaluations.

4.3 Combining SMA and Multibody Models

Finally, we shortly touch upon an extended simulation where the above shape memory model was treated as substructure for the dynamic motion of a robot finger composed of rigid bodies, cf. Pfeifer, Ulbrich and Schleich [11, 10]. Figure A.25 on page 338 shows the robot finger in open and closed position.

The coupling between the wire simulation and the multibody system was realized by computing the discrete stress $s(t)$ in the wire depending on the joint angles and their rates. This stress then was introduced as a momentum in the appropriate joint of the multibody system. Written in compact form, the overall system reads

$$M(p) \cdot \ddot{p} = f(p, \dot{p}, s) - G(p)^T \lambda, \qquad \text{(MBS)}$$
$$0 = g(p), \qquad \text{(joint constraints)}$$
$$s = h(\Xi), \qquad \text{(force of SMA wire)}$$
$$\dot{\Xi} = \Gamma(\Xi, p, \dot{p}), \qquad \text{(SMA model)}$$

with position coordinates $p(t)$ and Lagrange multipliers $\lambda(t)$.

After transformation of the rigid body equations to the stabilized formulation of index two, it is possible to solve the multibody equations simultaneously with the SMA equations by the Matlab integrator `ode15s`. The simulations performed have been visualized in a short movie [9] that shows how the finger is operated by changing the temperature of the SMA wire.

Acknowledgement. This work has been supported by the Deutsche Forschungsgemeinschaft under grant SI 756/1. The visualization of the robot finger has been implemented by our students Th. Schicketanz and A. Vuong.

Finally, on the occasion of his 75th birthday, the authors would like to thank Roland Bulirsch for being such an important and motivating source of inspiration. For many years, he has steadily paved the way for our beautiful discipline of applied and computational mathematics.

References

[1]   M. Arndt, M. Griebel, and T. Roubiček. Modelling and numerical simulation of martensitic transformation in shape memory alloys. *Continuum Mechanics and Thermodynamics*, 15(5):463–485, 2003.

[2]   C. Carstensen and P. Plechac. Numerical analysis of compatible phase transitions in elastic solids. *SIAM J. Num. Anal.*, 37(6):2061–2081, 2000.

[3]   W. Hackbusch. *Theorie und Numerik elliptischer Differentialgleichungen.* Teubner Studienbücher, Stuttgart, 1986.

[4]   D. Helm. *Formgedächtnislegierungen: Experimentelle Untersuchung, phänomenologische Modellierung und numerische Simulation der thermomechanischen Materialeigenschaften.* PhD thesis, Universität Gesamthochschule Kassel, 2001.

[5]   D. Helm and P. Haupt. Shape memory behaviour: modelling within continuum thermomechanics. *Intern. J. of Solids and Structures*, 40:827–849, 2003.

[6]   Y. Jung, P. Papadopoulos, and R.O. Ritchie. Constitutive modelling and numerical simulation of multivariant phase transformation in superelastic shape-memory alloys. *Int. J. Numer. Meth. Engng*, 60:429–460, 2004.

[7]   A. Mielke and T. Roubiček. A rate-independent model for inelastic behavior of shape-memory alloys. *MULTISCALE MODEL. SIMUL.*, 1(4):571–597.

[8]   O. Scherf and B. Simeon. Differential-algebraic equations in elasto-visco-plasticity. In K. Hutter and H. Baaser, editors, *Deformation and Failure in Metallic Materials*, volume 10 of *LN in Mechanics*, pages 31–50. Springer, 2003.

[9]   T. Schicketanz, A. Voung, G. Teichelmann, and B. Simeon. *Simulation eines Roboterfingers mit SMA Antrieb.* http://www-m2.ma.tum.de/~teichel/projekte/teichelmann/Finger.avi, 2006.

[10] M. Schleich. *Formgedächtnismaterialien als Fingerantrieb.* PhD thesis, Technische Universität München, 2003.

[11] M. Schleich and F. Pfeiffer. Modelling of shape memory alloys and experimental verification. *PAMM*, 2(1):294–295, 2003.

[12] L. F. Shampine and M. W. Reichelt. The matlab ode suite. *SIAM Journal on Scientific Computing*, 18:1–22, 1997.

[13] J.C. Simo and T.J.R. Hughes. *Computational Inelasticity.* Springer, 1998.

[14] E. Zeidler. *Nonlinear Functional Analysis and its Applications II/A.* Springer-Verlag, 1990.

[15] T.I. Zohdi and P. Wriggers. Introduction to computational micromechanics. volume 20 of *LN in Appl. and Comp. Mech.* Springer-Verlag, 2005.

A Appendix

A.1 Evolution Equations

Some details of the material's evolution in a point x are presented. There is a total of 6 possible phase changes between temperature induced martensite (TIM), stress induced martensite (SIM) and austenite (A). While the first four phase transitions are induced by the exertion of forces the last two are effected mainly by change in temperature.

$$
\begin{aligned}
\text{A} \to \text{SIM} \quad &: \quad \Delta\psi > 0,\ \Upsilon > 0,\ z_{\text{SIM}} < 1,\ \tau_{eff} > \|\mathbf{X}_\theta\|,\ \epsilon_{\text{p}}N \geq 0\,, \\
\text{SIM} \to \text{A} \quad &: \quad \Delta\psi > 0,\ \Upsilon > 0,\ z_{\text{SIM}} > 0,\ \epsilon_{\text{p}}N < 0\,, \\
\text{TIM} \to \text{SIM} \quad &: \quad \Upsilon > 0,\ z_{\text{TIM}} > 0,\ z_{\text{SIM}} < 1,\ \epsilon_{\text{p}}N \geq 0\,, \\
\text{SIM} \to \text{TIM} \quad &: \quad \Delta\psi < 0,\ \Upsilon > 0,\ z_{\text{TIM}} < 1,\ z_{\text{SIM}} > 0,\ \epsilon_{\text{p}}N < 0\,, \\
\text{A} \to \text{TIM} \quad &: \quad \Upsilon < 0,\ \dot{z}_{\text{SIM}} = 0,\ z < 1,\ \dot{\theta} < 0,\ \theta \leq M_s(\tau_{eff})\,, \\
\text{TIM} \to \text{A} \quad &: \quad \dot{z}_{\text{SIM}} = 0,\ z > 0,\ \dot{\theta} > 0,\ \theta \geq A_s(\tau_{eff})\,.
\end{aligned}
$$

Herein $N = (\sigma - X - X_\theta)/\|\sigma - X - X_\theta\|$ is the normal to the yield surface (see below) with the internal stress X and the temperature dependent stress X_θ. The quantity $\tau_{eff} = \|\sigma - X\|$ denotes the effective stress state, z_{SIM} is the fraction of stress induced martensite, and $\Delta\psi$ is the free energy difference at temperature θ. Now the evolution equation for the inelastic strain ϵ_{p} reads

$$
\dot{\epsilon}_{\text{p}} = \lambda \frac{\partial \Upsilon(\sigma, \theta, X)}{\partial \sigma} = \lambda N \ \text{ with yield function } \ \Upsilon(\sigma, \theta, X) = \|\sigma - X\| - \sqrt{\frac{2}{3}}k(\theta)
$$

and the inelastic multiplier $\lambda = \begin{cases} \Upsilon^m/\varpi & \text{if } \ \text{A} \leftrightarrow \text{SIM},\ \text{TIM} \leftrightarrow \text{SIM} \\ 0 & \text{otherwise} \end{cases}$

with the material parameter m. The fraction of temperature induced martensite behaves according to

$$\dot{z}_{\mathrm{TIM}} = \begin{cases} -\dfrac{|\dot{\theta}|}{M_s - M_f} \dfrac{(\Delta\psi)}{|(\Delta\psi)|} & A \rightarrow \mathrm{TIM} \\[2ex] -\dfrac{|\dot{\theta}|}{A_f - A_s} \dfrac{(\Delta\psi)}{|(\Delta\psi)|} & \text{if} \quad \mathrm{TIM} \rightarrow A \\[2ex] -\dot{z}_{\mathrm{SIM}} & \mathrm{TIM} \leftrightarrow \mathrm{SIM} \\[2ex] 0 & \text{otherwise} \end{cases} \quad \text{with} \quad \dot{z}_{\mathrm{SIM}} = \sqrt{\dfrac{2}{3}} \dfrac{\epsilon_p \cdot \dot{\epsilon}_p}{\gamma_d \|\epsilon_p\|}$$

The last two cases denote purely martensitic phase changes, where the sum of temperature induced and stress induced martensite $z = z_{\mathrm{TIM}} + z_{\mathrm{SIM}}$ is constant, while in the first two cases a phase transformation between martensite and austenite occurs. The internal stress X satisfies the equation

$$\dot{X} = c_1 \dot{\epsilon}_p + (\frac{dc_1}{c_1 d\theta}\dot{\theta} - c_1\varsigma(\theta, \dot{\epsilon}_p, z_{\mathrm{SIM}}))X.$$

A.2 Parameters

Symbol	Parameter	Value	Unit
c_0	specific heat capacity	460	$\mathrm{J/kg\,K}$
λ	specific heat conductance	80	$\mathrm{W/m\,K}$
θ_0	reference temp.		K
ρ	mass density	6400	$\mathrm{kg/m^3}$
k	yield value	50	MPa
$\Delta\eta_0$	entropy difference	-46875	$\mathrm{J/kg\,K}$
Δe_0	energy difference	-12375	J/kg
$\Delta\psi$	free energy difference	$\Delta e_0 - \theta\Delta\eta_0$	J/kg
κ	compression modulus	43000	MPa
μ	shear modulus	19800	MPa
α	linear expansion coefficient	1.06×10^{-5}	1/K
b	limitation term for X	700	-
c	shear modulus for X	14000	MPa
ϖ	viscosity	2×10^7	MPa/s
c_1	temperature dependent parameter		

A.3 The Heat Equation

The aim is to describe the heat conduction on domains, whose deformation varies in time by means of external and internal forces. The deformation function

$$\vartheta : [0, T] \times \Omega \rightarrow \hat{\Omega}(t) , \quad \vartheta(t, x) = x + u(t, x)$$

itself is thereby given by the structural mechanics simulation of the linear material by means of the displacement u. The large deformations namely take place in the internal variable ϵ_{p} while the elastic strain ϵ_e stays bounded. Since the deformation of the domain $\hat{\Omega}$ can be interpreted as a special case of convection we first name the heat equation

$$c_0\rho\frac{d}{dt}\,\hat{\theta}(t,\hat{x}) = -v(t,\hat{x})\,c_0\rho\nabla\hat{\theta}(t,\hat{x}) + \nabla\cdot(\lambda\nabla\hat{\theta}(t,\hat{x})) + \hat{f}(t,\hat{x})$$

$$\hat{\theta}(0,\hat{x}) = \hat{\theta}_0(x)\,, \quad \hat{x}\in\hat{\Omega}(0) \tag{16}$$

$$\hat{\theta}(t,\hat{x}) = \hat{g}(t,\hat{x})\,, \quad \hat{x}\in\partial\hat{\Omega}(t)\,, \; t\in[0,T]$$

with Dirichlet boundary values, that incorporates convective fluxes by the velocity vector v. The heat flux is hereby described by the Fourier model $Q = \lambda\nabla\theta$ with the heat conductivity λ. Apart from the user defined volume source r the source term (8) is provided by the material model. We identify the convection with the material deformation by simply replacing v by $\dot{\vartheta}$. Since we do our computations rather on the reference domain Ω in terms of the fixed referential coordinates than on the deformed domain $\hat{\Omega}$, we need a way to map the heat equation for $\hat{\theta}$ on $\hat{\Omega}$ to a differential equation for θ on Ω with the property

$$\hat{\theta}(t,\vartheta(t,x)) = \theta(t,x(t))\,,$$

adapting the Lagrangian formulation of structural mechanics. This can be accomplished by a transformation of differential operators. For simulating a one-dimensional wire it is sufficient to restrict the domain to one space dimension. The right hand side of (16) ends up with

$$c_0\rho\,\frac{\partial}{\partial t}\,\hat{\theta}(t,\hat{x}) = -\,\dot{\vartheta}(t,x)\,c_0\,\rho\,\frac{1}{\vartheta_x}\frac{\partial}{\partial x}\,\theta(t,x)$$
$$+\,\frac{1}{\vartheta_x}\cdot\frac{\partial}{\partial x}\left(\frac{\lambda}{\vartheta_x}\cdot\frac{\partial}{\partial x}\,\theta(t,x)\right) + \frac{1}{\vartheta_x}f(t,x)\,, \quad x\in\Omega\,. \tag{17}$$

By setting $\hat{f}(t,\vartheta(t,x)) = \frac{1}{\vartheta_x}f(t,x(t))$ and introducing (17) in

$$\frac{d}{dt}\,\theta(t,x) = \frac{d}{dt}\,\hat{\theta}(t,\vartheta(t,x)) = \frac{\partial}{\partial t}\,\hat{\theta}(t,\hat{x}) + \frac{\partial}{\partial\hat{x}}\,\hat{\theta}(t,\hat{x})\cdot\frac{d\vartheta(t,x)}{dt}$$

$$= \frac{\partial}{\partial t}\,\hat{\theta}(t,\hat{x}) + \frac{\dot{\vartheta}}{\vartheta_x}\frac{\partial}{\partial x}\,\theta(t,x)$$

the first term on the right hand side of (17) is canceled out. It follows

$$\rho c_0\cdot\frac{d}{dt}\,\theta = \frac{1}{\vartheta_x}\cdot\frac{\partial}{\partial x}\left(\frac{\lambda}{\vartheta_x}\cdot\frac{\partial}{\partial x}\,\theta\right) + \frac{1}{\vartheta_x}f \;\text{ with } \vartheta_x = 1 + u_x\,, \quad x\in\Omega$$

depending on the strain. In the special case of an undeformed region it holds $v_y = u_x = u_{xx} = 0$. This yields the standard heat conduction equation

$$\rho c_0\,\dot{\theta} = \lambda\cdot\theta_{xx} + f\,, \quad x\in\Omega$$

on the one dimensional domain under the assumption of isotropic material.

Part IV

Mathematics and Applications in Real World

Customer Tailored Derivatives: Simulation, Design and Optimization with the WARRANT-PRO-2 Software[*]

Michael H. Breitner

Institut für Wirtschaftsinformatik, Leibniz Universität Hannover,
Königsworther Platz 1, D-30167 Hannover,
breitner@iwi.uni-hannover.de, www.iwi.uni-hannover.de

Summary. Risk management is essential in a modern market economy. Financial markets enable firms and households to select an appropriate level of risk in their transactions. Risks can be redistributed towards others who are willing and able to assume them. Derivative instruments – derivatives, for short – like options or futures have a particular status. In the early 1970s Myron S. Scholes, Robert C. Merton and Fischer Black modeled an analytic pricing model for derivatives. This model is based on a continuous-time diffusion process (Ito process) for non-payout underlyings: The partial differential Black-Scholes equation. The WARRANT-PRO-2 software (Release 0.3) solves this equation with an adapted Crank-Nicholson scheme numerically. Arbitrary payments (boundary conditions) enable the design and optimization of customer tailored derivatives. WARRANT-PRO-2 computes derivative prices for given payments (simulation and expert design). But moreover this software can also optimize payments via parameterized boundary conditions of the Black-Scholes equation. The parameterized boundary conditions are optimized by nonlinear programming, i.e. an advanced SQP-method here. The deviation from a predefinable Δ of an option (performance index), e.g., can be minimized and the gradient can be computed highly accurate with automatic differentiation. A software quality and change management process for WARRANT-PRO-2, its comfortable and easy to use MATLAB-GUI (graphical user interface) and its portability to WINDOWS and LINUX operating systems is discussed. Optimized derivatives are very promising for both buyer and seller and can revolutionize modern financial markets: Examples like European double-barrier options are discussed.

---

[*] Prologue: Prof. Dr. Dr. h. c. mult. Roland Bulirsch has taught us from the first year of our studies that *Applied* Mathematics is *more* than definitions, lemmas, theorems, proofs, conclusions etc. An applied mathematician also has to solve relevant problems in every day's life, has to use reliable, accurate and fast algorithms on (high performance) computers and has to analyze, visualize and explain solutions, especially to non-mathematicians. Here definitions, theorems, proofs etc. are omitted, but referenced, to focus on the problem, the software and the exemplarily solved examples. And thus this paper is dedicated to Roland Bulirsch's credo on the occasion of his 75th birthday.

1 Introduction, Problem Relevance and Motivation

The paper is organized as follows:

- **Section 1** is dedicated to an overview and motivates the research and WARRANT-PRO-2 development. Some definitions and historical facts are given. The relevance of the problem *"How to design a derivative optimally?"* is addressed.
- **Section 2** shortly introduces the partial differential Black-Scholes equation which often enables the estimation of a derivative's value. Then the adapted Crank-Nicholson scheme for the numerical solution of the Black-Scholes equation is explained.
- **Section 3** gives an overview over automatic differentiation which allows to compute gradients of (almost) arbitrary functions very fast and without numerical discretization errors. Furthermore nonlinear programming and advanced SQP-methods are sketched. The usage of automatic differentiation and SQP-methods in WARRANT-PRO-2 are outlined.
- **Section 4** is dedicated to software quality and the WARRANT-PRO-2 software quality and change management process. A screenshot shows the most important windows of the GUI.
- After some examples for optimally designed derivatives in **Section 5** conclusions and an outlook are found in **Section 6**.
- The paper ends with an acknowledgment and the references.

Risk management is essential in a modern market economy. Financial markets enable firms and households to select an appropriate level of risk in their transactions. Risks can be redistributed towards others who are willing and able to assume them. Derivative instruments – **derivatives**, for short – like options or futures have a particular status. Futures allow agents to hedge against upcoming risks. These contracts promise future delivery of a certain, so called underlying at a certain, prespecified strike price. Options allow agents to hedge against one-sided risks. Options give the right, but not the obligation, to buy (call option) or sell (put option) a certain underlying at a prespecified strike price at expiration (European style option) or any time up to expiration (American style option). Common underlyings are, e. g., a certain amount of foreign currency, a certain number of bonds or shares or also a certain weight or volume of commodities, see [15], [24], [33] and [37] for an introduction. Today derivatives on various underlyings are worldwide traded around-the-clock. Computer systems and networks that worldwide link offices and markets permit immediate exchange of true market information (securities exchanges and "over the counter" (OTC) traders).

Two main **research questions** arise:

"What's the value of a derivative, e. g., a future or an option?"
"How to design a derivative optimally?"

The first question is answered by different approaches which model the value of a derivative dependent on its policy, i. e. essentially its guaranteed payments dependent on different scenarios, see [6], [7], [11], [15], [24], [31], [33], [34], [35] and [37] for various pricing approaches and models based on, e. g., Stochastics or Artificial Neural Networks. The second questions is also quite important and not well investigated, today. Usually derivative experts working for banks, investment banks, insurance companies and fund companies "manually" design derivatives tailored to their own companies' or their customers' needs. These tailored derivatives usually are traded "over the counter", i. e. not via exchanges or interbank security markets. Here we introduce a new software-based approach (WARRANT-PRO-2) using mathematical optimization for derivative design instead of intuition.

2 Black-Scholes Equation and Crank-Nicholson Scheme

In October 1997 the Swedish Nobel Prize in Economic Sciences was awarded to Myron S. Scholes, Stanford University, and Robert C. Merton, Harvard University. Inspired by the options and futures trading (Chicago Board Options Exchange) Scholes, Merton and Fischer Black, who died in 1995, studied European stock options' pricing since the early 1970s, see [3], [4], [5], [12], [13], [14], [27], [28], [29] and [30] for the most important historical papers. The analytic pricing model is based on a continuous-time diffusion process (Ito process) for non-payout underlyings. In the late 1970s John C. Cox, Stephen A. Ross and Mark Rubinstein used a numerical model to price also the more common American options. The diffusion process of the underlying price is simulated with a discretized time binomial tree (discrete-time stochastic process). Common to both models is that the correct option price is investigated by an option-underlying portfolio with risk-free profit, see [24] for an easily intelligible deduction. The theoretically fair **Black-Scholes-price** $p(t, u, r, \sigma)$ of an option depends on

- the time $t \in [0, T]$ with T the expiration time of the option,
- the underlying price $u \in [u_{\min}, u_{\max}]$ with $u_{\min} \geq 0$,
- the strike price $s \geq 0$ predefinable for $t = T$,
- the risk-free interest rate r independent from the maturity period $T - t$ and
- the so called future volatility σ of the underlying price u measured by its annualized standard deviation of percentage change in daily price independent from $T - t$, too.

During the last three decades both models have been improved in many ways, e. g., American options and put options can be valued and dividend yields of the underlying and variable interest rates can be handled, too.

The price movement of non-payout underlying assets is modeled by continuous-time (Ito process) or discrete-time stochastic processes (time binomial trees). The solution $p(t, u, r, \sigma)$ of the well known **Black-Scholes-equation**

$$\frac{\partial}{\partial t}p + r\,u\,\frac{\partial}{\partial u}p + \frac{1}{2}\sigma^2 u^2 \frac{\partial^2}{\partial u^2}p - r\,p = p_t + r\,u\,p_u + \frac{1}{2}\sigma^2 u^2 p_{uu} - r\,p \equiv 0, \quad (1)$$

which is a linear, homogeneous, parabolic partial differential equation of second order, often yields a sufficiently accurate approximation of option values, see [6], [11], [15], [24], [31], [33], [34], [35] and [37]. Note that mathematically t and u are the independent variables in Eq. (1) very similar to time and space in 1-dimensional diffusion equations, e. g., the heat equation. r and σ are constant parameters which are adaptable. Although developed for European options the Black-Scholes-equation easily can be used for the pricing of more general derivatives which will be discussed in the following.

The most important **derivative "greek" Delta** (symbol: Δ) is the first partial derivative p_u of $p(t, u, r, \sigma)$ w. r. t. u. A derivative's value $p(t, u, r, \sigma)$ is defined for all $t \in [0, T]$ and $u \in [u_{min}, u_{max}]$ if and only if the three boundary conditions at $u = u_{min}$, $t = T$ and $u = u_{max}$ are set. These **boundary conditions** correspond to the **cash settlements** at a derivative's expiration. W. l. o. g. a derivative expires if either one of the knock-out barriers is hit, i. e. $u = u_{max}$ at the upper barrier or $u = u_{min}$ at the lower barrier, or if the expiration date is reached, i. e. $t = T$.

In comparison to future contracts options have advantages and disadvantages. Exemplarily the German stock index DAX 30 is taken as underlying ("DAX" is used for the XETRA spot rate of the German stock index DAX 30). DAX-future contracts have an invariant profit/loss per DAX point (Δ independent from t and u), whereas the profit/loss per DAX point of common options varies undesirably with the DAX. But an advantage of buying options results from the immediate payment of the option premium. There are no further payments during the life of the option. Similarly margin requirements force a DAX-future buyer/seller to pay an initial margin plus a safety margin immediately. But in the event of adverse DAX movements the buyer/seller has to meet calls for substantial additional margin and may be forced to liquidate his position prematurely, see [8], [9], [10] and [25] for details. An idea is to combine the advantages of futures and options to an option with Δ (almost) independent from t and u. The only way is to optimize the cash settlements at a derivative's expiration, see [8], [9], [10] and [25] for details again. Mathematically an optimal control problem for a partial differential equation arises, comp. [17] or [23] for various, typical problems of that type.

For arbitrary boundary conditions the Black-Scholes-equation (1) is solvable numerically only. An **upgraded Crank-Nicholson scheme** is used which is a combination of an explicit and an implicit finite difference method, see [1], [10], [25], [34], [35], [36] and [38]. With a time discretization Δt and an underlying price discretization Δu the upgraded scheme is highly accurate: The absolute, global error of the numerical solution decreases fast by $\mathcal{O}(\Delta t^2) + \mathcal{O}(\Delta u^2)$ for $\max(\Delta t, \Delta u) \to 0$. The convergence of the upgraded Crank-Nicholson scheme for all relevant t, u, r and σ has been proven in [25].

3 Automatic Differentiation and SQP-Method

The optimization of the cash settlements at a derivative's expiration is done by minimization of an appropriate performance index. Different performance indeces for derivatives correspond to different needs of buyers and sellers of derivatives. e. g., an option with Δ (almost) independent from t and u is attractive for both buyers and sellers because hedging is easy. The gradient of a performance index always is needed in order to minimize it with an advanced nonlinear programming method numerically.

A gradient can be computed highly accurate, i. e. with the Crank-Nicholson scheme's accuracy, if **automatic differentiation** – also called algorithmic differentiation – instead of numerical differentiation is used, see [2], [9], [21], [22], [25], [39] and the historical book [32]. Automatic differentiation is concerned with the accurate and efficient evaluation of derivatives for functions defined by computer programs. No truncation errors are incurred and the resulting numerical derivative values can be used for all scientific computations. Chainrule techniques for evaluating derivatives of composite functions are used. Automatic differentiation today is applied to larger and larger programs, e. g. for optimum shape design. In many such applications modeling was restricted to simulations at various parameter settings. Today, with the help of automatic differentiation techniques, this trial and error approach can be replaced by a much more efficient optimization w. r. t. modes, design and control parameters, see [8], [9], [10], [16], [17], [23], [25] and the historical book [26]. Alternatively computer algebra systems, e. g. MAPLE, see www.maplesoft.com, can be used to calculate analytic derivatives for functions.

Here, based on the FORTRAN code for the computation of a performance index, efficient FORTRAN code for the gradient can be generated easily. For the code generation ADIFOR 2.0, see www.cs.rice.edu/~adifor, is used. In comparison to the first release WARRANT-PRO-2 (0.1) the gradient of a performance index can be computed not only highly accurate. The gradient also can be computed relatively cheap in comparison to numerical differentiation.

A performance index chosen can be minimized with the high end optimization software NPSOL 5.0, see [18] and [19] and also [20] for the large-scale sparse successor SNOPT 7.1. NPSOL is a general-purpose **sequential quadratic programming (SQP) method** to solve constrained nonlinear programming problems numerically. It minimizes a linear or nonlinear function subject to bounds on the variables and linear or nonlinear constraints. NPSOL finds solutions that are locally optimal. But very often these solutions are also globally optimal or almost as good as the global optimum. NPSOL can estimate gradient components automatically by numerical differentiation (not advisable, see above!). A quasi-Newton approximation to the Hessian of the Lagrangian (second order information) is made by BFGS updates applied after each so-called major iteration. In WARRANT-PRO-2 NPSOL converges very fast and reliable for most performance indices, underlyings and

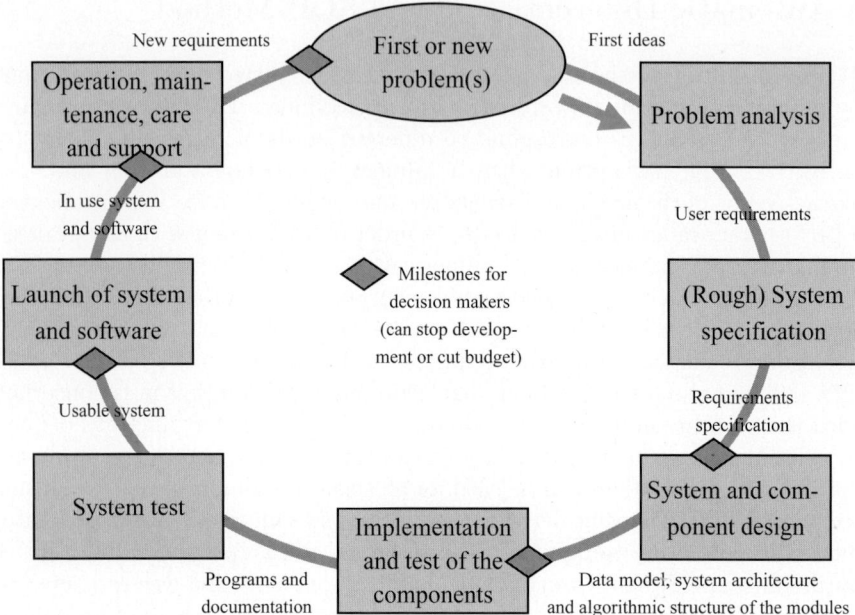

Fig. 1. Change management process for WARRANT-PRO-2: First ideas date back to 1999, Release (0.1) was finished in 2000, a reengineering was started in 2002 and finished with Release (0.2) in 2003 followed be a new reengineering in 2006 finished with actual Release (0.3) in 2007, see [8], [9], [10], [16] and [25].

times to expiration. In case of a convergence fail NPSOL reports likely reasons and difficulties and helps to reformulate the nonlinear programming problem. One important performance index is ready to use in WARRANT-PRO-2: The change of Δ as a function of t and u can be minimized to make Δ (almost) independent from t and u, see above.

4 Change Management Process and Software Quality

Today's **software-(re)engineering** is process oriented, i. e. structured in subsequent (or sometimes parallel) phases which reduce complexity. Milestones allow the management to stop software-(re)engineering processes, to cut down or raise budgets and/or resources and to change requirements and schedules, see Fig. 1. Today's software-(re)engineering is based on **software quality** principles according to ISO/IEC 9126 and DIN 66272, i. e. primarily

- suitable, correct, safe, secure and adequate functionality and interfaces for the integration into (often complex) IT-infrastructures,
- high stability and error tolerance and good restart ability,

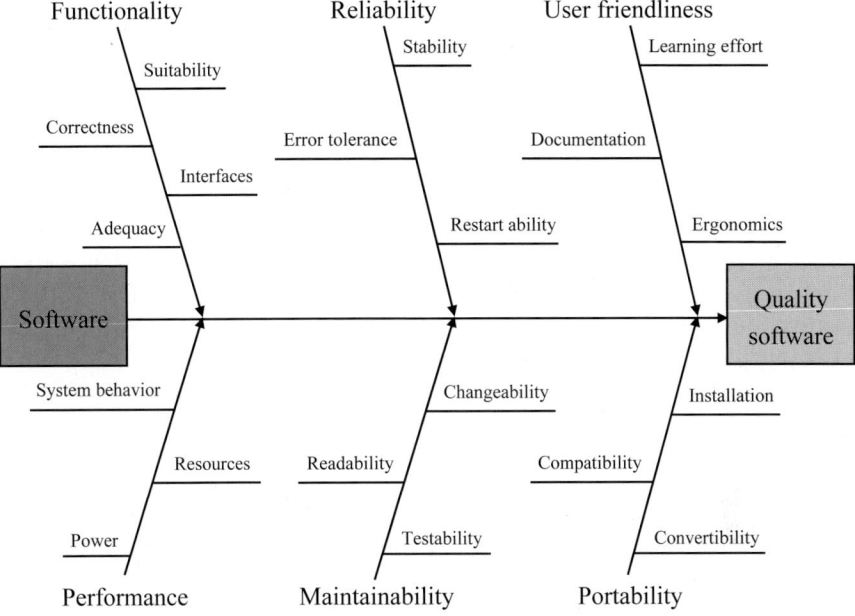

Fig. 2. Most important **software quality** aspects according to ISO/IEC 9126 and DIN 66272.

- user friendly documentation, easy self-learning, advanced ergonomics and high attractiveness,
- good overall performance and moderate allocation of resources, e. g. cache memory, RAM or computing times,
- good maintainability with readable, changeable and testable source code
- and high portability with easy installation, high compatibility with different hardware and operating systems and high convertibility,

see Fig. 2. In addition generally conformity with standards and also best practices is important.

Development of WARRANT-PRO-2 started in 1999 with the idea to optimize options numerically. Prototype Release 0.1 was completed at the end of 2000 by Tobias Burmester and the author at the Technische Universität Clausthal, see [8] and [10]. Primary goal was to optimize cash settlements for European double-barrier options and warrants. The boundary conditions of the Black-Scholes equation were parameterized. Then the Black-Scholes equation was solved with a *standard* Crank-Nicholson scheme (rectangular mesh) and the parameters could be optimized by NPSOL 5.0. As a performance index an option's deviation from a predefinable Δ_{opt} for a set of up to 8 volatilities was implemented. NPSOL estimated all gradient components of the performance index automatically by numerical differentiation. The NPSOL convergence was quite slow due to the numerical differentation and quite poor due to the low accuracy of the gradient. All computations were based on input and output files and an executable without GUI. Graphics were only possible via a MAPLE script. WARRANT-PRO-2 (0.1) was tested only on UNIX and LINUX computers. However, the prototype already enabled a glance at very interesting solutions and phenomena.

The **first reengineering** of WARRANT-PRO-2 (0.1) started in spring 2002. Release 0.2 was completed in March 2003 by Oliver Kubertin and the author at the Technische Universität Clausthal and the Leibniz Universität Hannover, see [9] and [25]. The kernel was still coded in ANSI FORTRAN to assure easy portability. With an **upgrade** of the **Crank-Nicholson scheme** it became highly accurate: The absolute, global error of the numerical solution decreases by $\mathcal{O}(\Delta t^2) + \mathcal{O}(\Delta u^2)$ for $\max(\Delta t, \Delta u) \to 0$. The convergence of the upgraded Crank-Nicholson scheme for all relevant t, u, r and σ has been proven in [25]. The **cache** and **main memory (RAM) allocation** was reduced by the choice of 9 different, precompiled modes for different grids/meshes. Dependent on the fineness of the Crank-Nicholson mesh and the fineness of the boundary conditions' parameterization the user could limit the memory allocation. For the parameterization 10 (rough), 19 (moderate) and 37 (fine) equidistant points were available. Different interpolation methods between the points were selectable: Linear, (cubic) spline and (cubic) Hermite-spline, too. For the gradient highly accurate **automatic differentiation**, i. e. with the Crank-Nicholson scheme's accuracy, was introduced. Summarized the new FORTRAN kernel was about 10 to 20 times faster compared to the kernel of Release 0.1. The robustness of the NPSOL convergence has improved significantly due to the significantly higher accuracy of the gradient. Smaller problems could run completely in level-2 or level-3 cache memories (≥ 4 MB). If so, an additional speed up of 2 to 3 was gained. Discussions to implement a GUI began. On the one hand programming a **user friendly GUI** is very important for modern software. On the other hand GUI programming usually is very time consuming. Often a mandatory GUI reduces portability. A well-designed GUI should be intuitive for the user. Providing an interface between user and application GUIs enable users to operate an application without knowing commands and/or formats required by a command line and/or file interface. Thus applications providing GUIs are easier to learn and easier to use. As a compromise MATLAB, see www.mathworks.com, was used for a semi-automatic setup of a WARRANT-PRO-2 GUI. MATLAB is a so called rapid application development (RAD) tool. For the WARRANT-PRO-2 (0.2) GUI the MATLAB GUIDE and MATLAB Compiler were used to generate C and C++ code. The C and C++ code was compiled and linked with the MATLAB run-time libraries. A stand-alone executable for LINUX PCs and laptops was built which called the compiled kernel program. The stand-alone application ran even if MATLAB is not installed on the end-user's system. Internal information flowed via the input and output files. The file concept had the important advantage that the compiled kernel program of Release 0.2 ran

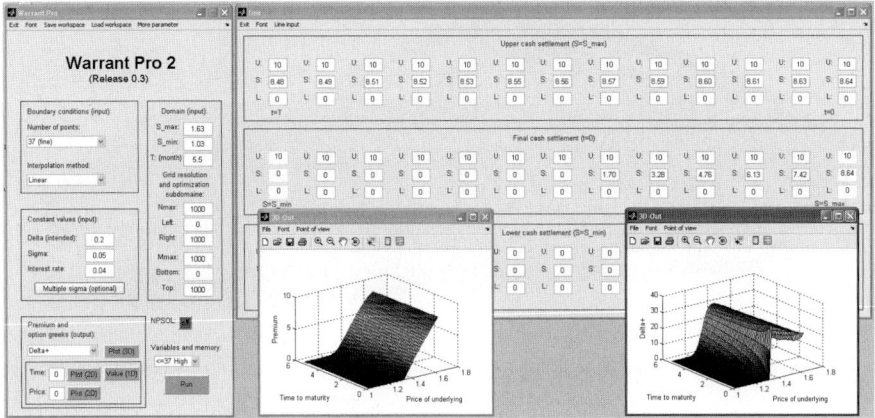

Fig. 3. WARRANT-PRO-2's **MATLAB GUI (graphical user interface)** helps to valuate and optimize arbitrary derivatives comfortably. The user defines two rectangular, independent meshes for the Crank-Nicholson scheme and the parameterization of the boundary conditions. The fineness of the meshes determines both accuracy of the solution and computing time. 3-dimensional, interactive MATLAB graphics enable to check the settings, to understand the solution and to grep a derivative's value easily, see also Fig. A.26 on page 339.

without GUI, too. Remote logins on LINUX compute servers were possible for time-consuming computations. The GUI had several windows: The main window, the boundary conditions' window, the multiple σ window and the 1-dimensional, 2-dimensional and 3-dimensional graphics windows, see Fig. 3 or also www.iwi.uni-hannover.de "Forschung" and "WARRANT-PRO-2".

The **second reengineering** of WARRANT-PRO-2 (0.2) started in fall 2006. The actual Release 0.3 was completed in March 2007 by Heiko Degro and the author at the Leibniz Universität Hannover, see [16]. **Full dynamic memory allocation** is achieved now. The FORTRAN kernel is again 5 to 10 times faster due to code optimization and optimal compiler settings. The MATLAB GUI of WARRANT-PRO-2 (0.3) is more intuitive and user friendly now. A **stand-alone application** for all new WINDOWS, LINUX and UNIX computers can be compiled and linked. WARRANT-PRO-2 (0.3) can be regarded as a tested beta-prototype and is available upon request from the author (free test for 30 days or commercial license).

5 Some Optimally Designed Derivatives

The Porsche AG, Stuttgart, sells about 40 – 50 % of its cars to non-Euro countries. Usually an importer pays in **foreign currency**, i. e. US or Canadian Dollar, Japanese Yen, GP Pounds or Swiss Franken. Development, production,

Fig. 4. Top: The **Porsche 911 Carrera S** is the most popular Porsche of the USA and Canada. About 50 % of all Porsche 911 built are delivered to this region (Source: Porsche AG WWW-pages). Bottom: **Exchange rate US Dollar (USD)/Euro** from mid 2006 to mid 2007 (Source: Comdirect WWW-pages, weekly candlestick analysis).

marketing and delivery costs arise in Euro almost 100 %. Thus Porsche has to face a high **risk** of **adverse exchange rate changes**. But, on the other hand, there are chances to make additional profit in the case of beneficial exchange rate movements. A customer tailored OTC option can insure against adverse changes *and* gives the opportunity to exploit beneficial movements.

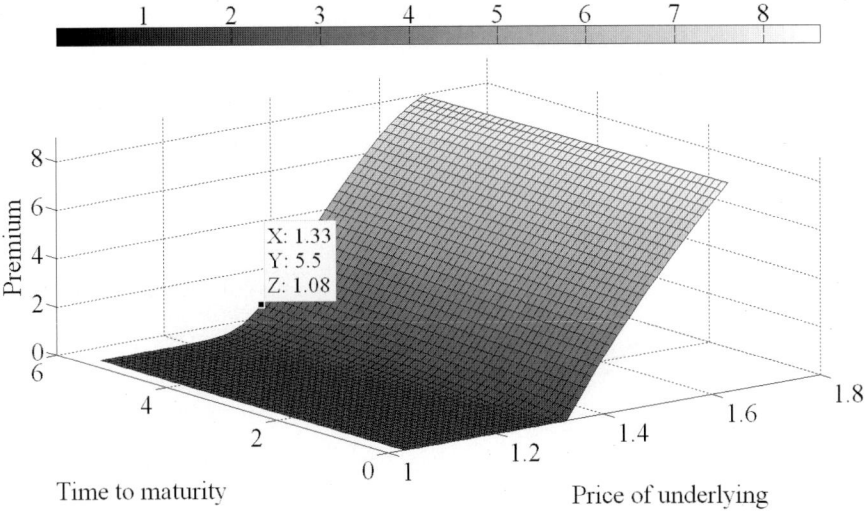

Fig. 5. The Porsche AG buys an **OTC option** with the right to sell 62.4 million USD, i. e. revenues from an US export, until December 3, 2007, for 46.917 million Euro. Thus the exchange rate will be better than 1.33 USD for 1 Euro, comp. all cash settlements, too. The option's premium is about 1.08 million Euro = 2.3 % about 5.5 month earlier in mid June 2007, see also Fig. A.27 on page 340.

Consider an US importer which orders 800 Porsche 911 Carrera S, see Fig. 4, for Christmas offers 2007 and the year 2008. After delivery to the USA end of November 2007 the importer will pay 78000 USD per car, i. e. total 62.4 million USD, at December 3, 2007. Mid June 2007 the exchange rate USD/Euro is 1.33 which results in a revenue of 46.917 million Euro. This revenue generates a satisfactory profit for Porsche. In case of a weaker USD, i. e. exchange rates of 1.40, 1.50 or even higher, the total costs exceed the revenue in Euro, see also Fig. 4. Thus Porsche buys an OTC option which gives the right – but no obligation – to sell 62.4 million USD until December 3, 2007, for 46.917 million Euro. The option will be executed if and only if the USD/Euro exchange rate is above 1.33. At maturity (December 3) a cash settlement for, e. g., 1.38 would be 1.700 million Euro and for 1.53 6.133 million Euro, see Fig. 3. With a future/implied volatility $\sigma = 0.05$, a common USD/Euro interest rate $r = 0.04$ and an initial time to maturity of 5.5 month WARRANT-PRO-2 (0.3) computes an option premium of about 1.08 million Euro (= 2.3 % of today's (potential) revenue of 46.917 million Euro), see Fig. 5. Including the premium Porsche has a guaranteed revenue of 45.837 million Euro. Note that in case of USD/Euro exchange rates below 1.33 in December 2007 Porsche benefits from additional revenues, e. g. of 1 million Euro for 1.302 and 2 million Euro for 1.276. Figure 6 shows the leverage effect of the OTC USD/Euro option. Asume an increase of the USD/Euro exchange

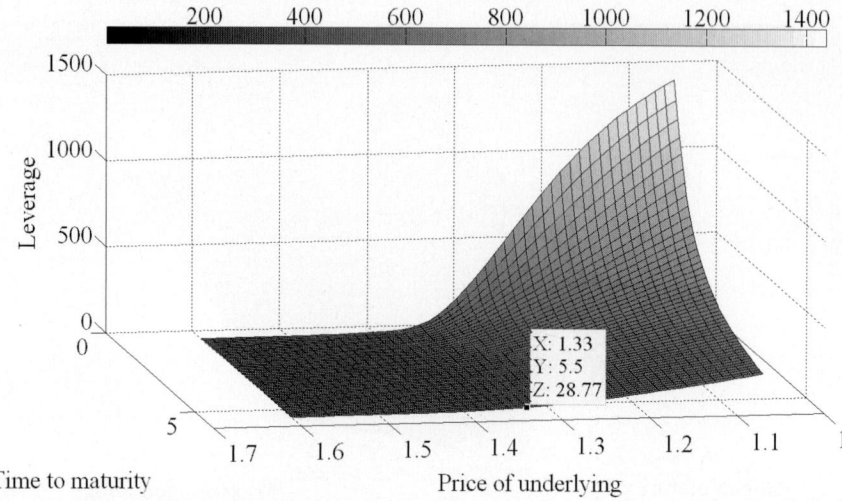

Fig. 6. Leverage of the **OTC option**, see Fig. 5, e.g. about 29 in mid June 2007 for an exchange rate of 1.33 USD for 1 Euro.

rate from 1.33 to 1.34. Then the value/price of the option increases from 1.08 to 1.37 million Euro (leverage of 29).

The Hannoversche Leben has a professional **asset management** for billion of Euro which are reserves for life insurance customers. An amount of 7.166 million TUI shares with 18.70 Euro each mid June, 2007, has a value of 134 million Euro. After a good performance in the first half of 2007 the **risk of a rebound** is quite high, see Fig. 7. But, on the other hand, a year end rallye up to 23 or even 25 Euro per TUI share has a likelihood of about 30 – 40 %.

The Hannoversche Leben/VHV buys *optimized* OTC put options with underlying TUI shares. Below 18.70 Euro per TUI share the option has an almost constant $\Delta \approx -1$, see Figs. 8 and 9. This allows an easy hedging for both the Hannoversche Leben and the option's issuer. For an immunization of all TUI shares the Hannoversche Leben must buy 7.166 million options. With a future/implied volatility $\sigma = 0.25$, an interest rate $r = 0.04$ and an initial time to maturity of 8 month WARRANT-PRO-2 (0.3) computes an option premium of 1.686 Euro, see Fig. 8 (= 9.0 % of the TUI share price in mid June 2007). After the payment of the option premium the Hannoversche Leben has a guaranteed value of about 17 Euro per TUI share. And the Hannoversche Leben can fully benefit from rising TUI share prices up to 23 or even 25 Euro. Note that the optimization of the cash settlement at maturity leads to an interessting phenomenon near 20 Euro. The "hook" enlarges the area with $\Delta \approx -1$ significantly. Note also that the optimized option shows a difficult behavior close to maturity, see Fig. 9. The initial time to maturity is chosen

Fig. 7. Quotation of TUI shares from mid 2006 to mid 2007 (Source: Comdirect WWW-pages, weekly candlestick analysis).

8 month to avoid hedging problems both for the Hannoversche Leben and the option's issuer.

6 Conclusions and Outlook

Unless originally designed to optimize cash settlements, WARRANT-PRO-2 (0.3) proved to be very useful and comfortable also for Black-Scholes value computations for derivatives *without* optimization. Based on MATLAB a stand-alone GUI provides interactive, 1-, 2- and 3-dimensional graphics of high quality. These graphics enable to evaluate and analyze a derivatives value and its most important option "greeks", see Figs. 3, 5, 6, 8 and 9. The upgraded Crank-Nicholson scheme is highly accurate and only reasonable computing time is needed. In the optimization mode the gradient computation with automatic differentiation is most important to enable fast and reliable convergence of the SQP method. Here the change of Δ as a function of t and u is minimized. Other performance indices can be implemented, too. ADIFOR 2.0 usage and moderate FORTRAN coding are necessary in this case.

The two real-life examples for customer tailored OTC derivatives show the power of the approach, see Sect. 5. With WARRANT-PRO-2 (0.3) a

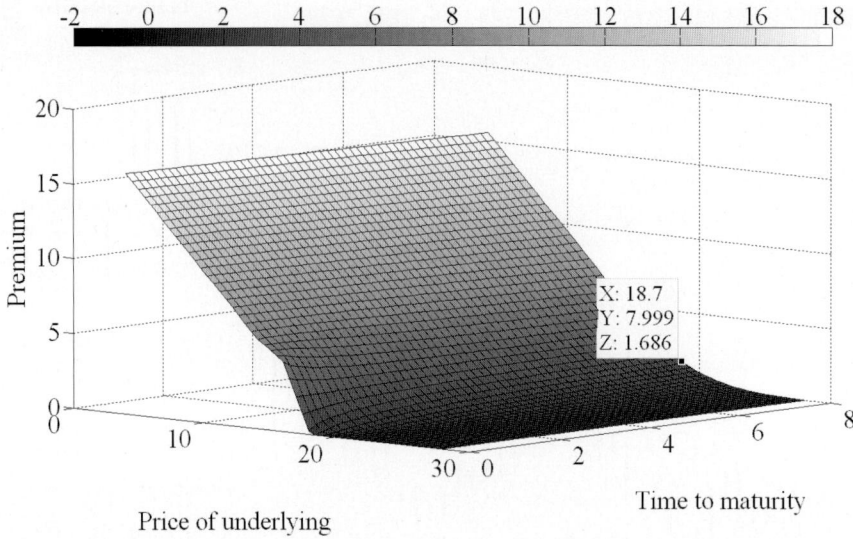

Fig. 8. The Hannoversche Leben buys **optimized OTC put options** with the right to sell 7.166 million TUI shares for 18.70 Euro per share. For share prices less than 18.70 Euro, i. e. the relevant scenarios, $\Delta \approx -1$ holds, see also Fig. 9. Mid June 2007 8 month is chosen for the time to maturity to cover a critical period until the end of 2007. The option's premium is about 1.686 Euro per share = 9.0 % in mid June 2007, see also Fig. A.27.

derivative designer concentrates on the customer's needs. The designer sets some or all of the cash settlements, the market parameters and an optimization criteria/performance index for the WARRANT-PRO-2 (0.3) computations. Customer tailored OTC derivatives allow to hedge against one-sided risks comparable to an insurance. The issuer of a derivative gets the (insurance) premium immediately comparable to an insurance company. An easy risk management essential in a modern market economy is possible.

WARRANT-PRO-2 (0.3) is a tested beta-prototype, see also www.iwi.uni-hannover.de "Forschung" and "WARRANT-PRO-2". The author concentrates on testing of the software and will establish a professional support in the near future. This support includes hotline and user/developer manual. Interested users can contact the author for a free test for 30 days or a commercial license (beta-testing), see first page.

Acknowledgement. The author gratefully appreciates joint work with his former graduate students Tobias Burmester (WARRANT-PRO-2 (0.1)), Oliver Kubertin (WARRANT-PRO-2 (0.2)) and Heiko Degro (WARRANT-PRO-2 (0.3)) and support by Prof. Dr. Philip E. Gill, University of California San Diego, providing excellent SQP optimization methods, and by Prof. Dr. Michael W. Fagan, Rice University, Houston, providing the excellent automatic differentiation package ADIFOR 2.0.

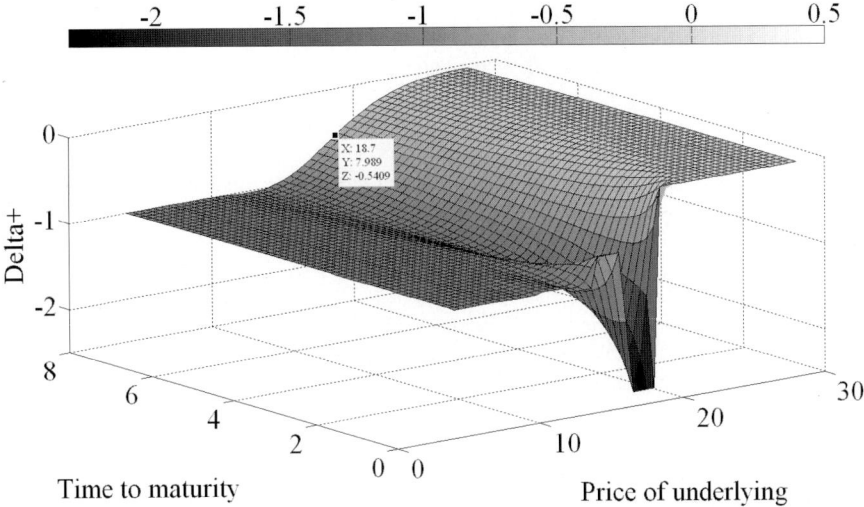

Fig. 9. Option "greek" $\Delta = p_u(t, u, r, \sigma)$ for the optimized OTC TUI put option, see also Fig. 8.

References

[1] Ames, W. F., *Numerical Methods for Partial Differential Equations.* Academic Press, San Diego, 1994 (3rd ed.).

[2] Bischof, C., Carle, A., Hovland, P., Khademi, P., and Mauer, A., *ADIFOR 2.0 User's Guide (Revision D).* Mathematics and Computer Science Division, Argonne National Laboratory, Technical Memorandum ANL/MCS-TM-192 and Center for Research on Parallel Computation, Rice University, Technical Report CRPC-TR95516-S, supported by the Department of Energy, Washington (D. C.), 1998. See also www-unix.mcs.anl.gov/autodiff/ADIFOR.

[3] Black, F., and Scholes, M., *The Valuation of Option Contracts and a Test of Market Efficiency.* Journal of Finance **27**, 1972.

[4] Black, F., and Scholes, M., *The Pricing of Options and Corporate Liabilities.* Journal of Political Economy **81**, 1973.

[5] Black, F., *How We came Up with the Option Formula.* Journal of Portfolio Management **15**, 1989.

[6] Bookstaber, R. M., *Option Pricing and Investment Strategies.* McGraw-Hill, London, 1991 (3rd ed.).

[7] Breitner, M. H., *Heuristic Option Pricing with Neural Networks and the Neuro-computer SYNAPSE 3*, Optimization **47**, pp. 319 – 333, 2000.

[8] Breitner, M. H., and Burmester, T., *Optimization of European Double-Barrier Options via Optimal Control of the Black-Scholes-Equation.* In Chamoni, P., et al. (Eds.), Operations Research Proceedings 2001 (Duisburg, September 3 – 5), pp. 167 – 174, Springer, Berlin, 2002.

[9] Breitner, M. H., and Kubertin, O., *WARRANT-PRO-2: A GUI-Software for Easy Evaluation, Design and Visualization of European Double-Barrier Options.* Presented at Operations Research 2003 (Heidelberg, September 3 – 5),

IWI Discussion Paper Series No. 5, Institut für Wirtschaftsinformatik, Leibniz Universität Hannover, 35 p., 2003.

[10] Burmester, T., *Optimale Steuerung der partiellen Black-Scholes-Gleichung zur Optimierung von Optionspreissensitivitäten für vorgegebene Zeit- und Basisobjektpreisintervalle.* Diploma thesis (supervised by the author), Technische Universität Clausthal, 2000.

[11] Chriss, N. A., *Black-Scholes and Beyond: Option Pricing Models.* McGraw-Hill, New York/London, 1997.

[12] Cox, J. C., and Ross, S. A., *The Valuation of Options for Alternative Stochastic Processes.* Journal of Financial Economics **3**, 1976.

[13] Cox, J. C., Ross, S. A., and Rubinstein, M., *Option Pricing: A Simplified Approach.* Journal of Financial Economics **7**, 1979.

[14] Cox, J. C., and Rubinstein, M., *Options Markets.* Prentice Hall, Upper Saddle River (New York), 2002 (Facsimile from 1985).

[15] Daigler, R. T., *Advanced Options Trading: The Analysis and Evaluation of Trading Strategies, Hedging Tactics and Pricing Models.* Probus Publisher, Chicago 1994.

[16] Degro, H., *Reengineering eines Softwarepakets zur Preisberechnung und Optimierung von Finanzderivaten.* Diploma thesis (supervised by the author), Leibniz Universität Hannover, 2007.

[17] Deutsche Forschungsgemeinschaft (DFG), *Schwerpunktprogramm 1253 Optimization with Partial Differential Equations,* see www.am.uni-erlangen.de/home/spp1253/wiki, launched 2006.

[18] Gill, P. E., Murray, W., and Wright, M. H., *Practical Optimization.* Elsevier/Academic Press, Amsterdam, 2004 (14th print, 1st print 1981).

[19] Gill, P. E., Murray, W., and Saunders, M. A., *Large-scale SQP methods and their Application in Trajectory Optimization.* In: Bulirsch, R., and Kraft, D., eds., Computational Optimal Control. Birkhäuser, International Series of Numerical Mathematics 115, Basel, 1994.

[20] Gill, P. E., Murray, W., and Saunders, M. A., *User's Guide for SNOPT Version 7: Software for Large-Scale Nonlinear Programming.* Report, downloadable from cam.ucsd.edu/~peg/, April 24, 2007.

[21] Griewank, A., *Evaluating Derivatives: Principles and Techniques of Algorithmic Differentiation.* SIAM, Philadelphia, 2000.

[22] Griewank, A., and Corliss, G. F. (eds.), *Automatic Differentiation of Algorithms: Theory, Implementation and Application (Proceedings of the First SIAM Workshop on Automatic Differentiation, Breckenridge, Colorado, January 6 – 8, 1991).* SIAM Proceedings, Philadelphia, 1991.

[23] Hoffmann, K.-H. (ed.), *Optimal Control of Partial Differential Equations (International Conference in Chemnitz, April 20 – 25, 1998).* Birkhäuser, Basel, 1999.

[24] Hull, J. C., *Options, Futures, and other Derivatives (w. CD-ROM).* Prentice Hall, Upper Saddle River (New York), 2006 (6th ed., 1st ed. 1989). In German: *Optionen, Futures und andere Derivate,* Pearson Studium, München, 2006.

[25] Kubertin, O., *Optimierung von Optionen durch effiziente Berechnung optimaler Anfangs-/Randbedingungen für die partielle Black-Scholes-Gleichung in WARRANT-PRO-2.* Diploma thesis (supervised by the author), Technische Universität Clausthal, 2003.

[26] Lions, J.-L., *Optimal Control of Systems Governed by Partial Differential Equations.* Springer, Berlin, 1971.

[27] Merton, R. C., *Optimum Consumption and Portfolio Rules in a Continuous Time Model*. Journal of Economic Theory **3**, 1971.

[28] Merton, R. C., *Theory of Rational Option Pricing*. Bell Journal of Economics and Management Science **4**, 1973.

[29] Merton, R. C., *An Intertemporal Capital Asset Pricing Model*. Econometrica **41**, 1973.

[30] Merton, R. C., *Option Pricing When Underlying Stock Returns Are Discontinuous*. Journal of Financial Economics **3**, 1976.

[31] Prisman, E. Z., *Pricing Derivative Securities: An Interactive Dynamic Environment with Maple V and Matlab (with CD-ROM)*. Academic Press, San Diego, 2001 (2nd print).

[32] Rall, L. B., *Automatic Differentiation — Techniques and Applications*. Springer Lecture Notes in Computer Science (LNCS) 120, 1981.

[33] Redhead, K., *Financial Derivatives — An Introduction to Futures, Forwards, Options and Swaps*. Prentice Hall, London, Herfordshire, 1997.

[34] Seydel, R. U., *Einführung in die numerische Berechnung von Finanz-Derivaten (Computational Finance)*. Springer, Berlin, 2000.

[35] Seydel, R. U., *Tools for Computational Finance*. Springer, Berlin, 2006 (3rd ed.).

[36] Stoer, J., and Bulirsch, R., *Introduction to Numerical Analysis*. Springer, Berlin, 2002 (3rd ed., 1st ed. 1980).

[37] Stoll, H. R., and Whaley, R. E., *Futures and Options: Theory and Applications*. South-Western Publications, Cincinnati 1994 (3rd print).

[38] Thomas, J. W., *Numerical Partial Differential Equations (Part 1): Finite Difference Methods*. Springer, New York, 1998 (2nd print).

[39] Wikipedia (The Free Encyclopedia), *Automatic Differentiation*, see en.wikipedia.org/wiki/Automatic_differentiation.

Epilogue

Prof. Dr. Rüdiger Seydel, Full Professor for Applied Mathematics and Computer Science at the Universität zu Köln, is first author of the book Seydel, R. U., and Bulirsch, R., *Vom Regenbogen zum Farbfernsehen: Höhere Mathematik in Fallstudien aus Natur und Technik*, Springer, Berlin, 1986, see Fig. A.3 on page 325. It inspired not only many students, but 20 years later also the editors of this book.

Flipside text:

"Fragt man bei einigen Gegenständen und Vorgängen des täglichen Lebens 'Wie funktioniert das?', so läßt sich eine Antwort mit sprachlichen Mitteln allein oftmals nicht geben. Solche Fragen sind etwa: 'Warum erscheint der Regenbogen gerade in dieser Höhe? Was passiert im Taschenrechner, wenn auf die SINUS-Taste gedrückt wird?' Hier ist die Mathematik zur Erklärung unentbehrlich. Das Buch erklärt in einer Reihe von Fallstudien solche Phänomene mit mathematischen Hilfsmitteln. Durch einen wiederholten Wechsel zwischen Motivation, Erklärung, Aufgaben und Lösungen versuchen die Autoren den Leser zum Mitdenken und Mitrechnen anzuregen. Die zahlreichen in den Text

eingearbeiteten Übungsaufgaben eignen sich vorzüglich, den Interessenten –
zu denen sicherlich nicht nur Mathematiker gehören – einsichtsreiche und un-
terhaltsame Stunden zu bereiten."

Translation:

"The question 'How does it work?' often arises for every day's things and
processes. The answers require more than lengthy explanations. Only Mathematics can answer questions like 'Why does a rainbow appear in a certain
altitude?' or 'How does a calculator compute the SINUS?'. In case studies this
book explains phenomenona like these with mathematical theory and calculations. Motivations, explanations, exercises and solutions alternate. The reader
is motivated to think and calculate together with the authors. The numerous
exercises are very interesting and informative, not only for mathematicians."

Complete the Correlation Matrix

C. Kahl[1] and M. Günther[2]

[1] Quantitative Analytics Group, ABN AMRO, 250 Bishopsgate, London EC2M
4AA, UK, christian.kahl@uk.abnamro.com
[2] Fachbereich Mathematik und Naturwissenschaften, Fachgruppe Mathematik und
Informatik, Lehrstuhl für Angewandte Mathematik / Numerische Analysis,
Bergische Universität Wuppertal, Gaußstraße 20, D-42119 Wuppertal, Germany,
guenther@mathematik.uni-wuppertal.de

Summary. In this article we discuss a method to complete the correlation matrix in
a multi-dimensional stochastic volatility model. We concentrate on the construction
of a positive definite correlation matrix. Furthermore we present a numerical integration scheme for this system of stochastic differential equations which improves
the approximation quality of the standard Euler-Maruyama method with minimal
additional computational effort.

1 Introduction

In stochastic models, especially in finance, often only some part of a correlation
matrix is given by measured data. This incomplete model may easily be completed by defining the respective correlations in a reasonable way. However,
the problem is to guarantee the positive definiteness of the matrix after completion. This paper develops and describes an efficient and feasible algorithm
to accomplish this task, which is based on combining Gaussian elimination
with arguments from graph theory.

In comparison to the results of Grone et al. [7] and Barrett et al. [1] our algorithm shows that it is possible to find a symmetric positive definite completion
of the correlation matrix under the additional restriction that all matrix en-
tries have to satisfy $|a_{(i,j)}| < 1$. Moreover we verify that our choice of the
unspecified entries leads to the unique determinant maximising completion
without the necessity of solving a sequence of optimisation problems.

The paper is organized as follows. Section 1 defines the problem of completing
a correlation matrix and introduces the model setup. The basic idea of the
algorithm is motivated by inspecting a small 2×2 example in Sect. 2, and
is applied to the general multi-dimensional case in the subsequent section.
Finally, we show how the completed correlation matrix can be incorporated
into numerical integration schemes for multi-dimensional volatility models.

2 Model setup

We consider the following $2n$-dimensional system of stochastic differential equations

$$dS_i = \mu_i S_i dt + f_i(V_i) S_i dW_{(S,i)}$$
$$dV_j = a_j(V_j) dt + b_j(V_j) dW_{(V,j)} \tag{1}$$

with $i, j = 1, \ldots, n$ and Brownian motions $W_{(S,i)}, W_{(V,j)}$. This article focuses on the question how to correlate the $2n$ Brownian motions. Furthermore, we assume that the diffusion of the underlying S_i is only directly coupled to one volatility V_i. If we concentrate on just one volatility-underlying-pair we obtain a typical stochastic volatility model to describe the non-flat implied volatility surface in European option's prices. This problem is intensively discussed in the literature in the recent years. More details can be found in the articles of Heston [8], Scott-Chesney [2] and Schöbel-Zhu [13].

We assume that we already know the correlation between the Wiener processes $W_{S,i}$ and $W_{V,i}$

$$dW_{S,i} dW_{V,i} = \eta_i dt ,$$

which in short-hand may be written as

$$W_{S,i} \cdot W_{V,i} \stackrel{\circ}{=} \eta_i .$$

Neglecting the stochastic volatility we also have to decide how to couple the underlyings S_i. This question is quite common in finance and we assume that we know these correlations, too:

$$W_{S,i} \cdot W_{S,j} \stackrel{\circ}{=} \rho_{(i,j)}.$$

This leads to the following structure for the correlation matrix

$$A = a(i,j)_{1 \leq i,j, \leq 2n} = \begin{pmatrix} \rho_{(1,1)} & \cdots & \rho_{(1,n)} & \eta_1 & & ? \\ \vdots & & \vdots & & \ddots & \\ \rho_{(n,1)} & \cdots & \rho_{(n,n)} & ? & & \eta_n \\ \eta_1 & & ? & 1 & & ? \\ & \ddots & & & \ddots & \\ ? & & \eta_n & ? & & 1 \end{pmatrix}, \tag{2}$$

with the undefined correlations marked with ?. The problem of completing a matrix A which is only specified on a given set of positions, the so called *pattern*, is directly related to the structure of the graph $G = (V, E)$ of A. In case of completing symmetric positive definite matrices, Grone et al. [7, Thm. 7] proved that the partial matrix A is completable if and only if the corresponding graph G is a chordal graph and each principal submatrix is positive semidefinite. Unfortunately this theorem is not applicable in our case as a correlation matrix has the further requirement that all entries are restricted by $|a_{(i,j)}| < 1$. For more informations on matrix completions problems we refer the reader to [11, 12].

The problem we have to deal with now is to define the yet unspecified correlations. To make this point clear we will discuss the simplest example of two underlyings and two stochastic volatility processes in the next section and we generalise this to the multi-dimensional case in Sect. 4.

Before we start with the discussion of a two-dimensional example we have to state the following general result which helps us to prove the positive definiteness of the correlation matrix.

Remark 1. A square matrix $A \in \mathbb{R}^{n \times n}$ is positive definite if the Gaussian algorithm[3] can be done with diagonal pivots p_i and if each pivot p_i is greater zero.

3 The 2 × 2-dimensional example

In this section we discuss how to complete the correlation matrix in the easiest case of two underlyings and two stochastic volatilities. We will refer to this as *two-dimensional* even though, strictly speaking, it would be more correct to call it 2×2-dimensional. The correlation matrix (2) is given by

$$A = \begin{pmatrix} 1 & \rho_{(1,2)} & \eta_1 & ? \\ \rho_{(2,1)} & 1 & ? & \eta_2 \\ \eta_1 & ? & 1 & ? \\ ? & \eta_2 & ? & 1 \end{pmatrix}. \tag{3}$$

We recognize that we have to specify the cross-correlations between $S_1 \sim V_2$ and $S_2 \sim V_1$, as well as the correlation between the different volatilities V_1 and V_2. This problem becomes more clear if we have a look at the corresponding graph[4] of the matrix (3) as shown in Fig. 1 . One ad-hoc and pragmatic way[5] to define the correlation between V_1 and S_2 is the product of the correlation between $V_1 \sim S_1$ and $S_1 \sim S_2$

$$a(3,2) \overset{\circ}{=} W_{V,1} \cdot W_{S,2} \overset{\circ}{=} (W_{V,1} \cdot W_{S,1})(W_{S,1} \cdot W_{S,2}) \overset{\circ}{=} \eta_1 \cdot \rho_{(1,2)}.$$

In the same way we are able define the correlation of the volatilities as the product of the correlation between $V_1 \sim S_1$, $S_1 \sim S_2$ and $S_2 \sim V_2$

$$a(4,3) \overset{\circ}{=} W_{V,1} \cdot W_{V,2} \overset{\circ}{=} (W_{V,1} \cdot W_{S,1})(W_{S,1} \cdot W_{S,2})(W_{S,2} \cdot W_{V,2})$$
$$\overset{\circ}{=} \eta_1 \cdot \rho_{(1,2)} \cdot \eta_2.$$

---

[3] The Gauss algorithm we are referring to is the standard Gaussian elimination in *kij-form* as in Duff, Erisman and Reid [3, Section 3.3]

[4] Graph theory and sparse matrices are closely linked topics. One can represent a symmetric matrix by an undirected graph. For more information see Golub and van Loan [6] or Frommer [4].

[5] This idea goes back to P. Jäckel [9].

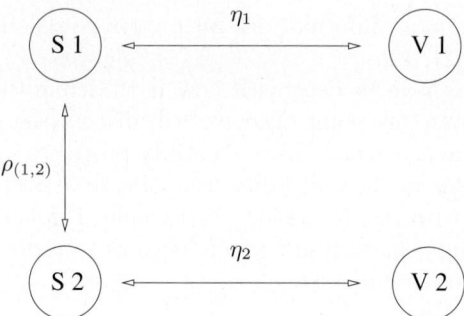

Fig. 1. Corresponding graph to matrix (3). The nodes (sometimes also called vertices) are referred by S_1, S_2, V_1 and V_2. This is not entirely consistent with the notation used in the literature as it would be correct to fill these nodes with the diagonal entries $a(i,i) = 1$. In the following we always choose the notation which provides the necessary information. The undirected edges are given by the non-diagonal entries of the matrix (3).

On the corresponding graph we just multiply the values of the edges on the path from V_1 to V_2. In the two-dimensional example there is just one possibility for this path but in the multi-dimensional case we have to choose the shortest one. The matrix now looks like

$$
A = \begin{pmatrix}
1 & \rho_{(1,2)} & \eta_1 & \eta_2 \cdot \rho_{(1,2)} \\
\rho_{(2,1)} & 1 & \eta_1 \cdot \rho_{(1,2)} & \eta_2 \\
\eta_1 & \eta_1 \cdot \rho_{(2,1)} & 1 & \eta_1 \cdot \rho_{(1,2)} \cdot \eta_2 \\
\eta_2 \cdot \rho_{(2,1)} & \eta_2 & \eta_1 \cdot \rho_{(2,1)} \cdot \eta_2 & 1
\end{pmatrix}. \tag{4}
$$

Figure 2 shows the corresponding graph to matrix (4). Next we have to verify that this choice of correlations leads to a positive definite matrix. In order to show this we use the Remark 1. In the kth step of the Gaussian elimination we have to choose the diagonal element $a(k,k)$ as the pivot p_k and we only operate on the elements $a(i,j)$ with $i,j \geq k$. To indicate the kth elimination step we

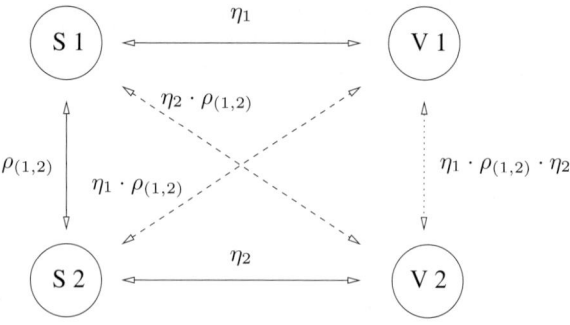

Fig. 2. Corresponding graph to matrix (4).

denote the matrix A as $A^{(k)}$ with entries $a(i,j)^{(k)}$ which will be updated via

$$a(i,j)^{(k+1)} = a(i,j)^{(k)} - \frac{a(i,k)^{(k)}a(k,j)^{(k)}}{a(k,k)^{(k)}}, \qquad i,j > k . \qquad (5)$$

The first pivot is $p_1 = a(1,1)^{(1)} = 1$ which is indeed greater zero and the remaining matrix looks as follows

$$A^{(2)} = \begin{pmatrix} 1 & \rho_{(1,2)} & \eta_1 & \eta_2 \cdot \rho_{(1,2)} \\ 0 & 1 - \rho_{(1,2)}^2 & 0 & \eta_2 \left(1 - \rho_{(1,2)}^2\right) \\ 0 & 0 & 1 - \eta_1^2 & 0 \\ 0 & \eta_2 \left(1 - \rho_{(1,2)}^2\right) & 0 & 1 - \left(\eta_2 \rho_{(1,2)}\right)^2 \end{pmatrix} .$$

After the first elimination step we can exclude the first row and first column from further consideration as they do not participate in the following calculations. Thus concentrating on the *active* part of the matrix we recognize that in the third row and third column only the diagonal element $a(3,3)^{(2)}$ is not zero. Therefore this node has lost any connection to other vertices in the corresponding graph, which means that in the following Gaussian elimination steps this whole row and column stays unmodified. In addition $a(3,3)^{(2)} = 1 - \eta_1^2 > 0$ hence we can choose this element as a positive pivot in elimination step $k = 3$. The next pivot is $a(2,2)^{(2)} = 1 - \rho_{(1,2)}^2 > 0$ and we obtain

$$A^{(3)} = \begin{pmatrix} 1 & \rho_{(1,2)} & \eta_1 & \eta_2 \cdot \rho_{(1,2)} \\ 0 & 1 - \rho_{(1,2)}^2 & 0 & \eta_2 \left(1 - \rho_{(1,2)}^2\right) \\ 0 & 0 & 1 - \eta_1^2 & 0 \\ 0 & 0 & 0 & 1 - \left(\eta_2 \rho_{(1,2)}\right)^2 - \frac{\eta_2^2 \left(1 - \rho_{(1,2)}^2\right)^2}{1 - \rho_{(1,2)}^2} \end{pmatrix}$$

$$= \begin{pmatrix} 1 & \rho_{(1,2)} & \eta_1 & \eta_2 \cdot \rho_{(1,2)} \\ 0 & 1 - \rho_{(1,2)}^2 & 0 & \eta_2 \left(1 - \rho_{(1,2)}^2\right) \\ 0 & 0 & 1 - \eta_1^2 & 0 \\ 0 & 0 & 0 & 1 - \eta_2^2 \end{pmatrix} .$$

The *active* part is now only the 2×2 submatrix containing all entries $a(i,j)^{(2)}$ with $i,j > 2$. In the last two elimination steps we can just choose the elements $a(3,3)^{(3)} = 1 - \eta_1^2 > 0$ and $a(4,4)^{(4)} = 1 - \eta_2^2 > 0$ as pivots p_3 and p_4 which proves that the original matrix A is positive definite. In the next section we show that the Gaussian algorithm is quite similar in the multi-dimensional case.

4 Multi-dimensional correlation

In this section we show how it is possible to complete the correlation matrix in the multi-dimensional setting in a very similar way as in the two-dimensional

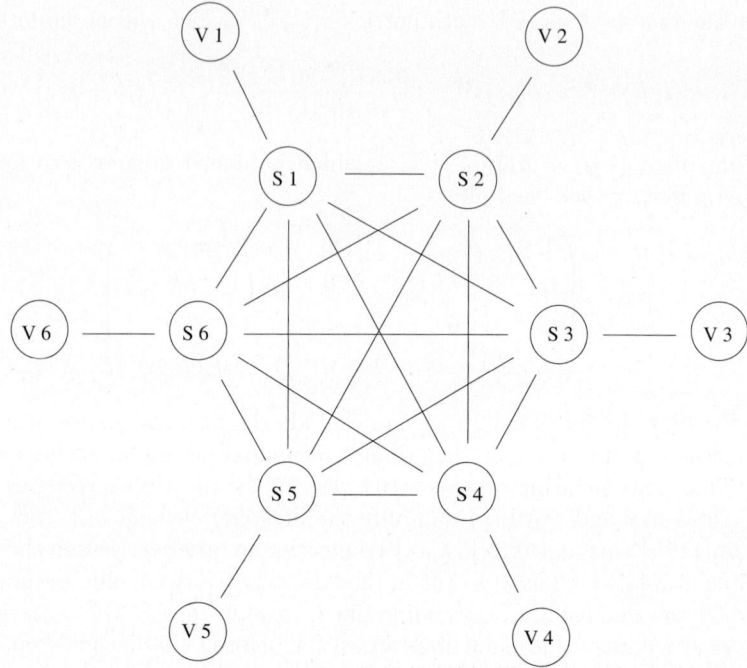

Fig. 3. Multi-dimensional correlation graph corresponding to matrix (2) with $n = 6$ where the unspecified correlations are interpreted as zero.

case. Moreover we verify that our choice leads to the determinant maximising completion. To get a first impression we draw the graph of six volatility-underlying-pairs and their corresponding correlations in Fig. 3. Within this figure we only show the fixed correlation between the underlyings S_i and the volatility-underlying correlation between S_j and V_j. As in the two-dimensional case we define the cross-correlation between the volatility V_i and the underlying S_j by the product of the correlation between $V_i \sim S_i$ and $S_i \sim S_j$

$$a(i + n, j) \overset{\circ}{=} W_{V,i} \cdot W_{S,j} \overset{\circ}{=} (W_{V,i} \cdot W_{S,i})(W_{S,i} \cdot W_{S,j}) \overset{\circ}{=} \eta_i \cdot \rho_{(i,j)}. \quad (6)$$

In the same way we define the correlation between two volatilities as

$$a(i + n, j + n) \overset{\circ}{=} W_{V,i} \cdot W_{V,j} \overset{\circ}{=} (W_{V,i} \cdot W_{S,i})(W_{S,i} \cdot W_{S,j})(W_{S,j} \cdot W_{V,j})$$
$$\overset{\circ}{=} \eta_i \cdot \rho_{(i,j)} \cdot \eta_j. \quad (7)$$

This corresponds to the shortest connection[6] in the graph between V_i and V_j. With this choice of the undefined correlations we are able to prove the following:

---

[6] In the shortest connection we only take into consideration the predefined paths between two underlyings S_i and S_j as well as the fixed underlying-volatility correlation between S_i and V_i

Theorem 1. *If the correlations between the underlyings*

$$W_{S,i} \cdot W_{S,j} = \rho_{(i,j)}$$

are given such that the correlation matrix

$$B = b(i,j) = \rho_{(i,j)} \tag{8}$$

is positive definite and if we choose the cross-correlations due to (6) and (7) then the whole correlation matrix $A = a(i,j)_{1 \le i,j \le n}$,

$$A = \begin{pmatrix} \rho_{(1,1)} & \cdots & \rho_{(1,n)} & \eta_1 & \cdots & \eta_n \cdot \rho_{(1,n)} \\ \vdots & \ddots & \vdots & \vdots & \ddots & \vdots \\ \rho_{(n,1)} & \cdots & \rho_{(n,n)} & \eta_1 \cdot \rho_{(n,1)} & \cdots & \eta_n \\ \eta_1 & \cdots & \eta_1 \cdot \rho_{(n,1)} & 1 & \cdots & \eta_1 \cdot \rho_{(1,n)} \cdot \eta_n \\ \vdots & \ddots & \vdots & \vdots & \ddots & \vdots \\ \eta_n \cdot \rho_{(1,n)} & \cdots & \eta_n & \eta_1 \cdot \rho_{(n,1)} \cdot \eta_n & \cdots & 1 \end{pmatrix}, \tag{9}$$

is positive definite.

Proof. In the two-dimensional setting we observed that the volatility V_1 given by the diagonal entry $a(1+n, 1+n)^{(2)}$, where n is the number of volatility-underlying-pairs, lost any connection in the corresponding graph after choosing $S_1 = a(1,1)^{(1)}$ as the first pivot. In the multi-dimensional setting this is equivalent to

$$a(n+1, j)^{(2)} = 0, \qquad j = 2, \ldots, n, n+2, \ldots, 2n . \tag{10}$$

In general we have to show that after selecting $a(k,k)^{(k)}$ as the kth pivot the corresponding volatility entry $a(k+n, k+n)^{(k+1)}$ is greater zero and has no further connection in the graph of $A^{(k)}$, which means that

$$a(n+k, j)^{(k+1)} = 0, \qquad j = k+1, \ldots, n+k-1, n+k+1, \ldots, 2n . \tag{11}$$

We will prove the positivity by showing that the following invariant holds

$$\begin{aligned} a(k+n, k+n)^{(k+1)} - \frac{\left(a(k, k+n)^{(k+1)}\right)^2}{a(k,k)^{(k+1)}} &= \\ a(k+n, k+n)^{(k)} - \frac{\left(a(k, k+n)^{(k)}\right)^2}{a(k,k)^{(k)}} &= \\ \ldots &= \\ a(k+n, k+n)^{(1)} - \frac{\left(a(k, k+n)^{(1)}\right)^2}{a(k,k)^{(1)}} &= 1 - \eta_k^2. \end{aligned} \tag{12}$$

We show this by induction. First we verify the following invariants:

$$a(i+n,j)^{(k)} = \frac{a(i+n,i)^{(k)}a(i,j)^{(k)}}{a(i,i)^{(k)}}, \tag{13}$$

$$a(i+n,j+n)^{(k)} = \frac{a(i+n,i)^{(k)}a(i,j)^{(k)}a(j+n,j)^{(k)}}{a(i,i)^{(k)}a(j,j)^{(k)}}, \tag{14}$$

with $i,j > k$. Before we start proving these statements, we explain the origin of the invariants (13) and (14). Having a look at the corresponding graph (see Fig. 4) in the elimination step k and bearing in mind that $a(i+n,j+n)^{(k)}$ describes the correlation between V_i and V_j then Eq. (14) is just the product over the values of the edges on the shortest path from V_i to V_j divided by the values of the vertices $S_i = a(i,i)^{(k)}$ and $S_j = a(j,j)^{(k)}$.

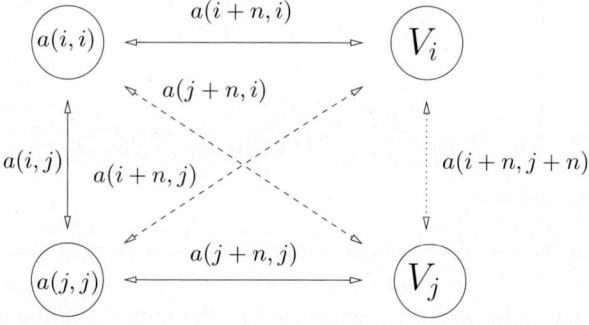

Fig. 4. Graph describing the correlation structure of the underlyings S_i and S_j as well as V_i and V_j in the kth step of the Gaussian algorithm. To simplify the notation we dropped all superscripts (k). The nodes on the left side belong to S_i and S_j which are here filled with the corresponding diagonal entries in the matrix $A^{(k)}$.

First we show by induction that (13) holds. The start of the induction is due to the construction (6)

$$a(i+n,j)^{(1)} = \frac{a(i+n,i)^{(1)}a(i,j)^{(1)}}{a(i,i)^{(1)}} = a(i+n,i) \cdot a(i,j) = \eta_i \cdot \rho_{(i,j)}.$$

Now we assume that (13) is valid until elimination step k and we want to verify that

$$a(i+n,j)^{(k+1)} = \frac{a(i+n,i)^{(k+1)}a(i,j)^{(k+1)}}{a(i,i)^{(k+1)}}, \qquad \text{for all } i,j > k+1.$$

Due to the Gaussian algorithm we know that[7]

$$a(i+n,j)^{(k+1)} = a(i+n,j)^{(k)} - \frac{a(i+n,k)^{(k)}a(k,j)^{(k)}}{a(k,k)^{(k)}}.$$

---

[7] In the following calculation we will not mention the index (k) indicating a variable in the elimination step k.

We thus calculate

$$a(i+n,j)^{(k+1)} = a(i+n,j)^{(k)} - \frac{a(i+n,k)^{(k)}a(k,j)^{(k)}}{a(k,k)^{(k)}}$$

$$= \frac{1}{a(k,k)}\left(a(k,k)a(i+n,j) - a(i+n,k)a(k,j)\right)$$

$$= \frac{1}{a(k,k)}\left(a(k,k)\frac{a(i+n,i)a(i,j)}{a(i,i)} - \frac{a(i+n,i)a(i,k)}{a(i,i)}a(k,j)\right)$$

$$= \frac{1}{a(k,k)}\frac{a(i+n,i)}{a(i,i)}\left(a(k,k)a(i,j) - a(i,k)a(k,j)\right)$$

$$= \frac{1}{a(k,k)}\left(a(k,k)a(i,j) - a(i,k)a(k,j)\right)$$

$$\cdot \frac{\frac{a(i+n,i)}{a(i,i)}\left(a(i,i)a(k,k) - a(i,k)a(i,k)\right)}{\left(a(i,i)a(k,k) - a(i,k)a(i,k)\right)}$$

$$= a(i,j)^{(k+1)}\frac{a(k,k)\left(a(i+n,i) - \frac{a(i,k)a(i,k)}{a(k,k)}\right)}{a(k,k)\left(a(i,i) - \frac{a(i,k)a(i,k)}{a(k,k)}\right)}$$

$$= \frac{a(i,j)^{(k+1)}a(i+n,i)^{(k+1)}}{a(i,i)^{(k+1)}} .$$

The proof of invariant (14) can be done in the same way. Next we show that Eq. (12) holds. Again the begin of the induction is valid due to construction (6). Assuming that the equation is valid up to elimination step $l \le k$, we obtain[8]

$$a(k+n,k+n)^{(l+1)} - \frac{a(k,k+n)^{(l+1)}a(k,k+n)^{(l+1)}}{a(k,k)^{(l+1)}} =$$

$$a(k+n,k+n) - \frac{a(k+n,l)a(k+n,l)}{a(l,l)} -$$

$$\frac{\left(a(l,l)a(k,k+n) - a(k,l)a(k+n,l)\right)^2}{a(l,l)\left(a(l,l)a(k,k) - a(k,l)a(k,l)\right)} =$$

$$a(k+n,k+n) - \frac{\left(a(k+n,l)a(k,l)\right)^2}{a(k,k)^2 a(l,l)} -$$

$$\frac{\left(a(l,l)a(k,k+n) - a(k,l)(a(k+n,k)a(k,l)/a(k,l))\right)^2}{a(l,l)\left(a(l,l)a(k,k) - a(k,l)a(k,l)\right)} =$$

---

[8] The first step is just the calculation rule of the Gaussian elimination and we drop again the superscript (l).

$$a(k+n, k+n) - \frac{(a(k+n, k)a(k, l))^2}{a(k, k)^2 a(l, l)} -$$

$$\frac{a(k+n, k)^2 (a(l, l)a(k, k) - a(k, l)a(k, l))}{a(k, k)^2 a(l, l)} =$$

$$a(k+n, k+n)^{(l)} - \frac{a(k+n, k)^{(l)} a(k+n, k)^{(l)}}{a(k, k)^{(l)}} .$$

Last we have to prove Eq. (11). Choosing $a(k, k)^{(k)}$ as the kth pivot leads to

$$a(k+n, j)^{(k+1)} = a(k+n, j) - \frac{a(k+n, k)a(k, j)}{a(k, k)} = 0 .$$

The same holds for

$$a(k+n, j+n)^{(k+1)} = a(k+n, j+n) - \frac{a(k+n, k)a(k, j+n)}{a(k, k)}$$

$$= a(k+n, j+n) - \frac{a(k+n, k)a(k, j)a(j, j+n)}{a(j, j)a(k, k)}$$

$$= 0 .$$

Let us summarize this proof. We know that we can choose the n underlying $S_i \sim a(i, i)$ as the first n pivots assuming that the coupling of the underlyings leads to a positive definite matrix. During these n steps all volatilities $V_i \sim a(i+n, i+n)$ are losing their connection in the corresponding graph and furthermore, we verified that $a(k+n, k+n)^{(k+1)} = 1 - \eta_k^2 > 0$. $\qquad\square$

Remark 2. The result can be generalised to the case where one underlying is not necessarily restricted to be directly coupled to one stochastic volatility process. It is also possible to have some underlyings without stochastic volatility and some with two or even more factors. Following the idea of the proof of Theorem 1 we just have to make sure that, if one underlying S_i is coupled to more than one volatility processes $V_i^{(m)}$, $m = 1, \ldots, M$, the matrix C_i with $c(k, l) \stackrel{\circ}{=} W_{V_i^{(k)}} \cdot W_{V_i^{(l)}}$, $c(k, M+1) \stackrel{\circ}{=} W_{V_i^{(k)}} \cdot W_{S,i}$, $c(M+1, M+1) \stackrel{\circ}{=} W_{S,i} \cdot W_{S,i} \stackrel{\circ}{=} 1$ is positive definite.

Next we verify that the choice of the cross-correlations due to (6) and (7) leads to the unique completion $\tilde{A}$ of the correlation matrix which maximises the determinant. Furthermore we know from [7, Thm. 2] that this is equivalent, that the inverse (9) contains zeros at each position which was previously unspecified.

Theorem 2. *Choosing the cross-correlations due to (6) and (7) leads to the unique determinant maximising positive completion.*

Proof. First we write the matrix A given by Eq. (9) as

$$A = \begin{pmatrix} B & D^{\top} \\ D & C \end{pmatrix}$$

with square matrices B, D and C. The entries of B are already specified. Next we introduce the term $\tilde{C} = C - DB^{-1}D^{\top}$ which is the well known Schur-complement. Using this notation we can formally write the inverse of A as

$$A^{-1} = \begin{pmatrix} B^{-1}\left(I + D^T\tilde{C}^{-1}DB^{-1}\right) & -B^{-1}D^T\tilde{C}^{-1} \\ -\tilde{C}^{-1}D^T B^{-1} & \tilde{C}^{-1} \end{pmatrix}.$$

Thus we have to show that $\tilde{C}$ and $B^{-1}D^T$ are diagonal. Since the Gaussian elimination on B coincides with calculating its inverse, Eq. (11) verifies that $\tilde{C}$ is diagonal. Moreover (11) also shows that $B^{-1}D^T$ only contains zeros below the diagonal. As this matrix is symmetric, caused by the diagonality of B and D, its diagonal. Hence the inverse of A contains zeros at each previously unspecified position which is equivalent with finding the determinant maximising completion due to Grone et al. [7]. □

Now we are able to complete the correlation matrix such that we obtain a symmetric positive definite matrix. Next we are confronted with the problem of integrating this system of stochastic differential equations. In case of one volatility-underlying-pair Kahl and Jäckel [10] compared the efficiency of various numerical integration methods. In the next section we show that these results are also applicable in the multidimensional setting.

5 Numerical tests for the multidimensional stochastic volatility model

In this section we discuss suitable numerical integration schemes for the multidimensional stochastic volatility model (1). Without stochastic volatility this problem is comparatively easy to solve as the n underlyings S_i are n-dimensional lognormal distributed. Thus in case of European options we do not have to discretise the time to maturity at all. The situation becomes much more complicated when stochastic volatility comes into play. As we do not know the distribution density we have to apply numerical integration schemes for stochastic differential equations. In the standard model with only one underlying and one stochastic volatility Kahl and Jäckel [10] discussed different integration methods with special regard to the numerical efficiency. It turned out that higher order methods, i.e. the Milstein scheme are inappropriate due to the fact that we have to generate additional random numbers. The finally most efficient integration scheme is referred to IJK[9]

---

[9] The name IJK refers to the originators' of this scheme.

$$\ln S_{(m+1)} = \ln S_{(m)} + \left(\mu - \tfrac{1}{4}\left(f^2\left(V_{(m)}\right) + f^2\left(V_{(m+1)}\right)\right)\right)\Delta t$$
$$+\rho f\left(V_{(m)}\right)\Delta W_{(V,m)}$$
$$+\tfrac{1}{2}\left(f\left(V_{(m)}\right) + f\left(V_{(m+1)}\right)\right)\left(\Delta W_{(S,m)} - \rho\Delta W_{(V,m)}\right)$$
$$+\tfrac{1}{2}\sqrt{1-\rho^2}f'\left(V_{(m)}\right)b\left(V_{(m)}\right)\left(\left(\Delta W_{(V,m)}\right)^2 - \Delta t\right)$$

with correlation $dW_S \cdot dW_V = \rho dt$. As one underlying S_i is only directly coupled to one stochastic volatility process V_i we can generalise this integration scheme straightforward to the multidimensional case

$$\ln S_{(i,m+1)} = \ln S_{(i,m)} + \left(\mu_i - \tfrac{1}{4}\left(f^2\left(V_{(i,m)}\right) + f^2\left(V_{(i,m+1)}\right)\right)\right)\Delta t$$
$$+\rho f\left(V_{(i,m)}\right)\Delta W_{(V,i,m)}$$
$$+\tfrac{1}{2}\left(f\left(V_{(i,m)}\right) + f\left(V_{(i,m+1)}\right)\right)\left(\Delta W_{(S,i,m)} - \rho\Delta W_{(V,i,m)}\right)$$
$$+\tfrac{1}{2}\sqrt{1-\rho^2}f'\left(V_{(i,m)}\right)b\left(V_{(i,m)}\right)\left(\left(\Delta W_{(V,i,m)}\right)^2 - \Delta t\right)$$

where $S_{i,m+1}$ denotes the $(m+1)$th step of the ith underlying. For the *IJK* scheme we assume that we already know the numerical approximation of the whole path of the process V_t. This path has to be computed with a suitable numerical integration scheme depending on the stochastic differential equation for the stochastic volatility. This problem is intensively discussed in [10, Section 3]. The benchmark scheme for the *multidimensional IJK* scheme is the standard *Euler-Maruyama* method

$$\ln S_{(i,m+1)} = \ln S_{(i,m)} + \left(\mu_i - \tfrac{1}{2}f_i\left(V_{(i,m)}\right)\right)\Delta t_m$$
$$+f_i\left(V_{(i,m)}\right)\Delta W_{(S,i,m)} \tag{15}$$
$$V_{(j,m+1)} = a_j\left(V_{(j,m)}\right)\Delta t_m + b_j\left(V_{(j,m)}\right)\Delta W_{(V,j,m)}.$$

Here one has to bear in mind that if the stochastic volatility V_j is given by a mean-reverting process the Euler scheme is not able to preserve numerical positivity. Thus if the financial derivative is sensitive to the dynamic of the variance of the underlying we recommend more advanced numerical integration schemes to preserve positivity.

Next we set up a 4×4-dimensional benchmark model to obtain a first impression on the numerical efficiency of both integration schemes. The stochastic volatility $\sigma_t = f(V_t)$ is described by a hyperbolic transformed Ornstein-Uhlenbeck process

$$dy_t = -\kappa y_t dt + \alpha\sqrt{2\kappa}dW_V ,$$

with transformation function $\sigma_t = \sigma_0\left(y_t + \sqrt{y_t^2 + 1}\right)$ which was introduced and discussed in [10]. We choose the following parameter configuration $\kappa = 1$, $\alpha = 0.35$, $\sigma_0 = 0.25$ and $y_0 = 0$ throughout the whole section and for all volatility processes. The initial value of the four underlyings is set to $S_0 = 100$. The decisive point for these tests is the correlation structure of the Wiener processes. The underlying correlation matrix (8) is chosen as follows

$$B = \begin{pmatrix} 1.0 & 0.2 & 0.0 & 0.5 \\ 0.2 & 1.0 & 0.4 & 0.0 \\ 0.0 & 0.4 & 1.0 & 0.6 \\ 0.5 & 0.0 & 0.6 & 1.0 \end{pmatrix}.$$

For the underlying-volatility correlation we assume a highly negative correlation corresponding to a downward sloping implied volatility surface in European vanilla option markets

$$v = \begin{pmatrix} -0.7 \\ -0.8 \\ -0.9 \\ -0.8 \end{pmatrix}. \tag{16}$$

This directly leads to the following correlation matrix completed due to (6) and (7)

$$A = \begin{pmatrix} 1.000 & 0.200 & 0.000 & 0.500 & -0.700 & -0.160 & 0.000 & -0.400 \\ 0.200 & 1.000 & 0.400 & 0.000 & -0.140 & -0.800 & -0.360 & 0.000 \\ 0.000 & 0.400 & 1.000 & 0.600 & 0.000 & -0.320 & -0.900 & -0.480 \\ 0.500 & 0.000 & 0.600 & 1.000 & -0.350 & 0.000 & -0.540 & -0.800 \\ -0.700 & -0.140 & 0.000 & -0.350 & 1.000 & 0.112 & 0.000 & 0.280 \\ -0.160 & -0.800 & -0.320 & 0.000 & 0.112 & 1.000 & 0.288 & 0.000 \\ 0.000 & -0.360 & -0.900 & -0.540 & 0.000 & 0.288 & 1.000 & 0.432 \\ -0.400 & 0.000 & -0.480 & -0.800 & 0.280 & 0.000 & 0.432 & 1.000 \end{pmatrix}.$$

The first numerical test compares the evaluation of a basket option. Thereby we consider the payoff function to be the mean of the four underlying assets. Thus the fair value of this option is given by

$$C(T, K) = e\left(\frac{1}{4} \sum_{i=1}^{4} S_i(T) - K \right)^{+}.$$

The numerical results are compared with a numerical reference solution computed with the *Euler-Maruyama* scheme (15) and a stepsize of $\Delta t_{\text{exact}} = 2^{-10}$. The prices are calculated for a whole range of strikes $K = \{75, \ldots, 133.3\}$ and a termstructure of maturities $T = \{0.5, 1, 1.5, \ldots, 4\}$. As there is a great difference in prices of at-the-money options compared to in-the-money and out-of-money options we compute the implied Black volatility, denoted as $IV(C, S, K, r, T)$, where C is the given option price, S the initial value and r the risk free interest rate, to get a fair error measure. The biggest advantage of the implied volatility is that the error-size throughout the whole level of strikes and maturities becomes comparable. The first Fig. 5 (A) shows the numerical reference solutions where we recognize the strongly downward sloping skew structure of the implied volatility surface as a consequence of the negative correlation between underlyings and volatilities (16).

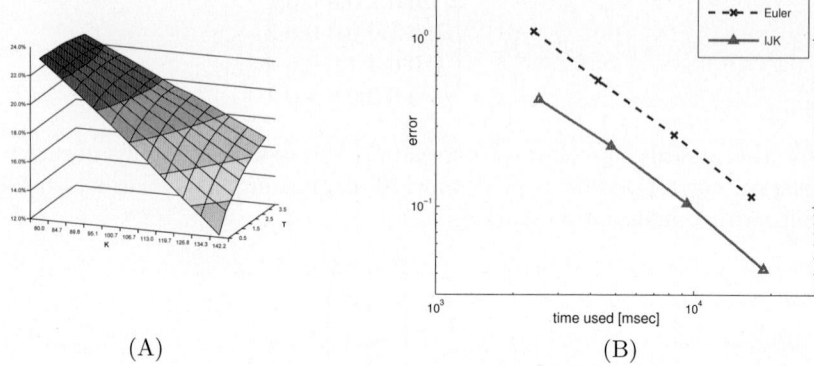

(A)   (B)

Fig. 5. (A): Implied volatility surface of the reference solution with a stepsize of $\Delta t_{\mathrm{exact}} = 2^{-10}$ and 32767 paths . (B): Weak approximation error (17) as a function of CPU time [in msec] for the simulation of 32767 paths. The number generator in (A) and (B) was Sobol's method and the paths were constructed via the Brownian bridge.

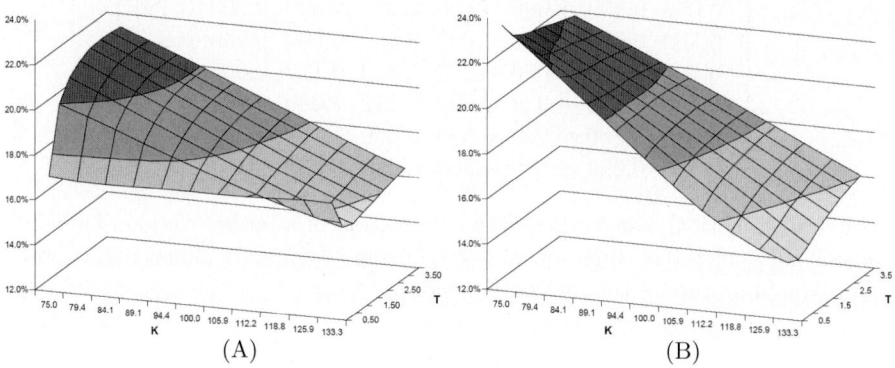

(A)   (B)

Fig. 6. Implied volatility surface calculated with the (A): *Euler* method and the (B): *multidimensional IJK* method. The stepsize was $\Delta t = 2^{-1}$ and the prices were averaged over 32767 paths . The number generator was the Sobol's method and the paths were constructed via the Brownian bridge.

In Fig. 6 we compare the numerical results of the *Euler* scheme with the *multidimensional IJK* scheme where we integrate along the maturities with a stepsize of $\Delta t = 0.5$. The superiority of the *IJK* scheme is at its most impressive for the first maturity $T = 0.5$ as we obtain a skew within one integration step in comparison to the flat implied-volatility of the *Euler* for this maturity.

To underscore this result we also compare the error and the computational effort of both integration schemes for different stepsizes. The error is computed as the sum over the absolute value of the differences between the implied volatility of the reference solution $C(t_i, K_j)$ and the numerical approximation $\tilde{C}(t_i, K_j)$

$$\text{Error} = \sum_{i,j} \left| IV(C(t_i, K_j), S, K_j, r, t_i) - IV(\tilde{C}(t_i, K_j), S, K_j, r, t_i) \right| . \quad (17)$$

In Fig. 5 (B) we see that the *IJK* leads to better results comparing the relation between approximation quality and computational effort. The decisive point is that the computational effort of the *IJK* scheme is only slightly higher than the *Euler* scheme since we do not have to draw any additional random number.

6 Summary

Based on combining Gaussian elimination and graph theory, we have introduced an algorithm to complete the correlation matrix, if only an incomplete set of measured data is available, which does not allow to define all correlations a unique way. Compared to the results of Grone et al. [7] and Barrett et al. [1], our algorithm preserves that all entries are bounded by $|a_{(i,j)}| < 1$, and avoids the costly computation of optimisation problems. From an application-oriented point of view, our algorithm can be implemented within pricing tools based on simulating multidimensional stochastic volatility models.

Acknowledgement. This work has been supported by ABN AMRO, London and the authors like to thank Peter Jäckel for fruitful discussions. Moreover thanks go to Cathrin van Emmerich, Bruno Lang and Roland Pulch from the University of Wuppertal for valuable hints.

References

[1]   Barett, W.W. and Johnson, C.R. and Lundquist, M.: Determinantal Formulae for Matrix Completions Associated with Chordal Graphs. Linear Algebra and its Applications, 121, 265–289 (1989).

[2]   Chesney, M. and Scott, L.: Pricing European Currency Options: A comparison of the modified Black-Scholes model and a random variance model. Journal of Financial and Quantitative Analysis, 24, 267–284 (1989).

[3]   Duff, I. S. and Erisman, A. M. and Reid, J. K.: Direct Methods for Sparse Matrices. Oxford Science Publications, Oxford, 1986.

[4]   Frommer, A.: Dünn besetzte Matrizen. Manuscript, University of Wuppertal, 2003.

[5]   A. George and J. W. H. Liu: Computer solution of large sparse positive-definite systems. Prentice-Hall, New-Jersey, 1981.

[6]   G. Golub and C. van Loan: Matrix computations. The Johns Hopkins University Press, 1996.

[7]   Grone, R. and Johnson, C.R. and Sá, E.M. and Wolkowicz, H.: Positive Definite Completions of Partial Hermitian Matrices. Linear Algebra and its Applications, 58, 109–124 (1984).

[8]   Heston, S.: A closed-form solution for options with stochastic volatility with application to bond and currency options. Rev. Financ. Studies, 6, 327–343 (1993)

[9]   Jäckel, P.: Private communication, 2005.

[10]   Kahl, C. and Jäckel, P.: Fast strong approximation Monte-Carlo schemes for stochastic volatility models. Submitted to Journal of Quantitative Finance.

[11]   Laurent, M.: Matrix completion problems. The Encyclopedia of Optimization, 3, 221–229 (2001).

[12]   Johnson, C.: Matrix completion problems: a survey. Matrix Theory and Applications, 44, 171–198 (1990).

[13]   Schöbel, R. and Zhu, J.: Stochastic Volatility with an Ornstein-Uhlenbeck Process: An Extension. European Finance Review, 3, 23–46 (1999).

Accelerating the Distributed Multiplication Protocol with Applications to the Distributed Miller-Rabin Primality Test

P. Lory

Institut für Wirtschaftsinformatik, Universität Regensburg, D-93040 Regensburg, Germany, Peter.Lory@wiwi.uni-regensburg.de

Summary. In the light of information security it is highly desirable to avoid a "single point of failure" because this would be an attractive target for attackers. Cryptographic protocols for distributed computations are important techniques in pursuing this goal. An essential module in this context is the secure multiparty multiplication of two polynomially shared values over $\mathbb{Z}_q$ with a public prime number q. The multiplication protocol of Gennaro, Rabin and Rabin (1998) is considered as the best protocol for this purpose. It requires a complexity of $O(n^2 k \log n + nk^2)$ bit operations per player, where k is the bit size of the prime q and n is the number of players. The present paper reduces this complexity to $O(n^2 k + nk^2)$ with unaltered communication and round complexities. This improvement is possible by a loan from the field of numerical analysis, namely by the use of Newton's classical interpolation formula. The distributed version of the famous probabilistic primality test of Miller and Rabin is built of several modules, which depend on distributed multiplications. Applications of the new method to these modules is studied and its importance for distributed signatures is outlined.

1 Introduction

Under the title *Where is the Football?* the *Time Magazine* wrote in its issue dated May 4, 1992: *Anytime President Bush or Russian President Boris Yeltsin travels, an aide tags along carrying the briefcase of electronic controls that Americans call the nuclear football – the ignition key, in effect, for nuclear war. The former Soviet Union has three operational sets of such devices: Yeltsin has one, which can be used only in conjunction with another set controlled by Defense Minister Yevgeni Shaposhnikov. A third system is usually held by the Defense Ministry and can replace either of the other two. But after last year's aborted coup, Western intelligence lost sight of the third football, and officials were forced to ponder the implications of a nuclear fumble. Now the intelligence boys have cleared up the mystery: the third football is safe in the hands of the Defense Ministry chief of staff. Civilian power may be in flux, but at least the nuclear authority has not changed hands.*

Obviously, the Russian system is an example for a two-out-of-three access mechanism. Situations like this motivate the following definition: Let t be a nonnegative and n a positive integer with $t < n$. A $(t+1)$-*out-of-n threshold scheme* is a method of sharing a secret key a among a set of n participants (denoted by $\mathcal{P}$), in such a way that any $t+1$ participants can compute the value of a, but no group of t or less participants can do so.

The value of a is chosen randomly by a special entity called the *dealer* that is not a member of $\mathcal{P}$ but has to be trusted by its members. When the dealer wants to share the secret a among the participants in $\mathcal{P}$, he gives each participant some partial information called a *share*. The shares should be distributed secretly, so no participant knows the share given to another participant. At a later time, a subset of members of $\mathcal{P}$ will pool their shares in an attempt to compute the secret a. If this subset consists of $t+1$ or more members, then they should be able to compute the value of a as a function of the shares; otherwise they should not be able to compute a. A survey of threshold schemes of this type can be found in the textbook of Stinson [17].

In [16] A. Shamir has proposed a seminal $(t+1)$-out-of-n threshold scheme. A comprehensive presentation can be found in Catalano [4]. Let q be a public prime number with $q > n$ and let $\mathbb{Z}_q = \{0, 1, \ldots, q-1\}$ denote the field of residues modulo q. To share a secret a, the dealer randomly chooses t elements $b_i \in \mathbb{Z}_q$ and sets $f(z)$ as the polynomial

$$f(z) = a + \sum_{i=1}^{t} b_i z^i \bmod q.$$

Then he sends the value $f(i)$ to player P_i ($i = 1, 2, \ldots n$). The Lagrange interpolation formula (e.g. Stoer and Bulirsch [18]) allows one to retrieve the unique polynomial f whose degree does not exceed t from $t+1$ support ordinates. Let $S = \{P_{i_1}, \ldots, P_{i_{t+1}}\}$ be any subset of $t+1$ players. The formula for the free term of the polynomial is

$$f(0) = \sum_{j=1}^{t+1} f(i_j) \prod_{\substack{1 \le k \le t+1 \\ k \ne j}} \frac{i_k}{i_k - i_j} \bmod q.$$

Defining the Lagrange interpolation coefficients by

$$\lambda_{i_j} = \prod_{\substack{1 \le k \le t+1 \\ k \ne j}} \frac{i_k}{i_k - i_j} \bmod q,$$

the shared secret a can be computed by the formula

$$a = f(0) = \sum_{j=1}^{t+1} \lambda_{i_j} f(i_j) \bmod q. \tag{1}$$

The values of the Lagrange interpolation coefficients λ_{i_j} depend on q but are independent from the specific polynomial one wants to interpolate. For this reason they can be precomputed and their values do not need to be kept secret.

Whereas the classical approach of Shamir [16] assumes the existence of a trusted dealer, more recent approaches want to avoid the problem of such a *single point of failure* for two reasons: 1) A dealer has full control of the system. If he is dishonest, he could misuse this power. 2) The server of the dealer is an attractive target for malicious adversaries and has to be protected with high costs.

It is the aim of *secure multiparty compuations* to avoid the need for a trusted dealer. The last two decades have seen an exciting development of techniques of this type. Classical theoretical results [2, 6, 10, 19] show that any multiparty computation can be performed securely, if the number of corrupted participants does not exceed certain bounds. However already Gennaro, Rabin and Rabin [9] point out, that these generic secure circuit techniques are too inefficient in the area of practical feasibility, which might render them impractical. Thus, it is a high priority to optimize such techniques.

Indeed, under the "honest-but-curious" model significant gains in efficiency are possible. This model assumes that all players follow the protocols honestly, but it is guaranteed that even if a minority of players "pool" their information they cannot learn anything that they were not "supposed" to.

First, the question arises how in this model the members of $\mathcal{P}$ can generate shares of a secret chosen jointly and at random in $\mathbb{Z}_q$ without a trusted dealer. For this purpose each player P_i chooses a random value $r_i \in \mathbb{Z}_q$ and shares it according to the above described secret sharing scheme. Then he sends the obtained shares to the remaining players of $\mathcal{P}$. At this point each player sums up (modulo q) all the received values and sets the obtained value as his share of the jointly chosen random value. Please note, that the latter is the sum of the r_i and is never generated explicitly.

To bring this idea to its full potential it is necessary to perform computations with two or more jointly shared secrets. For example, in several cryptographic protocols a number of participants is required to have an RSA [15] modulus $N = pq$ (p and q primes) for which none of them knows the factorization (see [1] and [4] for surveys on this). An essential step in the distributed establishment of such a modulus is of course a primality test. Algesheimer, Camenisch and Shoup [1] have presented a distributed version of the famous probabilistic Miller-Rabin [12, 14] primality test. Their protocol is modularly built of several (both elementary and rather complex) subprotocols (see also Sect. 4 of the present paper).

Please note again, that these computations must be accomplished without the use of a trusted dealer in a completely distributed manner. For elementary operations, this is an easy task:

1. *Multiplication or addition of a constant (public) value and a polynomially shared secret*: This is done by having each player multiply (or add) his share to the constant. This works because, if $f(i)$ is a share of a, then $f(i) + c$ will be a share of $a + c$ and $c \cdot f(i)$ one of $c \cdot a$.

2. *Addition of two polynomially shared values*: This is done by having the players locally add their own shares. Indeed, if $f(i)$ is a share of a secret a and $g(i)$ a share of a secret b, the value $f(i) + g(i)$ is actually a share of the sum $a + b$.

The *muliplication of two polynomially shared values* is more complicated. At the beginning of the multiplication protocol each player P_i holds as input the function values $f_\alpha(i)$ and $f_\beta(i)$ of two polynomials f_α and f_β with degree at most t and $\alpha = f_\alpha(0), \beta = f_\beta(0)$. At the end of the protocol each player owns the function value $H(i)$ of a polynomial H with degree at most t as his share of the product $\alpha\beta = H(0)$. A first multiplication protocol of the this type has been presented by Ben-Or, Goldwasser and Wigderson [2]. A considerable improvement was proposed by Gennaro, Rabin and Rabin [9]. Presently, their approach is considered as the most efficient protocol (see [1, 4]). It requires $O(k^2 n + kn^2 \log n)$ bit operations per player. Here, k is the bit size of the prime q and n is the number of players; $\log x$ denotes the logarithm of x to the base 2. In the present paper, this complexity is reduced to $O(k^2 n + kn^2)$. Remarkably, the key idea for this success is the application of a rather old technique, namely Newton's classical interpolation formula [13] (*Methodus Differentialis*, 1676).

The paper is organized as follows: Section 2 describes the the protocol of Gennaro, Rabin and Rabin [9] for the reader's convenience and investigates its complexity in detail. Section 3 presents the new protocol and examines its complexity. In Sect. 4 the impact of the new technique on the distributed Miller-Rabin primality test is studied. In the last section the results of the previous sections are viewed in the context of distibuted digital signatures.

2 The Protocol of Gennaro, Rabin and Rabin

The protocol in [9] assumes that two secrets α and β are shared by polynomials $f_\alpha(x)$ and $f_\beta(x)$ respectively and the players would like to compute the product $\alpha\beta$. Both polynomials are of degree at most t. Denote by $f_\alpha(i)$ and $f_\beta(i)$ the shares of player P_i on $f_\alpha(x)$ and $f_\beta(x)$ respectively. The product of these two polynomials is

$$f_\alpha(x)f_\beta(x) = \gamma_{2t}x^{2t} + \ldots \gamma_1 x + \alpha\beta \overset{def}{=} f_{\alpha\beta}(x).$$

Because of Eq. (1)

$$\alpha\beta = \lambda_1 f_{\alpha\beta}(1) + \ldots + \lambda_{2t+1} f_{\alpha\beta}(2t + 1)$$

with known non-zero constants λ_i. Let $h_1(x), \ldots, h_{2t+1}(x)$ be polynomials of degree at most t which satisfy that $h_i(0) = f_{\alpha\beta}(i)$ for $1 \leq i \leq 2t + 1$. Define

$$H(x) \overset{def}{=} \sum_{i=1}^{2t+1} \lambda_i h_i(x).$$

Then this function is a polynomial of degree at most t with the property

$$H(0) = \lambda_1 f_{\alpha\beta}(1) + \ldots + \lambda_{2t+1} f_{\alpha\beta}(2t + 1) = \alpha\beta.$$

Clearly, $H(j) = \lambda_1 h_1(j) + \lambda_2 h_2(j) + \ldots + \lambda_{2t+1} h_{2t+1}(j)$. Thus, if each of the players P_i $(1 \leq i \leq 2t+1)$ shares his share $f_{\alpha\beta}(i)$ with the other participants using a polynomial $h_i(x)$ with the properties as defined above, then the product $\alpha\beta$ is shared by the polynomial $H(x)$ of degree at most t. This idea is the basis of the protocol given in Fig. 1.

Input of player P_i: The values $f_\alpha(i)$ and $f_\beta(i)$.

1. Player P_i $(1 \leq i \leq 2t + 1)$ computes $f_\alpha(i) f_\beta(i)$ and shares this value by choosing a random polynomial $h_i(x)$ of degree at most t, such that

$$h_i(0) = f_\alpha(i) f_\beta(i).$$

He gives player P_j $(1 \leq j \leq n)$ the value $h_i(j)$.
2. Each player P_j $(1 \leq j \leq n)$ computes his share of $\alpha\beta$ via a random polynomial H, i.e. the value $H(j)$, by locally computing the linear combination

$$H(j) = \lambda_1 h_1(j) + \lambda_2 h_2(j) + \ldots + \lambda_{2t+1} h_{2t+1}(j).$$

Fig. 1. The multiplication protocol of Gennaro, Rabin and Rabin

Please note, that the protocol implicitly assumes that the number of players n obeys $n \geq 2t + 1$.

For the investigation of the complexity the basic assumption is made, that the bit-complexity of a multiplication of a k-bit-integer and an l-bit-integer is $O(kl)$. This is a reasonable estimate for realistic values (e.g. $k = l = 1024$). Step 1 of the protocol of Fig. 1 requires n evaluations of the polynomial $h_i(x)$ of degree t. If Horner's scheme is used for this purpose, one evaluation requires t multiplications of a k-bit integer and an integer with at most $\text{entier}(\log n) + 1$ bits. In step 2 of the protocol each player has to compute $2t + 1$ multiplications of two k-bit numbers. Taking into account that $t < 2t + 1 \leq n$, a complexity of $O(n^2 k \log n + nk^2)$ bit operations per player follows. This is consistent with the corresponding propositions in Algesheimer, Camenish and Shoup [1] and Catalano [4].

3 The New Protocol and its Complexity

3.1 The New Protocol

The key for reducing the complexity in the multiplication protocol of Gennaro, Rabin and Rabin [9] is the observation that in Step 1 of the protocol in Fig. 1 each of the players P_i $(1 \leq i \leq 2t+1)$ chooses a random polynomial of degree at most t

$$h_i(x) = a_t x^t + a_{t-1} x^{t-1} + \ldots + a_1 x + a_0$$

with $a_0 = f_\alpha(i) f_\beta(i)$ and then has to evaluate this polynomial at n different points. The present paper suggests that instead of choosing the coefficients a_j $(1 \leq j \leq t)$, each of the players P_i $(1 \leq i \leq 2t+1)$ randomly picks t support ordinates f_j for the t abscissas $x_j = j$ $(1 \leq j \leq t)$. Together with the condition

$$h_i(0) = f_\alpha(i) f_\beta(i)$$

this implicitly defines the unique interpolation polynomial $h_i(x)$ of degree at most t. Then player P_i has to evaluate this polynomial for $x_j = j$ $(t+1 \leq j \leq n)$. Using Newton's scheme of divided differences [13] (see e.g. Stoer and Bulirsch [18]) these computations can be performed very efficiently. The details are given in Fig. 2 with $f_j = h_i(j)$ $(0 \leq j \leq n)$. For readability reasons the index i is omitted.

A few remarks are in place:

1. As in [9] the support abscissas for the interpolating polynomial $h_i(x)$ are chosen as $x_j = j$ for $0 \leq j \leq n$.
2. The zeros in the columns $l = t+1, t+2, \ldots, n$ of Fig. 2 are not computed. Instead, they are prescribed and force the interpolating polynomial to be of degree at most t.
3. The first $t+1$ ascending diagonal rows are computed from left to right starting from the prescribed support ordinate $f_0 = h_i(0) = f_\alpha(i) f_\beta(i)$ and the randomly chosen support ordinates $f_1 = h_i(1), \ldots, f_t = h_i(t)$. The following diagonal rows are computed from right to left starting from the prescribed zeros and ending in the computed support ordinates

$$f_{t+1} = h_i(t+1), \ldots, f_n = h_i(n).$$

4. Instead of calculating the divided differences according to the usual definition (c.f. Stoer and Bulirsch [18])

$$f_{i_0, i_1, \ldots, i_l} \stackrel{def}{=} \frac{f_{i_1, i_2, \ldots, i_l} - f_{i_0, i_1, \ldots, i_{l-1}}}{x_{i_l} - x_{i_0}},$$

the numbers

$$f_{i_0, i_1, \ldots, i_l} \cdot (x_{i_l} - x_{i_0}) = f_{i_1, i_1, \ldots, i_l} - f_{i_0, i_1, \ldots, i_{l-1}}.$$

are computed. This modification is the reason for the factors $l!$ in column l of the scheme of Fig. 2 and avoids superfluous arithmetic operations.

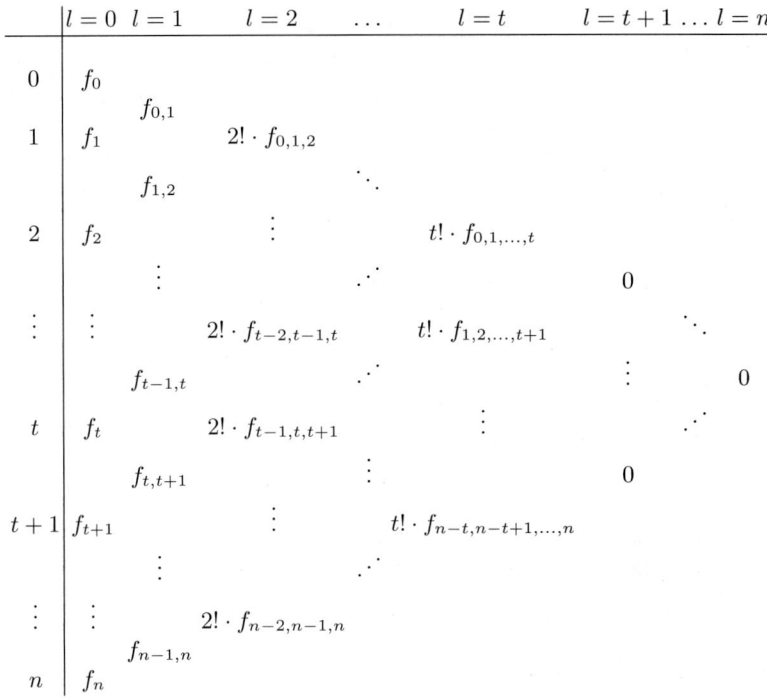

Fig. 2. Newton's divided-difference scheme for the new protocol

These ideas are the basis of the new protocol given in Fig. 3, where all operations take place in $\mathbb{Z}_q$ with a public prime number q (see Sect. 1).

A few comments are in place:

1. Step 1(a) of the protocol in Fig. 3 calculates the upper left corner in Newton's diveded difference scheme of Fig. 2.
2. Step 1(b) of this protocol calculates the following t ascending diagonal rows from left to right. Here, the index k is running downwards for storage efficiency reasons.
3. Step 1(c) of this protocol calculates the following $n-t$ ascending diagonal rows from right to left.

Apart from technical details in the calculations, the protocol of Gennaro, Rabin and Rabin [9] (cf. Fig. 1) and the new protocol of Fig. 3 differ in only one respect: In the protocol of Gennaro, Rabin and Rabin each player P_i randomly chooses a polynomial $h_i(x)$ of degree at most t by choosing its coefficients of $x^1, x^2, \ldots, x^t$. In the new protocol the same player P_i randomly chooses t support ordinates for this polynomial:

$$f_1 = h_i(1), \ldots, f_t = h_i(t).$$

Input of player P_i: The values $f_\alpha(i)$ and $f_\beta(i)$.

1. Player P_i $(1 \leq i \leq 2t + 1)$ computes $f_\alpha(i)f_\beta(i)$ and shares this value by randomly choosing t support ordinates $f_1, f_2, \ldots, f_t$ and executing the following steps:

   a)
   $$g_0 := f_\alpha(i)f_\beta(i) \,.$$

   b) For $j = 1, 2, \ldots, t$:
   $$g_j := f_j \,,$$
   for $k = j - 1, j - 2, \ldots, 0$:
   $$g_k := g_{k+1} - g_k \,.$$
   c) For $j = t + 1, t + 2, \ldots, n$:
   for $k = 0, 1, \ldots, t - 1$:
   $$g_{k+1} := g_{k+1} + g_k \,.$$
   $$f_j := g_t \,.$$
   He gives player P_j $(1 \leq j \leq n)$ the value

   $$h_i(j) := f_j.$$

2. This step is identical to Step 2 in the protocol of Gennaro, Rabin and Rabin (Fig. 1).

Fig. 3. The new multiplication protocol

This does not affect the randomness of the chosen polynomial $h_i(x)$. Therefore, the proof of Theorem 3 in [9] applies to the new protocol as well and the following theorem follows:

Theorem 1. *The protocol of Fig. 3 is a secure multiplication protocol in the presence of a passive adversary computationally unbounded.*

3.2 Complexity of the New Protocol

Step 1(b) of the new protocol (Fig. 3) needs $t(t + 1)/2$ additions of two k-bit numbers, where k is the bit size of the prime q and $t+1$ is the threshold. Step 1(c) of the same protocol requires $(n - t)t$ additions, where n is the number of players. Clearly, the complexity for the addition of two k-bit numbers is $O(k)$. Taking into account that $t < 2t + 1 \leq n$, a complexity of $O(n^2k)$ bit operations per player for step 1 of the new protocol follows. Step 2 requires $O(nk^2)$ bit operations (see Sect. 2). So the following theorem is proven:

Theorem 2. *The multiplication protocol of Fig. 3 requires $O(n^2k + nk^2)$ bit operations per player.*

This result has to be compared with the bit-complexity of $O(n^2k \log n + nk^2)$ for the multiplication protocol of Gennaro, Rabin and Rabin [9]. Please note, that obviously the communication complexity and the round complexity (for definitions see e.g. [4]) are not affected by a shift from the old to the new protocol.

4 Impact on the Distributed Miller-Rabin Primality Test

Algesheimer, Camenisch and Shoup [1] and Catalano [4] have pointed out the attractiveness of modularity in the construction of protocols for distributed computations. Simple protocols can be combined to address more complicated tasks. In the following, MUL_{old} denotes the distributed multiplication protocol of Gennaro, Rabin and Rabin (see Sect. 2), whereas MUL_{new} means the new protocol given in Sect. 3 of the present paper.

Algesheimer, Camenisch and Shoup [1] have presented an efficient protocol for the distributed Miller-Rabin primality test. It is built of several subprotocols. The following list gives their complexities. For details the reader is referred to [1] or [4].

1. The protocol I2Q-BIT transforms additive shares over $\mathbb{Z}$ of a k bit secret b into polynomial shares of the bits of b over $\mathbb{Z}_q$. Unfortunately, the only known solution to perform this task is not very efficient, because it requires one to resort to general multiparty computation protocols. Its bit-complexity per player is

$$O(\gamma k n^3 \log n + \gamma^2 k n^2 + k^2 n^2 \log n),$$

   where $\gamma = O(\rho + \log n)$ with ρ as a security parameter. This protocol does not depend on distributed multiplications.
2. The protocol SI2SQ converts additive shares over $\mathbb{Z}$ into additive shares over $\mathbb{Z}_q$. The protocol SQ2PQ converts the latter shares into polynomial shares modulo q. The bit-complexities of these conversions are lower than those of the other protocols and need not be considered further.
3. The protocol APPINV generates polynomial shares modulo q of an approximation to $1/p$, where the polynomial shares of p are given. The cost of this protocol is dominated by the cost of the distributed multiplication protocol, which has to be repeated $2[\log(t - 3 - \log(n + 1))] \approx O(\log t)$ times, where $t > l + 5 + \log(n + 1)$ with an accuracy parameter l. Consequently, the cost of APPINV in terms of bit operations per player is

$$O((k^2 n + k n^2 \log n) \log t), \quad \text{if } MUL_{old} \text{ is used,}$$
$$O((k^2 n + k n^2) \log t), \quad \text{if } MUL_{new} \text{ is used.}$$

4. The TRUNC-protocol takes as input polynomial shares of a secret a and a parameter κ and returns as output polynomial shares of b such that $|b - a/2^\kappa| \le n + 1$. All shares are modulo q. The protocol does not rely on distributed multiplication and its bit-complexity per player is $O(k^2 n + k n^2 \log n)$.
5. The protocol MOD takes as input polynomial shares (over $\mathbb{Z}_q$) of secrets c and p and the output of the APPINV-protocol and computes (roughly speaking) polynomial shares (again over $\mathbb{Z}_q$) of $d \equiv c \bmod p$ in a distributed manner. The cost of this protocol is dominated by two invocations of the MUL-protocol and two invocations of the TRUNC-protocol.

Let MOD$_{\text{MUL}}$ and MOD$_{\text{TRUNC}}$ denote those parts of the MOD-protocol that are dominated by the MUL- and TRUNC-protocols respectively. The bit-complexity per player of MOD$_{\text{MUL}}$ is

$$O(k^2 n + k n^2 \log n), \quad \text{if MUL}_{old} \text{ is used,}$$
$$O(k^2 n + k n^2), \quad \text{if MUL}_{new} \text{ is used,}$$

whereas the bit-complexity of MOD$_{\text{TRUNC}}$ is always $O(k^2 n + k n^2 \log n)$.

6. The protocol EXPMOD assumes that the players want to compute polynomial shares (over $\mathbb{Z}_q$) of $c \equiv a^b \bmod p$, where a, b and p are shared secrets. The protocol uses the shared secrets of an approximation to $1/p$ as computed by the APPINV-protocol. It also assumes that the players hold shares of the bits of b as computed by the protocol I2Q-BIT. The EXPMOD-protocol does about $3k$ distributed multiplications and about $2k$ invocations of the MOD-protocol. Consequently, it separates into a fraction EXPMOD$_{\text{MUL}}$, that is dominated by the distributed multiplication protocol, and into a fraction EXPMOD$_{\text{TRUNC}}$, that is dominated by the TRUNC-protocol. For the former the bit-complexity per player is

$$O(k^3 n + k^2 n^2 \log n), \quad \text{if MUL}_{old} \text{ is used,}$$
$$O(k^3 n + k^2 n^2), \quad \text{if MUL}_{new} \text{ is used,}$$

whereas the bit-complexity for fraction EXPMOD$_{\text{TRUNC}}$ is always $O(k^3 n + k^2 n^2 \log n)$.

7. The protocol SETMEM allows the players to check whether a shared secret a equals one of of the shared secrets $\beta_1, \beta_2, \ldots, \beta_m$ modulo a shared secret p. All shares are polynomial over $\mathbb{Z}_q$. In the distributed primality test SETMEM is applied for $\beta_1 = +1$ and $\beta_2 = -1$, i.e. $m = 2$. In this case the cost of the protocol is dominated by approximately $6n$ invocations of the distributed multiplication protocol. Consequently, it requires

$$O(k^2 n^2 + k n^3 \log n), \quad \text{if MUL}_{old} \text{ is used,}$$
$$O(k^2 n^2 + k n^3), \quad \text{if MUL}_{new} \text{ is used}$$

bit operations per player.

The Miller-Rabin [12, 14] primality test is a probabilistic algorithm that takes as input a candidate p for a prime and investigates this number in a certain number (say ζ) of rounds. Each round gives at least a $3/4$ probability of finding a witness to the compositeness of p, if n is composite. These probabilites are independent, so if the test is run with $\zeta = 50$ rounds, then the probability that p is composite and the test never finds a witness to this compositeness is at most 2^{-100}. This pobability is sufficient for practical purposes in cryptography.

Let p_j be the additive shares (over the integers) of a candidate p for a prime. The structure of the distributed version of the Miller-Rabin test as developed by Algesheimer, Camenisch and Shoup [1] is as follows:

1. Set $b_1 = (p_1 - 1)/2$ and $b_j = p_j/2$ for $j \geq 2$.
2. One invocation of I2Q-BIT.
3. One invocation of SI2SQ, SQ2PQ and APPINV.
4. The following steps are repeated ζ times:
   a) Choose the shares of an additive sharing over the integers uniformly and at random from an appropriate set.
   b) One invocation of SI2SQ, SQ2PQ and MOD.
   c) One invoation of EXPMOD.
   d) One invocation of SETMEM (with $m = 2$). If the output of this protocol is `failure`, then stop and output `composite`.
5. Output `probably prime`.

A simple analysis based on the results presented above shows that the bit-complexity of the distributed Miller-Rabin test is dominated by the bit-complexity of the I2Q-BIT-protocol and ζ times the bit-complexities for the EXPMOD- and SETMEM-protocols. The cost of the first is unaffected by a shift from MUL_{old} to MUL_{new}, whereas the cost of the second and the last are advantageously reduced by such a shift. Nevertheless, for a reduction of the global order of the bit-complexity of the distributed Miller-Rabin test improvements of further subprotocols are necessary.

5 Remarks on Distributed Signatures

Using the algorithms described so far, it is possible to generate in a distributed manner a composite integer $N = pq$ with p and q primes or even safe primes. In other words a shared RSA modulus can be computed for which none of the players knows its factorization.

In many situations however, the parties are required to efficiently generate not only the modulus but also shares of the private exponent. This task is much less computationally involved than distributively generating the modulus N. In particular, Boneh and Franklin [3] and Catalano, Gennaro and Halevi [5] have presented efficient protocols to accomplish this. One of the main applications of these results is the construction of theshold variants of signature schemes. In such a scheme n parties hold a $(t + 1)$-out-of-n sharing of the secret key. Only when at least $t + 1$ of them cooperate they can sign a given message. The reader is referred to [5], where two such signature schemes are constructed. The first is an appropriate variant of the signature scheme of Gennaro, Halevi and Rabin [8]; the second relies on the signature scheme of Cramer and Shoup [7].

References

[1]   J. Algesheimer, J. Camenisch, and V. Shoup. Efficient computation modulo a shared secret with application to the generation of shared safe-prime products. In *Advances in Cryptology – CRYPTO 2002*, Lecture Notes in Computer Science 2442:417–432, Springer, Berlin, 2002.

[2]   M. Ben-Or, S. Goldwasser, and A. Wigderson. Completeness theorems for non-cryptographic fault-tolerant distributed computation. In *Proceedings of 20th Annual Symposium on Theory of Computing (STOC'88)*, 1–10, ACM Press, 1988.

[3]   D. Boneh and M. Franklin. Efficient generation of shared RSA keys. In *Advances in Cryptology – CRYPTO 1997*, Lecture Notes in Computer Science 1294:425–439, Springer, Berlin, 1997.

[4]   D. Catalano. Efficient distributed computation modulo a shared secret. In D. Catalano, R. Cramer, I. Damgård, G. Di Crescenco, D. Pointcheval, and T. Takagi (eds.) *Contemporary Cryptology*, Advanced Courses in Mathematics CRM Barcelona, 1–39, Birkhäuser, Basel, 2005.

[5]   D. Catalano, R. Gennaro, and S. Halevi. Computing inverses over a shared secret. In *Proceedings of EUROCRYPT 2000*, Lecture Notes in Computer Science 1807:190–206, Springer, Berlin, 2000.

[6]   D. Chaum, C. Crépeau, and I. Damgård. Multiparty unconditionally secure protocols. In *Proceedings of 20th Annual Symposium on Theory of Computing (STOC'88)*, 11–19, ACM Press, 1988.

[7]   R. Cramer and V. Shoup. Signature schemes based on the Strong RSA Assumption. *ACM Transactions on Information and System Security (ACM TISSEC)*, 3(3):161-185, 2000.

[8]   R. Gennaro, S. Halevi, and T. Rabin. Secure hash-and-sign signatures without the random oracle. In *Advances in Cryptology – CRYPTO 1999*, Lecture Notes in Computer Science 1592:123–139, Springer, Berlin, 1999.

[9]   R. Gennaro, M. O. Rabin, and T. Rabin. Simplified VSS and fast-track multyparty computations with applications to threshold cryptography. In *Proceedings of the 17th ACM Symposium on Principles of Distributed Computing (PODC'98)*, 1998.

[10]   O. Goldreich, S. Micali, and A. Wigderson. How to play any mental game. In *Proceedings of 19th Annual Symposium on Theory of Computing (STOC'87)*, 218–229, ACM Press, 1987.

[11]   P. Lory. Reducing the complexity in the distributed multiplication protocol of two polynomially shared values. In *Proceedings of 3rd IEEE International Symosium on Security in Networks and Distributed Systems, May 21-23, 2007*, IEEE Computer Society, submitted.

[12]   G. L. Miller. Riemann's hypothesis and tests for primality. *Journal of Computers and System Sciences*, 13:30–317, 1976.

[13]   I. Newton. *Methodus differentialis*, manuscript mentioned in a letter of Oct. 1676. Published in *Analysis per quantitatum series, fluxiones, ac differentias*, W. Jones, London, 1711.

[14]   M. O. Rabin. Probabilistic algorithms for testing primality. *Journal of Number Theory*, 12:128–138, 1980.

[15]   R. Rivest, A. Shamir, and L. Adleman. A method for obtaining digital signatures and public key cryptosystems. *Communications of the ACM*, 21(2):120–126, 1978.

[16] A. Shamir. How to share a secret. *Communications of the ACM*, 22(11):612–613, 1979.

[17] D. R. Stinson. *Cryptography – Theory and Practice*, Chapman & Hall/CRC, Boca Raton (FL), 2006.

[18] J. Stoer and R. Bulirsch. *Introduction to Numerical Analysis*, Springer, Berlin, 2002.

[19] A. C. Yao. How to generate and exchange secrets. In *Proceedings of 27th IEEE Symposium on Foundations of Computer Science (FOCS'86)*, 162–167, IEEE Computer Society, 1986.

Mathematics and Applications in Space

Techniques and Applications in Space

Optimal Control of Free-Floating Spin-Stabilized Space Robotic Systems

R. Callies[1] and Ch. Sonner[2]

[1] Zentrum Mathematik M2, Technische Universität München, Boltzmannstr. 3, 85748 Garching, Germany, `callies@ma.tum.de`
[2] Kuka Roboter GmbH, Zugspitzstr. 140, 86165 Augsburg, Germany, `christian.sonner@web.de`

Summary. Future robotic manipulators mounted on small satellites are expected to perform important tasks in space, like servicing other satellites or automating extravehicular activities on a space station. In a free-floating space manipulator, the motion of the manipulator affects the carrier satellite's position and attitude. Spin-stabilization improves the system performance, but further complicates the dynamics. The combination of satellite and multi-link manipulator is modeled as a rigid multi-body system. A Maximum Principle based approach is used to calculate optimal reference trajectories with high precision. To convert the optimal control problem into a nonlinear multi-point boundary value problem, complicated adjoint differential equations have to be formulated and the full apparatus of optimal control theory has to be applied. For that, an accurate and efficient access to first- and higher-order derivatives is crucial. The special modeling approach described in this paper allows it to generate all the derivative information in a structured and efficient way. Nonlinear state and control constraints are treated without simplification by transforming them into linear equations; they do not have to be calculated analytically. By these means, the modeling of the complete optimal control problem and the accompanying boundary value problem is automated to a great extent. The fast numerical solution is by the advanced multiple shooting method *JANUS*.

1 Introduction

In early 2007 satellite hopes ride on "Orbital Express". It is a flight experiment seeking to demonstrate the feasibility of extending the lives of spacecraft already in orbit by refueling or even upgrading them in space by free-flying space robotic systems. In such a system, and while the manipulator is active, the position and attitude of the carrier satellite is actively three-axes controlled by thrusters. In contrast, there is no active position and attitude control during manipulator activity in a free-floating system to conserve fuel, to avoid contamination with fuel exhausted and to keep system costs low. Additional attitude control by spin-stabilization will help to make future free-floating service systems more robust. For the model configuration of a rigid body manipulator mounted on a despun platform on top of a small spin-stabilized satellite the combined dynamic model is formulated in this paper in a structured way. Based on this model, optimal example motions are calculated by indirect methods with high accuracy.

Reviews of main approaches to the dynamic modeling of space robotic systems are given in [11, 17]. It is well-known in classical mechanics, that the motion of a rigid body system can be split up into the motion of the center of mass of the total system (CM) and the motion around the CM; this principle is widespread used in space applications. It proved to be especially efficient in combination with the Newton-Euler formalism [16, 23]. The approach was further developed by completely decoupling the translational motion of the CM [26]. The problem of dynamic singularities is treated in [22].

Optimal control applications focus on different major tasks. One is to fix the spacecraft's attitude and position by help of thrusters [12]. This requires a relatively large amount of fuel. A second task is only to control the attitude, e.g. by use of reaction wheels [20]. Another class of problems is the optimal control of free-floating systems, which is addressed by direct methods in [24] for a Shuttle mounted robot and in [14] for a more refined model including collision checking. Free-flying systems, which reach the desired location and orientation in space by help of thrusters, are treated in [25].

2 Modeling the Space Robotic System

An overall control-oriented modeling approach is presented combining elements from [17, 23, 22, 26] in a way well-suited for an efficient numerical calculation of the dynamic equations of motion and their partial derivatives.

2.1 Coordinate System

The satellite-mounted manipulator is modeled as a chain of n rigid links connected by revolute joints. The links are numbered starting from the satellite base (link 1). For every $i \in \{1, \ldots, n\}$ a reference frame $\{i\}$ is rigidly attached to link i according to the following convention: The z-axis of frame $\{i\}$ is coincident with the joint axis i, the x-axis is perpendicular to the z-axes of $\{i\}$ and $\{i+1\}$ and the y-axis is defined by the right hand rule to complete frame $\{i\}$. Using the Denavit-Hartenberg notation [9, 10] the transition from $\{i-1\}$ to $\{i\}$ is carried out by a rotation around x_{i-1} counterclockwise with the link twist α_i (angle between z_{i-1} and z_i), a translation along x_{i-1} by the link length a_i (distance from z_{i-1} to z_i), another counterclockwise rotation around z_i with the joint angle θ_i (angle between x_{i-1} and x_i) and a translation along z_i by the link offset d_i (distance from x_{i-1} to x_i).

Each revolute joint i is described via θ_i. The leading superscript $\langle j \rangle$ refers to frame $\{j\}$, so e.g. $^{\langle j \rangle}\mathbf{x}$ denotes the vector $\mathbf{x} \in \mathbb{R}^3$ expressed in an inertial frame which instantaneously has the same orientation as frame $\{j\}$. With

$$
{}^{\langle j-1\rangle}R_{\langle j\rangle} := \begin{pmatrix} \cos\theta_j & -\sin\theta_j & 0 \\ \cos\alpha_j\sin\theta_j & \cos\alpha_j\cos\theta_j & -\sin\alpha_j \\ \sin\alpha_j\sin\theta_j & \sin\alpha_j\cos\theta_j & \cos\alpha_j \end{pmatrix} \in \mathbb{R}^{3\times3}
$$

and the translation vector ${}^{\langle j-1\rangle}\mathbf{p}_j := (a_j, -d_j\sin\alpha_j, d_j\cos\alpha_j)^T \in \mathbb{R}^3$, the following transformation formula holds for $j \in \{2,\ldots,n\}$

$$
{}^{\langle j-1\rangle}\mathbf{x} = {}^{\langle j-1\rangle}R_{\langle j\rangle}\,{}^{\langle j\rangle}\mathbf{x} + {}^{\langle j-1\rangle}\mathbf{p}_j\,, \quad {}^{\langle j-1\rangle}R_{\langle j\rangle} = {}^{\langle j-1\rangle}R_{\langle j\rangle}(\theta_j) = {}^{\langle j\rangle}R_{\langle j-1\rangle}{}^T(\theta_j)\,.
$$

All quantities with the leading superscript $\langle 0\rangle$ or without a leading superscript are expressed in a frame with the same orientation as the inertial frame.

2.2 Geometrical Description

Let m_i be the mass and $\boldsymbol{\rho}_i$ the position vector of the center of mass of link i (CM i) with respect to an inertial system, $i = 1,\ldots,n$. Let $\boldsymbol{\rho}_c$ denote the position of the CM of the total system and $\mathbf{r}_i$ the difference vector between $\boldsymbol{\rho}_i$ and $\boldsymbol{\rho}_c$

$$
\boldsymbol{\rho}_c = \frac{1}{M}\sum_{i=1}^{n} m_i\boldsymbol{\rho}_i \;\Rightarrow\; \sum_{i=1}^{n} m_i\mathbf{r}_i = \mathbf{0}; \quad M := \sum_{i=1}^{n} m_i\,, \quad \mathbf{r}_i := \boldsymbol{\rho}_i - \boldsymbol{\rho}_c\,. \tag{1}
$$

With $\mathbf{r}_k^-$ and $\mathbf{r}_k^+$ defined as the vectors from $\boldsymbol{\rho}_k$ to joint k and $k+1$, respectively

$$
\mathbf{r}_i = \mathbf{r}_1 + \sum_{k=1}^{i-1}\left(\mathbf{r}_k^+ - \mathbf{r}_{k+1}^-\right)\,, \quad \mathbf{r}_i - \mathbf{r}_{i-1} = \mathbf{r}_{i-1}^+ - \mathbf{r}_i^-\,, \quad i = 2,\ldots,n\,.
$$

The relative masses μ_i^- and μ_i^+ connected to link i via joint i and $i+1$ respectively are

$$
\mu_i^- := \frac{1}{M}\sum_{k=1}^{i-1} m_k\,, \quad \mu_i^+ := \frac{1}{M}\sum_{k=i+1}^{n} m_k\,.
$$

Using (1), one gets with the i-th link barycenter $\mathbf{r}_i^b := \mu_i^-\mathbf{r}_i^- + \mu_i^+\mathbf{r}_i^+$

$$
\mathbf{r}_i = \sum_{k=1}^{i-1}\left(\mathbf{r}_k^+ - \mathbf{r}_k^b\right) - \mathbf{r}_i^b + \sum_{k=i+1}^{n}\left(\mathbf{r}_k^- - \mathbf{r}_k^b\right)\,.
$$

Introducing $\mathbf{y}_{k,i}$, this expression can be written more compactly

$$
r_i = \sum_{k=1}^{n}\mathbf{y}_{k,i}\,, \quad \mathbf{y}_{k,i} := \begin{cases} \mathbf{r}_k^+ - \mathbf{r}_k^b & \text{for } k < i\,, \\ -\mathbf{r}_i^b & \text{for } k = i\,, \\ \mathbf{r}_k^- - \mathbf{r}_k^b & \text{for } k > i\,. \end{cases} \tag{2}
$$

The following relations hold: $\sum_{j=1}^{n} m_j\,\mathbf{y}_{k,j} = 0\,, \quad k \in \{1,\ldots,n\}\,.$

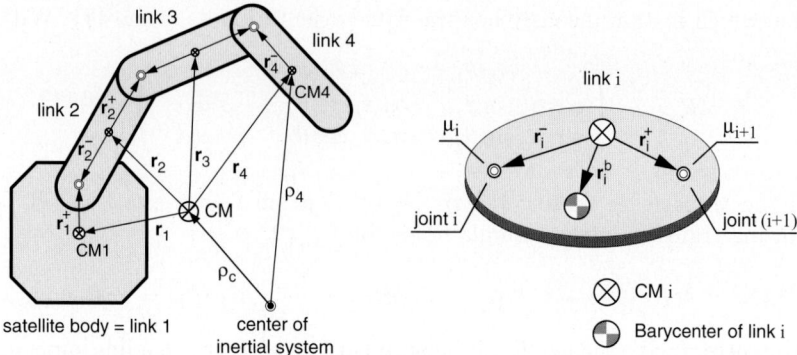

Fig. 1. Schematic view of the geometry and definitions of vectors.

2.3 Manipulator Dynamics in Space

Applying Newton's law to the i-th link yields for the linear motion

$$m_i \ddot{\boldsymbol{\rho}}_i = \mathbf{F}_{i,ex} + \mathbf{f}_i - \mathbf{f}_{i+1}, \quad i = 1, \dots, n, \quad \mathbf{f}_{n+1} = \mathbf{0}. \tag{3}$$

Time $t \in [t_0, t_f]$ is the independent variable. Dots indicate time derivatives. $\mathbf{f}_i$ denotes the force exerted on link i by link $i-1$ and $\mathbf{F}_{i,ex}$ the external force acting on CM i. For the linear motion of the CM, one gets from Newton's law

$$M \ddot{\boldsymbol{\rho}}_c = \sum_{i=1}^n \mathbf{F}_{i,ex} =: \mathbf{F}_{ex}. \tag{4}$$

¿From (1) one obtains $\sum_{i=1}^n m_i \ddot{\mathbf{r}}_i = 0$. With this expression and $\ddot{\mathbf{r}}_i = \ddot{\boldsymbol{\rho}}_i - \ddot{\boldsymbol{\rho}}_c$, summing up the Eqs. (3) and adding (4) yields $\mathbf{f}_1 = \mathbf{0}$. This is equivalent to

$$\mathbf{f}_i = \sum_{k=1}^{i-1} \left(\mathbf{F}_{k,ex} - \frac{m_k}{M} \mathbf{F}_{ex} - m_k \ddot{\mathbf{r}}_k \right) \tag{5}$$

$$= -\sum_{k=i}^n \left(\mathbf{F}_{k,ex} - \frac{m_k}{M} \mathbf{F}_{ex} - m_k \ddot{\mathbf{r}}_k \right). \tag{6}$$

The rotational motion of the i-th link is described by Euler's equation

$$I_i \dot{\boldsymbol{\omega}}_i + \boldsymbol{\omega}_i \times I_i \boldsymbol{\omega}_i = \mathbf{N}_{i,ex} + \tilde{\mathbf{n}}_i - \tilde{\mathbf{n}}_{i+1} + \left(\mathbf{r}_i^- \times \mathbf{f}_i - \mathbf{r}_i^+ \times \mathbf{f}_{i+1} \right), i = 1, \dots, n. \tag{7}$$

According to the definition of the geometry, $\tilde{\mathbf{n}}_1 = \mathbf{0}$ and $\tilde{\mathbf{n}}_{n+1} = \mathbf{0}$. $\boldsymbol{\omega}_i$ is the angular velocity of frame $\{i\}$ and I_i the tensor of inertia of link i with respect to CM i. $\tilde{\mathbf{n}}_i$ is the torque exerted on link i by link $i-1$ and $\mathbf{N}_{i,ex}$ the external torque acting on CM i.

(5) is substituted for $\mathbf{f}_i$ and (6) for $\mathbf{f}_{i+1}$ in (7). Then the $\ddot{\mathbf{r}}_k$ are eliminated using (2) and $\dot{\mathbf{y}}_{k,j} = \boldsymbol{\omega}_k \times \mathbf{y}_{k,j}$. Reordering the resulting terms in (7) with respect to $\dot{\boldsymbol{\omega}}_i$ and $\boldsymbol{\omega}_i$ finally yields

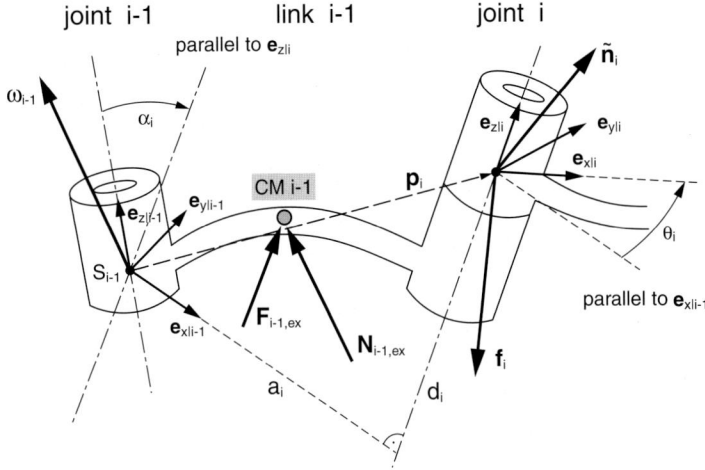

Fig. 2. Revolute joints: link frames, link parameters, force and torque vectors. $\mathbf{e}_{x|i}, \mathbf{e}_{y|i}, \mathbf{e}_{z|i}$ are the unit vectors in x, y, z-direction of frame $\{i\}$.

$$\sum_{k=1}^{n} \mathcal{I}_{i,k} \dot{\boldsymbol{\omega}}_k = \mathbf{N}_{i,ex} + \tilde{\mathbf{n}}_i - \tilde{\mathbf{n}}_{i+1} + \sum_{k=1}^{n} \left(\mathbf{y}_{i,k} \times \mathbf{F}_{k,ex} - \boldsymbol{\omega}_k \times \mathcal{I}_{i,k} \boldsymbol{\omega}_k \right), \quad (8)$$

$$\mathcal{I}_{i,k} := \begin{cases} I_i + \sum_{j=1}^{n} m_j \left(\mathbf{y}_{i,j}^T \mathbf{y}_{i,j} \mathbf{E}_3 - \mathbf{y}_{i,j} \mathbf{y}_{i,j}^T \right), & i = k, \\ -M \left(\mathbf{y}_{i,k}^T \mathbf{y}_{k,i} \mathbf{E}_3 - \mathbf{y}_{k,i} \mathbf{y}_{i,k}^T \right), & i \neq k, \end{cases} \quad (9)$$

for $i = 1, \ldots, n$. $\mathbf{E}_k = \mathrm{diag}\,(1, \ldots, 1) \in \mathbb{R}^{k \times k}$ is the k-dimensional unit matrix. The $\boldsymbol{\omega}_i$ are expressed by

$$\boldsymbol{\omega}_i = \boldsymbol{\omega}_1 + \sum_{k=2}^{i} \dot{\theta}_k \mathbf{g}_k \,. \quad (10)$$

$\dot{\theta}_i$ denotes the rotational velocity at the revolute joint i and $\mathbf{g}_i$ with $\|\mathbf{g}_i\|_2 = 1$ the normalized direction of the respective rotational axis. Insertion into (8) with $\mathbf{z}_k := \sum_{j=2}^{k} \dot{\theta}_j \dot{\mathbf{g}}_j = \sum_{j=2}^{k} \boldsymbol{\omega}_{j-1} \times \dot{\theta}_j \mathbf{g}_j$ leads to

$$\sum_{k=1}^{n} \mathcal{I}_{i,k} \left(\dot{\boldsymbol{\omega}}_1 + \sum_{j=2}^{k} \ddot{\theta}_j \mathbf{g}_j \right) = \mathbf{N}_{i,ex} + \tilde{\mathbf{n}}_i - \tilde{\mathbf{n}}_{i+1} + \sum_{k=1}^{n} \left(\mathbf{y}_{i,k} \times \mathbf{F}_{k,ex} \right)$$
$$- \sum_{k=2}^{n} \mathcal{I}_{i,k} \mathbf{z}_k - \sum_{k=1}^{n} \left(\boldsymbol{\omega}_k \times \mathcal{I}_{i,k} \boldsymbol{\omega}_k \right) \,. \quad (11)$$

The coupling torques $\tilde{\mathbf{n}}_i$ can be split up into two orthogonal components: The driving torque $u_i \mathbf{g}_i$ around the rotational axis and the torsion force $\boldsymbol{\kappa}_i$

$$\tilde{\mathbf{n}}_i = u_i \mathbf{g}_i + \boldsymbol{\kappa}_i \,, \quad \mathbf{g}_i^T \cdot \boldsymbol{\kappa}_i = 0 \,.$$

The manipulator motion is controlled via the motor torques u_i, $i = 2, \ldots, n$. Summing up (11) from $i = 1$ to $i = n$ yields

$$a_{1,1}\dot{\boldsymbol{\omega}}_1 + \sum_{k=2}^{n} \ddot{\theta}_k \mathbf{a}_{1,k} = \sum_{i=1}^{n} \boldsymbol{\mathcal{N}}_i. \tag{12}$$

Here

$$\boldsymbol{\mathcal{N}}_i := \mathbf{N}_{i,ex} + \sum_{k=1}^{n} \left(\mathbf{y}_{i,k} \times \mathbf{F}_{k,ex} - \boldsymbol{\omega}_k \times \mathcal{I}_{i,k}\boldsymbol{\omega}_k \right) - \sum_{k=2}^{n} \mathcal{I}_{i,k}\mathbf{z}_k \tag{13}$$

and $a_{1,1} := b_{1,1}$, $\mathbf{a}_{1,k} := \mathbf{b}_{1,k}$ with

$$b_{j,1} := \sum_{i=j}^{n} \sum_{l=1}^{n} \mathcal{I}_{i,l} \in \mathbb{R}^{3 \times 3}, \quad \mathbf{b}_{j,k} := \left(\sum_{i=j}^{n} \sum_{l=k}^{n} \mathcal{I}_{i,l} \right) \mathbf{g}_k. \tag{14}$$

Analogously summing up (11) from $i = j$ to $i = n$ for $j \in \{2, \ldots, n\}$ and multiplying the result with $\mathbf{g}_j^T$ from the left results in

$$\mathbf{a}_{j,1}^T \dot{\boldsymbol{\omega}}_1 + \sum_{k=2}^{n} \ddot{\theta}_k a_{j,k} = u_j + \mathbf{g}_j^T \sum_{i=j}^{n} \boldsymbol{\mathcal{N}}_i, \quad \mathbf{a}_{j,1}^T := \mathbf{g}_j^T b_{j,1}, \, a_{j,k} := \mathbf{g}_j^T \mathbf{b}_{j,k}. \tag{15}$$

The relative orientation of the first body of the space robotic system with respect to the inertial reference frame is defined in terms of the three Eulerian angles yaw ψ, pitch ϑ and roll ϕ. One transforms from the inertial frame to the body fixed frame by rotating about the vertical z-axis by ψ, then rotating about the node vector (= new y-axis) by ϑ, and finally rotating about the x-axis of the body fixed frame by ϕ [19]

$$\boldsymbol{\omega}_1 = {}^{\langle 0 \rangle}\boldsymbol{\omega}_1 = \begin{pmatrix} \omega_{1,x} \\ \omega_{1,y} \\ \omega_{1,z} \end{pmatrix} = \Omega \begin{pmatrix} \dot{\psi} \\ \dot{\vartheta} \\ \dot{\phi} \end{pmatrix}, \, \Omega := \begin{pmatrix} 0 & -\sin(\psi) & \sin(\psi)\cos(\vartheta) \\ 0 & \cos(\psi) & \cos(\psi)\cos(\vartheta) \\ 1 & 0 & -\sin(\vartheta) \end{pmatrix}. \tag{16}$$

${}^{\langle 0 \rangle}R_{\langle 1 \rangle}$ can be written as the product of three elementary rotation matrices

$${}^{\langle 0 \rangle}R_{\langle 1 \rangle} = \begin{pmatrix} \cos\psi & -\sin\psi & 0 \\ \sin\psi & \cos\psi & 0 \\ 0 & 0 & 1 \end{pmatrix} \begin{pmatrix} \cos\vartheta & 0 & \sin\vartheta \\ 0 & 1 & 0 \\ -\sin\vartheta & 0 & \cos\vartheta \end{pmatrix} \begin{pmatrix} 1 & 0 & 0 \\ 0 & \cos\phi & -\sin\phi \\ 0 & \sin\phi & \cos\phi \end{pmatrix}. \tag{17}$$

The quantities $\mathbf{y}_{i,k}$, I_i and $\mathbf{g}_i$ are constant in the body-fixed system $\{i\}$. For a transformation to the inertial system, they are multiplied by

$${}^{\langle 0 \rangle}R_{\langle i \rangle} := {}^{\langle 0 \rangle}R_{\langle 1 \rangle} \cdot \prod_{j=2}^{i} {}^{\langle j-1 \rangle}R_{\langle j \rangle}. \tag{18}$$

Using this strategy, (9,10) and (12,15) can be rewritten to allow an efficient numerical computation

$$\mathcal{I}_{i,i} = {}^{\langle 0 \rangle}R_{\langle i \rangle}\left({}^{\langle i \rangle}I_i + \sum_{j=1}^{n} m_j \left({}^{\langle i \rangle}\mathbf{y}_{i,j}^T {}^{\langle i \rangle}\mathbf{y}_{i,j}\mathbf{E}_3 - {}^{\langle i \rangle}\mathbf{y}_{i,j} {}^{\langle i \rangle}\mathbf{y}_{i,j}^T \right) \right) {}^{\langle 0 \rangle}R_{\langle i \rangle}^T, \quad (19)$$

$$\mathcal{I}_{i,k} = -M\left(\mathbf{y}_{i,k}^T \mathbf{y}_{k,i}\mathbf{E}_3 - \mathbf{y}_{k,i}\mathbf{y}_{i,k}^T \right), \quad \mathbf{y}_{i,k} = {}^{\langle 0 \rangle}R_{\langle i \rangle} {}^{\langle i \rangle}\mathbf{y}_{i,k}. \quad (20)$$

The expression in parenthesis in (19) can be precomputed. (20) is reduced to vector operations. With (2), only $(2n-2)$ matrix-vector multiplications ${}^{\langle 0 \rangle}R_{\langle i \rangle} {}^{\langle i \rangle}\mathbf{y}_{i,k}$ have to be performed in (20). In addition, $\mathcal{I}_{i,k} = \mathcal{I}_{k,i}^T$.

Inserting the results into (14) and defining ${}^{\langle k \rangle}\mathbf{e}_{z|k} := (0,0,1)^T$ yields

$$b_{j,1} := \sum_{i=j}^{n}\sum_{l=1}^{n}\mathcal{I}_{i,l}, \quad \mathbf{b}_{j,k} := \left(\sum_{i=j}^{n}\sum_{l=k}^{n}\mathcal{I}_{i,l} \right) \left({}^{\langle 0 \rangle}R_{\langle k \rangle} {}^{\langle k \rangle}\mathbf{e}_{z|k} \right), \quad (21)$$

$$\mathbf{a}_{j,1}^T = \left({}^{\langle 0 \rangle}R_{\langle j \rangle} {}^{\langle j \rangle}\mathbf{e}_{z|j} \right)^T b_{j,1}, \quad a_{j,k} = \left({}^{\langle 0 \rangle}R_{\langle j \rangle} {}^{\langle j \rangle}\mathbf{e}_{z|j} \right)^T \mathbf{b}_{j,k}, \quad j = 2, \dots, n.$$

$\boldsymbol{\omega}_i$ and $\mathbf{z}_i$ – required for $\mathcal{N}_i$ in (13) – are computed recursively

$$\text{Initial values:} \quad \boldsymbol{\omega}_1 := {}^{\langle 0 \rangle}R_{\langle 1 \rangle} {}^{\langle 1 \rangle}\boldsymbol{\omega}_1, \quad \mathbf{z}_1 := 0.$$

$$i: 2 \rightarrow n: \quad \boldsymbol{\omega}_i := \boldsymbol{\omega}_{i-1} + \left(\dot{\theta}_i {}^{\langle 0 \rangle}R_{\langle i \rangle} {}^{\langle i \rangle}\mathbf{e}_{z|i} \right)$$

$$\mathbf{z}_i := \mathbf{z}_{i-1} + \boldsymbol{\omega}_{i-1} \times \left(\dot{\theta}_i {}^{\langle 0 \rangle}R_{\langle i \rangle} {}^{\langle i \rangle}\mathbf{e}_{z|i} \right)$$

Analogously, a backward recursion can be applied for the calculation of the $\left(\sum_{i=j}^{n}\sum_{l=k}^{n}\mathcal{I}_{i,l} \right)$ in $\mathbf{b}_{j,k}$, starting from $\mathcal{I}_{n,n}$ $(j=k=n)$.

With these quantities and the equations of the linear motion of the CM, the equations of motion of the total space robotic system are obtained

$$\mathcal{M}\dot{\mathbf{x}} = \begin{pmatrix} \tilde{\mathbf{x}}_2 \\ \tilde{\mathcal{N}} \end{pmatrix} + \begin{pmatrix} \mathbf{0} \\ \tilde{\mathbf{u}} \end{pmatrix}, \quad \mathcal{M} \in \mathbb{R}^{(2n+10)\times(2n+10)}. \quad (22)$$

Due to the special approach chosen, the linear motion and the rotational motion are completely decoupled.

The following definitions have been used for the state vectors

$$\tilde{\mathbf{x}}_1 := (\boldsymbol{\rho}_c, \psi, \vartheta, \phi, \theta_2, \dots, \theta_n)^T, \quad \tilde{\mathbf{x}}_2 := \left(\dot{\boldsymbol{\rho}}_c, \boldsymbol{\omega}_1, \dot{\theta}_2, \dots, \dot{\theta}_n \right)^T, \quad \mathbf{x} := (\tilde{\mathbf{x}}_1, \tilde{\mathbf{x}}_2)^T,$$

the controls and the force- and torque-dependent terms

$$\tilde{\mathbf{u}} := \begin{pmatrix} \mathbf{0} \\ \mathbf{0} \\ u_2 \\ \vdots \\ u_n \end{pmatrix}, \quad \tilde{\mathcal{N}} := \begin{pmatrix} \sum_{k=1}^{n}\mathbf{F}_{k,ex} \\ \sum_{k=1}^{n}\mathcal{N}_k \\ \left({}^{\langle 0 \rangle}R_{\langle 2 \rangle} {}^{\langle 2 \rangle}\mathbf{e}_{z|2} \right)^T \sum_{k=2}^{n}\mathcal{N}_k \\ \vdots \\ \left({}^{\langle 0 \rangle}R_{\langle n \rangle} {}^{\langle n \rangle}\mathbf{e}_{z|n} \right)^T \mathcal{N}_n \end{pmatrix} \quad (23)$$

and the extended manipulator joint space inertia matrix

$$
\mathcal{M} := \begin{pmatrix} \mathbf{E}_3 & & & \\ & \Omega & & \\ & & \mathbf{E}_{n-1} & \\ \hline & & M\mathbf{E}_3 & \\ & & & A \end{pmatrix} , \quad A := \begin{pmatrix} a_{11} & \mathbf{a}_{12} \cdots \mathbf{a}_{1n} \\ \hline \mathbf{a}_{21}^T & \\ \vdots & (a_{jk})_{2 \le j,k \le n} \\ \mathbf{a}_{n1}^T & \end{pmatrix} . \tag{24}
$$

2.4 6-DOF Example System

Fig. 3 shows a free-floating space robot consisting of a small spin-stabilized satellite (body 1) and a 3-DOF manipulator (bodies 2,3,4) with revolute joints, mounted on a despun platform on top of the satellite. The rotational axes $\mathbf{e}_{z|3}$ and $\mathbf{e}_{z|4}$ are parallel and perpendicular to $\mathbf{e}_{z|2}$. Table 1 summarizes the data for the example manipulator.

Motor torques are limited to

$$
u_i \in [-u_{i,max}, u_{i,max}], \quad u_{2,max} = 0.5 \text{ Nm}, \quad u_{3,max} = u_{4,max} = 1.0 \text{ Nm}.
$$

The structured modeling approach allows it easily to incorporate spin-stabilization with ω_0 [rpm] via the initial values

$$
{}^{\langle 1 \rangle}\boldsymbol{\omega}_1(\tau_0) \rightarrow {}^{\langle 1 \rangle}\boldsymbol{\omega}_1(\tau_0) + (0, 0, \omega_0/60)^T, \quad \dot{\theta}_2(\tau_0) \rightarrow \dot{\theta}_2(\tau_0) - \omega_0/60.
$$

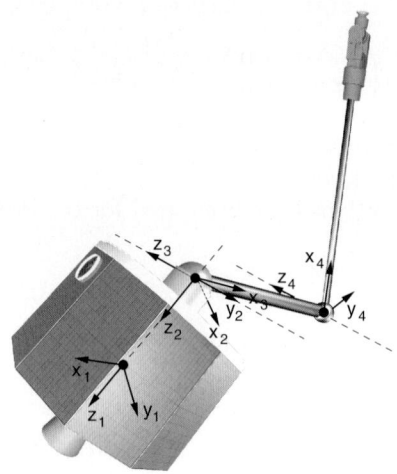

Fig. 3. The 6-DOF example system with body-fixed coordinate systems.

Table 1. Data for the example manipulator.

	Unit	Variable	Body 1	Link 2	Link 3	Link 4
Mass	[kg]	m	90.00	5.00	5.00	10.00
		r_x^+	0.00	0.00	0.50	1.00
Center of mass	[m]	r_y^+	0.00	0.00	0.00	0.00
$(\mathbf{r}_i^+ = -\mathbf{r}_i^-)$		r_z^+	−0.65	0.00	0.00	0.00
		I_{xx}	15.00	0.12	0.31	0.62
Moments of inertia	[kg m^2]	I_{yy}	15.00	0.12	0.57	3.60
$(I_{xy}=I_{xz}=I_{yz}=0.00)$		I_{zz}	18.00	0.10	0.57	3.60
	[m]	a		0.00	0.00	1.00
Denavit-Hartenberg	[rad]	α		0.00	$\pi/2$	0.00
parameters	[m]	d		−0.65	0.00	0.00
	[rad]	θ		θ_2	θ_3	$\theta_4 + \pi$

3 Calculation of Derivatives

3.1 Structural Considerations

Optimal control applications require a direct and efficient access to first and higher order derivatives. For that, structural information is used to a great extent in the present paper: The additive substructure of the individual parts of the model (22) – achieved by the special modeling technique – complements the typical linear basic structure of the equations of motion (22)

$$\mathcal{M}(\mathbf{x})\dot{\mathbf{x}} = \mathbf{h}(\mathbf{x}), \quad \mathbf{h}(\mathbf{x}) := \begin{pmatrix} \tilde{\mathbf{x}}_2 \\ \tilde{\mathcal{N}} \end{pmatrix} + \begin{pmatrix} \mathbf{0} \\ \tilde{\mathbf{u}} \end{pmatrix}. \tag{25}$$

Repeated partial differentiations of (25) result in the following recursive sequence of linear systems for the determination of $\dot{\mathbf{x}}$ and its derivatives [6]

$$\mathcal{M}(\mathbf{x})\,\dot{\mathbf{x}} = \mathbf{h}(\mathbf{x})\,,$$
$$\mathcal{M}(\mathbf{x})\frac{\partial \dot{\mathbf{x}}}{\partial x_j} = \frac{\partial \mathbf{h}(\mathbf{x})}{\partial x_j} - \frac{\partial \mathcal{M}(\mathbf{x})}{\partial x_j}\dot{\mathbf{x}}\,,$$
$$\mathcal{M}(\mathbf{x})\frac{\partial^2 \dot{\mathbf{x}}}{\partial x_j \partial x_k} = \frac{\partial^2 \mathbf{h}(\mathbf{x})}{\partial x_j \partial x_k} - \frac{\partial^2 \mathcal{M}(\mathbf{x})}{\partial x_j \partial x_k}\dot{\mathbf{x}} - \frac{\partial \mathcal{M}(\mathbf{x})}{\partial x_j}\frac{\partial \dot{\mathbf{x}}}{\partial x_k} - \frac{\partial \mathcal{M}(\mathbf{x})}{\partial x_k}\frac{\partial \dot{\mathbf{x}}}{\partial x_j}\,, \tag{26}$$

$$\ldots$$

The decomposition of $\mathcal{M}(\mathbf{x})$ has to be done only once, the subsequent steps are reduced to forward and backward substitutions. The special structure of $\mathcal{M}(\mathbf{x})$ is exploited to decrease the effort for the matrix decomposition by more than one order of magnitude.

3.2 Derivatives of Basic Building Blocks

For the rotational matrices $^{\langle 0\rangle}R_{\langle i\rangle}$ the first partial derivatives are

$$\frac{\partial^{\langle 0\rangle}R_{\langle i\rangle}}{\partial x_j} = \begin{cases} \left(\dfrac{\partial^{\langle 0\rangle}R_{\langle 1\rangle}}{\partial x_j}\right) \cdot \displaystyle\prod_{k=2}^{i} {}^{\langle k-1\rangle}R_{\langle k\rangle} & \text{, if } j = 4,5,6 \,, \\[4mm] \displaystyle\prod_{k=1}^{j-6} {}^{\langle k-1\rangle}R_{\langle k\rangle} \left(\dfrac{\partial^{\langle j-6\rangle}R_{\langle j-5\rangle}}{\partial x_j}\right) \displaystyle\prod_{k=j-4}^{i} {}^{\langle k-1\rangle}R_{\langle k\rangle} & \text{, if } j = 7,\ldots,n+5 \\ & \quad \wedge\; j \leq i+5 \,, \\[4mm] \mathbf{0}_3 & \text{, else.} \end{cases}$$

$\mathbf{0}_k \in \mathbb{R}^{k\times k}$ denotes the k-dimensional null matrix. The extra effort for the calculation of the derivatives is reduced by forming the $\partial^{\langle 0\rangle}R_{\langle i\rangle}/\partial x_j$ in parallel and simultaneously with $^{\langle 0\rangle}R_{\langle i\rangle}$ and splitting up every $^{\langle j-1\rangle}R_{\langle j\rangle}$ into two elementary rotations [13].

For the $\mathcal{I}_{i,i}$, $i = 1,\ldots,n$, the derivatives can be written

$$\frac{\partial \mathcal{I}_{i,i}}{\partial x_j} = \begin{cases} \tilde{\mathcal{I}} + \tilde{\mathcal{I}}^T, \ \tilde{\mathcal{I}} := \dfrac{\partial^{\langle 0\rangle}R_{\langle i\rangle}}{\partial x_j} {}^{\langle 0\rangle}R_{\langle i\rangle}^T \mathcal{I}_{i,i} & \begin{array}{l}\text{, if } j = 4,5,6 \text{ or} \\ (j = 7,\ldots,n+5 \wedge j \leq i+5), \end{array} \\[4mm] \mathbf{0}_3 & \text{, else.} \end{cases}$$

The partial derivatives of $\mathcal{I}_{i,k}$ with respect to x_j are non-zero for $j = 4,5,6$ or $(j \in \{7,\ldots,n+5\}) \wedge ((j \leq i+5) \vee (j \leq k+5))$ and are calculated similar to (20) by vector operations only

$$\mathbf{w}_a := \frac{\partial^{\langle 0\rangle}R_{\langle i\rangle}}{\partial x_j} {}^{\langle i\rangle}\mathbf{y}_{i,k}, \quad \mathbf{w}_b := \frac{\partial^{\langle 0\rangle}R_{\langle k\rangle}}{\partial x_j} {}^{\langle k\rangle}\mathbf{y}_{k,i}, \quad \mathbf{y}_{i,k} = {}^{\langle 0\rangle}R_{\langle i\rangle} {}^{\langle i\rangle}\mathbf{y}_{i,k},$$

$$\frac{\partial \mathcal{I}_{i,k}}{\partial x_j} = -M \left(\mathbf{w}_a^T \mathbf{y}_{k,i} + \mathbf{y}_{i,k}^T \mathbf{w}_b\right) \mathbf{E}_3 + M \left(\mathbf{w}_b \mathbf{y}_{i,k}^T + \mathbf{y}_{k,i} \mathbf{w}_a^T\right).$$

The partial derivatives of $\boldsymbol{\omega}_i$ and $\mathbf{z}_i$ are computed recursively by differentiating the recursion for $\boldsymbol{\omega}_i$ and $\mathbf{z}_i$. The new recursion reads as

$$\text{Initial values:} \quad \frac{\partial \boldsymbol{\omega}_1}{\partial x_j} := \frac{\partial \left({}^{\langle 0\rangle}R_{\langle 1\rangle} {}^{\langle 1\rangle}\boldsymbol{\omega}_1\right)}{\partial x_j}, \quad \frac{\partial \mathbf{z}_1}{\partial x_j} := \mathbf{0} \,.$$

$$i : 2 \to n : \quad \frac{\partial \boldsymbol{\omega}_i}{\partial x_j} := \frac{\partial \boldsymbol{\omega}_{i-1}}{\partial x_j} + \frac{\partial(\dot{\theta}_i {}^{\langle 0\rangle}R_{\langle i\rangle})}{\partial x_j} {}^{\langle i\rangle}\mathbf{e}_{z|i}$$

$$\frac{\partial \mathbf{z}_i}{\partial x_j} := \frac{\partial \mathbf{z}_{i-1}}{\partial x_j} + \frac{\partial \boldsymbol{\omega}_{i-1}}{\partial x_j} \times \left(\dot{\theta}_i {}^{\langle 0\rangle}R_{\langle i\rangle} {}^{\langle i\rangle}\mathbf{e}_{z|i}\right)$$

$$+ \boldsymbol{\omega}_{i-1} \times \frac{\partial(\dot{\theta}_i {}^{\langle 0\rangle}R_{\langle i\rangle})}{\partial x_j} {}^{\langle i\rangle}\mathbf{e}_{z|i}$$

Exploiting the special structure of the recursion for the various x_j further reduces the effort by a factor of greater than 3.

Higher order partial derivatives of the basic building blocks are derived analogously.

3.3 Derivatives of Macro Expressions

The partial derivatives of the macro expressions $\mathcal{M}(\tilde{\mathbf{x}}_1)$ and $\mathbf{h}(\mathbf{x})$ in (25) can be efficiently formed using the partial derivatives of the basic building blocks

$$
\frac{\partial \mathcal{M}}{\partial x_j} := \left(
\begin{array}{c|c}
\begin{array}{c} \mathbf{0}_3 \\ \dfrac{\partial \Omega}{\partial x_j} \\[4pt] \mathbf{0}_{n-1} \end{array} & \\ \hline
 & \begin{array}{c} \mathbf{0}_3 \\ \dfrac{\partial A}{\partial x_j} \end{array}
\end{array}
\right) , \quad j = 4, \ldots, n .
\tag{27}
$$

$\partial \Omega / \partial x_j$ is non-zero only for $j \in 4, 5$, the elements of $\partial A / \partial x_j$ are obtained from sums of the $\mathcal{I}_{i,k}$ and their derivatives, multiplied by vectors $^{\langle 0 \rangle}R_{\langle i \rangle}{}^{\langle i \rangle}\mathbf{e}_{z|i}$ and/or their derivatives, if necessary (cf. (21) and the following equations).

The main effort in calculating the derivatives of $\mathbf{h}(\mathbf{x})$ is the calculation of the partial derivatives of $\mathcal{N}_i$ in (23). From (13) one obtains

$$
\frac{\partial \mathcal{N}_i}{\partial x_j} := \sum_{k=1}^{n} \left(\frac{\partial {}^{\langle 0 \rangle}R_{\langle i \rangle}}{\partial x_j}{}^{\langle i \rangle}\mathbf{y}_{i,k} \times \mathbf{F}_{k,ex} - \frac{\partial \boldsymbol{\omega}_k}{\partial x_j} \times \mathcal{I}_{i,k}\boldsymbol{\omega}_k - \boldsymbol{\omega}_k \times \frac{\partial \mathcal{I}_{i,k}}{\partial x_j}\boldsymbol{\omega}_k \right.
$$
$$
\left. - \boldsymbol{\omega}_k \times \mathcal{I}_{i,k}\frac{\partial \boldsymbol{\omega}_k}{\partial x_j} \right) - \sum_{k=2}^{n} \left(\frac{\partial \mathcal{I}_{i,k}}{\partial x_j}\mathbf{z}_k + \mathcal{I}_{i,k}\frac{\partial \mathbf{z}_k}{\partial x_j} \right)
\tag{28}
$$

Again using the results from Sect. 3.2 significantly simplifies the procedure.

4 Optimal Control of the Space Robotic System

4.1 General Formulation

The equations of motion derived in Sect. 2 form a key element of the optimal control problem of the n-body space robotic system (here: $n = 4$). In a more abstract way the optimal control problem can be stated as follows:

Find a state function $\mathbf{x} : [\tau_0, \tau_F] \longrightarrow \mathbb{R}^{2n+10}$ and a control function $\mathbf{u} : [\tau_0, \tau_F] \longrightarrow U \subset \mathbb{R}^{n-1}$, which minimize the objective function $\int_{\tau_0}^{\tau_F} L(\mathbf{x}, \mathbf{u})\, dt$ subject to

$$
\dot{\mathbf{x}} = \mathbf{f}_i(\mathbf{x}, \mathbf{u}) \quad \text{for } t \in [\tau_i, \tau_{i+1}[, \; \tau_{\tilde{m}+1} =: \tau_F, \; i = 0, 1, \ldots, \tilde{m},
$$
$$
0 = \tilde{\mathbf{r}}(\tau_0, \mathbf{x}(\tau_0), \tau_F, \mathbf{x}(\tau_F)) \in \mathbb{R}^k, \; k \le 4n + 20
$$
$$
0 = \mathbf{q}_i(\tau_i, \mathbf{x}(\tau_i), \mathbf{x}(\tau_i^-), \mathbf{x}(\tau_i^+)) \in \mathbb{R}^{l_i},
$$
$$
0 \le C_{ij}(\mathbf{x}, \mathbf{u}), \; 1 \le j \le k_i; \quad \tilde{\mathbf{r}}, \mathbf{q}_i \text{ sufficiently smooth}; \; k, l_i, k_i \in \mathbb{N}.
$$

$\mathbf{x}$ is assumed to be an element of space $\mathcal{W}^{1,\infty}([\tau_0, \tau_F], \mathbb{R}^{2n+10})$ of the uniformly Lipschitz-continuous functions under the norm $\|\mathbf{x}\|_{1,\infty} := \|\mathbf{x}(\tau_0)\|_2 + \|\dot{\mathbf{x}}\|_\infty$;

$\|\cdot\|_2$ denotes the Euclidean norm and $\|\dot{\mathbf{x}}\|_\infty := \sup\{\|\dot{\mathbf{x}}(t)\|_2 \,|\, t \in [\tau_0, \tau_F]\}$. $\mathbf{u} := (u_2, \ldots, u_n)^T$ is assumed to be an element of space $L^\infty([\tau_0, \tau_F], \mathbb{R}^{n-1})$ of the bounded functions under the norm $\|\mathbf{u}\|_\infty$. $(\mathcal{W}^{1,\infty}([\tau_0, \tau_F], \mathbb{R}^{2n+10}), \|\cdot\|_{1,\infty})$ and $(L^\infty([\tau_0, \tau_F], \mathbb{R}^{n-1}), \|\cdot\|_\infty)$ are Banach spaces.

τ_0 is the initial time, τ_F the final time and the τ_i are intermediate times with interior point conditions. $\bar{U}_i := [\tau_i, \tau_{i+1}] \times \mathbb{R}^{2n+10} \times \mathbb{R}^{n-1}$ and $\mathbf{f}_i \in \mathcal{C}^3(\bar{U}_i, \mathbb{R}^{2n+10})$, $C_{ij} \in \mathcal{C}^{N_2}(\bar{U}_i, \mathbb{R})$, $N_2 \in \mathbb{N}$ sufficiently large. As usual we define $\mathbf{x}(\tau_j^\pm) := \lim_{\varepsilon \to 0, \varepsilon > 0} \mathbf{x}(\tau_j \pm \varepsilon)$.

4.2 Control-Constrained Problem

In a well-known manner (see e.g. [2, 21]) the problem of optimal control defined above is transformed into a multi-point boundary value problem.

In case of an unconstrained system, the following $(4n + 20)$-dimensional system of coupled nonlinear differential equations results

$$\dot{\mathbf{x}} = \mathbf{f}(\mathbf{x}, \mathbf{u}), \quad \dot{\boldsymbol{\lambda}} = -\mathbf{H}_{\mathbf{x}}(\mathbf{x}, \boldsymbol{\lambda}, \mathbf{u}). \tag{29}$$

$\boldsymbol{\lambda}$ denotes the vector of the adjoint variables, $\mathbf{f}$ the right-hand side of the system of differential equations of motion and

$$H(\mathbf{x}, \mathbf{u}, \boldsymbol{\lambda}) = L(\mathbf{x}, \mathbf{u}) + \boldsymbol{\lambda}^T \mathbf{f}(\mathbf{x}, \mathbf{u}) = L(\mathbf{x}, \mathbf{u}) + \boldsymbol{\lambda}^T \dot{\mathbf{x}} \tag{30}$$

the Hamiltonian. The controls $\mathbf{u}$ are derived from

$$\mathbf{H}_{\mathbf{u}}(\mathbf{x}, \boldsymbol{\lambda}, \mathbf{u}) = 0 \quad \wedge \quad H_{uu}(\mathbf{x}, \boldsymbol{\lambda}, \mathbf{u}) \quad \text{pos. semidefinite.}$$

The explicit form (29) of the equations of motion is achieved by solving the linear system (22) for $\dot{\mathbf{x}}$. The equations for the adjoint variables read

$$\dot{\lambda}_{x_j} = -\frac{\partial H}{\partial x_j} = -\frac{\partial L}{\partial x_j} - \boldsymbol{\lambda}^T \frac{\partial \dot{\mathbf{x}}}{\partial x_j}, \quad j = 1, \ldots, 2n + 10. \tag{31}$$

The subscript of each adjoint variable refers to the respective state variable. The partial derivatives $\partial \dot{\mathbf{x}} / \partial x_j$ are obtained from the recursion (26).

For control constraints of type $u_i \in [u_{i,min}, u_{i,max}]$, $i = 2, \ldots, n$, – for which the set of admissible controls is a convex polyhedron – one gets

$$u_i = \begin{cases} u_{i,min} & , \text{if } S_i > 0 \\ u_{i,max} & , \text{if } S_i < 0 \end{cases}.$$

The switching function S_i is defined by $S_i := H_{u_i} = \partial L / \partial u_i - \boldsymbol{\lambda}^T \partial \dot{\mathbf{x}} / \partial u_i$, the zeros of S_i are called the switching points. $\partial \dot{\mathbf{x}} / \partial u_i$ is obtained from the special linear system with the matrix $\mathcal{M}(\mathbf{x})$ already decomposed

$$\mathcal{M}(\mathbf{x}) \frac{\partial \dot{\mathbf{x}}}{\partial u_i} = \tilde{\mathbf{h}}_i, \quad \left(\tilde{\mathbf{h}}_i\right)_j = \begin{cases} 1 & , \text{if } j = n + 10 + i, \\ 0 & , \text{else.} \end{cases} \tag{32}$$

For $S_i(t) \equiv 0 \quad \forall t \in [t_1, t_2]$, $\tau_0 < t_1 < t_2 < \tau_F$, $i \in \{2, \ldots, n\}$, singular control exists [21]. The numerical treatment is similar to that of state constraints.

4.3 State-Constrained Problem

In addition to control constraints, state constraints can be active. In case of one active scalar constraint

$$0 = C(\mathbf{x}(t)) \in \mathbb{R} \quad \forall\, t \in [t_1, t_2] \subset\,]\tau_0, \tau_F[\,,$$

the solution procedure is briefly outlined. Total time derivatives $C^{(j)} :=$ $\mathrm{d}^j C/\mathrm{d}t^j$ have to be calculated, until at $j = k$

$$\exists\, i \in \{2, \dots, n\} \ni \frac{\partial}{\partial u_i} C^{(k)} \neq 0$$

$$\wedge \quad \frac{\partial}{\partial u_\nu} C^{(j)} = 0 \quad \forall\, j \in \{0, 1, \dots, k-1\}, \ \forall\, \nu \in \{2, \dots, n\}\,.$$

One component of $\mathbf{u}$ is then determined from $C^{(k)} \equiv 0$; without loss of generality, let this component be u_2. k is called the order of the state constraint.

General constraints are coupled to the Hamiltonian of the unconstrained problem by Lagrangian multiplier functions (see e.g. [2, 18]) to define the augmented Hamiltonian $\tilde{H}$. One way to define $\tilde{H}$ is

$$\tilde{H}(\mathbf{x}, \mathbf{u}, \boldsymbol{\lambda}, \mu) := H(\mathbf{x}, \mathbf{u}, \boldsymbol{\lambda}) + \mu C^{(k)}(\mathbf{x}) \tag{33}$$

with the multiplier function $\mu(t)$.

Appropriate interior point conditions have to be added at the beginning and the end of a constrained arc, sign-conditions concerning the multiplier functions have to be checked (for a survey see e.g. [15]).

Optimal control problems with singular controls or state constraints can be regarded as differential-algebraic systems [8] and at least partly reformulated into minimum coordinates [1, 5, 7].

The core element in the determination of u_2 from the constraint is the efficient calculation of the total time derivatives. In case of $k = 1$ one gets

$$0 = \frac{\mathrm{d}}{\mathrm{d}t} C(\mathbf{x}) = \sum_{i=1}^{2n+10} \frac{\partial C(\mathbf{x})}{\partial x_i} \dot{x}_i \quad \wedge \quad \frac{\partial}{\partial u_2}\left(\frac{\mathrm{d}}{\mathrm{d}t} C(\mathbf{x})\right) \neq 0\,. \tag{34}$$

The unknowns $(\dot{\mathbf{x}}, u_2) \in \mathbb{R}^{2n+11}$ are the solution of the inhomogeneous linear system formed by (22) and (34). An explicit analytical expression for u_2 is not necessary. For the complete control formalism, $\tilde{H}$ from (33) is differentiated to obtain the differential equations for $\boldsymbol{\lambda}$ and the information on $\mu(t)$. The following derivatives have to be calculated in addition

$$\frac{\partial}{\partial x_j}\left(\frac{\mathrm{d}}{\mathrm{d}t} C(\mathbf{x})\right) = \left(\frac{\partial C(\mathbf{x})}{\partial x_1}, \dots, \frac{\partial C(\mathbf{x})}{\partial x_{2n+10}}\right) \cdot \frac{\partial \dot{\mathbf{x}}}{\partial x_j} + \sum_{i=1}^{2n+10} \frac{\partial^2 C(\mathbf{x})}{\partial x_i \partial x_j} \dot{x}_i\,, \tag{35}$$

$$\frac{\partial}{\partial u_k}\left(\frac{\mathrm{d}}{\mathrm{d}t} C(\mathbf{x})\right) = \left(\frac{\partial C(\mathbf{x})}{\partial x_1}, \dots, \frac{\partial C(\mathbf{x})}{\partial x_{2n+10}}\right) \cdot \frac{\partial \dot{\mathbf{x}}}{\partial u_k}\,. \tag{36}$$

For $\partial \dot{\mathbf{x}}/\partial x_j$ and $\partial \dot{\mathbf{x}}/\partial u_k$ the values from (26,32) are inserted. Higher order state constraints are calculated analogously [6].

5 Numerical Results

For the 6-DOF example system specified in Sect. 2.4 time-optimal point-to-point trajectories are calculated. Initial spin-stabilization is with 5 rpm, the manipulator platform is completely despun at τ_0.

Besides the control constraints, state constraints of order 1 have been imposed on the joint velocities and become active on various subarcs

$$C_i(\mathbf{x}(t)) := \dot{\theta}_i^2 - 0.4^2 \le 0, \quad i = 2, 3, 4. \tag{37}$$

In case of $C_i(\mathbf{x}(t))$ active, (34) can be written

$$2\dot{\theta}_i\ddot{\theta}_i = 0 \quad \Leftrightarrow \quad \ddot{\theta}_i = 0 \quad \text{for} \quad \dot{\theta}_i \ne 0.$$

The simplification is possible, because the constraint can be active only for $\dot{\theta}_i \ne 0$. The expressions in (35,36) decrease to

$$\frac{\partial}{\partial x_j}\left(\frac{\mathrm{d}}{\mathrm{d}t}C_i(\mathbf{x})\right) = 2\dot{\theta}_i\frac{\partial\ddot{\theta}_i}{\partial x_j}, \quad \frac{\partial}{\partial u_k}\left(\frac{\mathrm{d}}{\mathrm{d}t}C_i(\mathbf{x})\right) = 2\dot{\theta}_i\frac{\partial\ddot{\theta}_i}{\partial u_k}.$$

With the augmented Hamiltonian (33) the switching and jump conditions for a constrained arc $[t_1, t_2] \in [\tau_0, \tau_F]$ read as

$$C_i(\mathbf{x})\big|_{t_1^-} = 0, \quad \tilde{H}\Big|_{t_1^-} = \tilde{H}\Big|_{t_1^+}, \quad \lambda_{x_j}(t_1^-) = \lambda_{x_j}(t_1^+) \; \forall j \wedge (j \ne n + 10 + i),$$

$$\tilde{H}\Big|_{t_2^-} = \tilde{H}\Big|_{t_2^+}, \quad \boldsymbol{\lambda}(t_2^-) = \boldsymbol{\lambda}(t_2^+).$$

The following sign conditions have to be fulfilled

$$\mu(t) \ge 0, \quad \text{sign}\left(\dot{\theta}_i(t_1^-)\right)\left(\lambda_{x_{n+10+i}}(t_1^-) - \lambda_{x_{n+10+i}}(t_1^+)\right) \ge 0.$$

The numerical treatment of the multi-point boundary value problem is by the advanced version JANUS [4] of the multiple shooting method [3]. It provides an improved stability of the solution process and an increased rate of convergence. A detailed description is given in [4].

The results for an example trajectory are depicted in Fig. 4. The behaviour of the following quantities vs. the normalized time $\xi := t/(\tau_F - \tau_0) = t/\tau_F$ is shown: satellite orientation (ψ, ϑ, ϕ) (upper left), optimal controls (upper right), joint angles (lower left) and joint velocities (lower right). Three constrained arcs are active. Total operation time $\tau_F - \tau_0 = 8.7355$ s. The Hamiltonian has to be constant; this is fulfilled with a rel. tolerance of 10^{-9}.

Summary

Maximum Principle based methods are used to calculate optimal trajectories with high accuracy for free-floating space robotic systems. Significant drawbacks of the indirect approach have been resolved. By the special approach

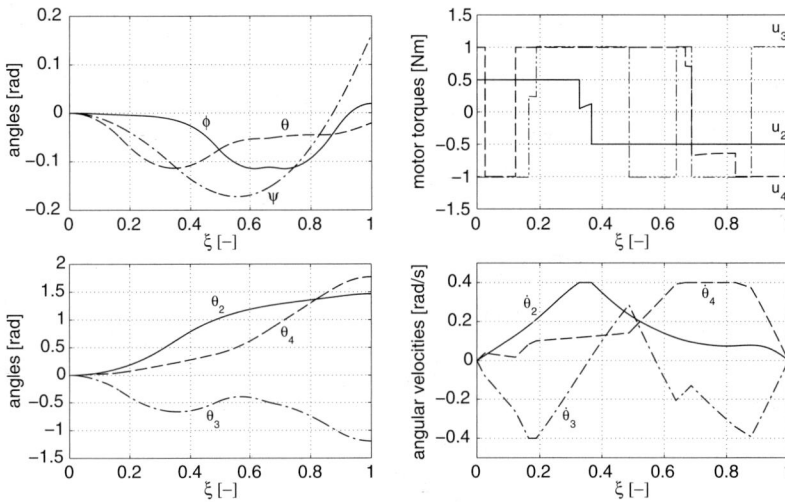

Fig. 4. Example trajectory: Satellite orientation, optimal controls, joint angles and joint velocities vs. normalized time $\xi := t/(\tau_F - \tau_0)$.

described in this paper, the model structure is revealed, additional effects like spin-stabilization can be directly included. Derivative information is generated in a structured and efficient way. State and control constraints are transformed into linear equations and do not have to be calculated analytically. Optimal motion trajectories are calculated for a 6-DOF example system.

References

[1] S. Breun and R. Callies. Redundant optimal control of manipulators along specified paths. In C.A. Mota Soares et al., editor, *Proc. of the III. European Conference on Computational Mechanics: Solids, Structures and Coupled Problems in Engineering*, Lissabon, Portugal, 2006.

[2] A. E. Bryson and Y.-C. Ho. *Applied Optimal Control*. Revised Printing, Hemisphere Publishing Corp., Washington D.C., 1975.

[3] R. Bulirsch. Die Mehrzielmethode zur numerischen Lösung von nichtlinearen Randwertproblemen und Aufgaben der optimalen Steuerung. Report, Carl-Cranz-Gesellschaft e.V., Oberpfaffenhofen, 1971.

[4] R. Callies. *Entwurfsoptimierung und optimale Steuerung. Differential-algebraische Systeme, Mehrgitter-Mehrzielansätze und numerische Realisierung*. Habilitationsschrift, Zentrum Mathematik, Technische Universität München, 2000.

[5] R. Callies and R. Bulirsch. *3D Trajectory Optimization of a Single-Stage VTVL System*. Paper AIAA-96-3903, San Diego, 1996.

[6] R. Callies and T. Schenk. *Recursive Modeling of Optimal Control Problems for Multi-Link Manipulators*. Submitted for publication, available as preprint TUM-NUM13, Munich University of Technology, 2005.

[7]   K. Chudej. *Effiziente Lösung zustandsbeschränkter Optimalsteuerungsaufgaben.* Habilitationsschrift, Universität Bayreuth, 2001.

[8]   K. Chudej and M. Günther. Global state space approach for the efficient numerical solution of state-constrained trajectory optimization problems. *JOTA*, 103(1):75–93, 1999.

[9]   J. J. Craig. *Introduction to Robotics.* Addison-Wesley, Reading, MA, 1986.

[10] J. Denavit and R.S. Hartenberg. A kinematic notation for lower-pair mechanisms based on matrices. *ASME J. Appl. Mech.*, pp. 215–221, 1955.

[11] S. Dubowsky and E. Papadopoulos. The kinematics, dynamics, and control of free-flying and free-floating space robotic systems. *IEEE Trans. on Robotics and Automation*, 9(5):531–543, 1993.

[12] S. Dubowsky, E. E. Vance, and M. A. Torres. The control of space manipulators subject to spacecraft attitude control saturation limits. In *Proc. NASA Conf Space Telerobotics, Vol. 4*, pp. 409–418, Pasadena, 1989.

[13] R. Featherstone. *Robot Dynamics Algorithms.* Kluwer, Boston, 1987.

[14] M. Glocker, M. Vögel, and O. von Stryk. Trajectory optimization of a shuttle mounted robot. In G. Sachs, editor, *Optimalsteuerungsprobleme von Hyperschall-Flugsystemen*, pp. 71–82, Greifswald, 2000.

[15] R. F. Hartl, S. P. Sethi, and R. G. Vickson. A survey of the maximum principles for optimal control problems with constraints. *SIAM Review*, 37:181–218, 1995.

[16] W. W. Hooker and G. Margulies. The dynamical attitude equations for an *n*-body satellite. *J. Astronautical Sciences*, 12(4):123–128, 1965.

[17] P. C. Hughes. *Spacecraft Attitude Dynamics.* Wiley, New York, 1986.

[18] D. H. Jacobson, M. M. Lele, and J. L. Speyer. New necessary conditions of optimality for control problems with state- variable inequality constraints. *J. Math. Anal. Appl.*, 35:255–284, 1971.

[19] J. L. Junkins and J. D. Turner. *Optimal Spacecraft Rotational Maneuvers.* Elsevier Science Publ. Comp., New York, 1986.

[20] R. Longman, R. Lindberg, and M. Zedd. Satellite-mounted robot manipulators - new kinematics and reaction moment compensation. *Int. J. Robotics Res.*, 6 (3):87–103, 1987.

[21] H. J. Oberle. *Numerische Behandlung singulärer Steuerungen mit der Mehrzielmethode am Beispiel der Klimatisierung von Sonnenhäusern.* Habilitationsschrift, Technische Universität München, 1982.

[22] E. Papadopoulos and S. Dubowsky. Dynamic singularities in the control of free-floating space manipulators. *ASME J. Dynamic Systems, Measurement and Control*, 115:44–52, 1993.

[23] R. Robertson and J. Wittenberg. A dynamical formalism for an arbitrary number of interconnected bodies, with reference to the problem of satellite attitude control. In *Proc. of the IFAC Congress 1966*, pp. 46D.1–46D.9, London, 1968. Butterworth.

[24] V. H. Schulz, H. G. Bock, and R. W. Longman. Optimal path planning for satellite mounted robot manipulators. *Advances in the Astronautical Sciences*, 82:311–329, 1993.

[25] J. R. Spofford and D. L. Akin. Redundancy control of a free-flying telerobot. *Proc. AIAA Guid., Nav. and Control Conf.*, Minneapolis, 1988.

[26] Z. Vafa and S. Dubbowsky. The kinematics and dynamics of space manipulators: The virtual manipulator approach. *Int. J. Robotics Res.*, 9(4):3–21, 1990.

Computing the Earth Gravity Field with Spherical Harmonics

Michael Gerstl

Deutsches Geodätisches Forschungsinstitut, Bayerische Akademie der Wissenschaften, Alfons-Goppel-Str. 11, D-80539 München

Summary. The expensive evaluation of the spherical-harmonic series expansion of the earth gravity field is optimised by transition to 4-dimensional coordinates. That makes disappear square roots and trigonometric functions. The singularity at the poles inherent in spherical coordinates is removed by the increase of dimension. Instead of the associated Legendre functions we obtain a basis of hypergeometric Jacobi polynomials that reproduces under derivation. Thus, the calculation of their derivatives cancels in the Forsythe summation technique; for the Clenshaw summation, the recursions of function value and derivatives are decoupled.

1 Introduction

Geodesy, the science of surveying the earth, is one of the fields of personal interest of Prof. Bulirsch. Geodesy today deals with the determination of an **earth model** consisting of a time-dependent surface – represented by position and velocity coordinates of marked points on the earth surface –, physical parameters – describing the gravitational field, the ocean surface, or the refractivity of the atmosphere, e.g. –, and last but not least the time series of the earth's origin and orientation with respect to the celestial inertial system – more precisely, origin and orientation of the conventional triplet of unit vectors underlying that model. Suppose all these model parameters being combined in a vector $p \in \mathbb{R}^n$, the vector of the parameters solved for. Hereat n can reach a magnitude of 600 000.

The current geodetic measurement techniques use satellites or extragalactic radio sources. If, in satellite geodesy, the orbit coordinates of one or more satellites are collected in a vector $x(t)$, the problem arises to integrate the differential equation

$$\ddot{x}(t) = F(t, x(t), \dot{x}(t), p)$$

together with its variational equations with respect to the dynamic parameters within p. For it, multistep integration methods are chosen, because the numerical cost of the integrator is negligible compared to the expense to evaluate the right hand side F.

The perturbing acceleration F is a sum of all the forces acting on the satellite (divided by the mass of the satellite). F includes gravitational acceleration due to a lunar gravity field and due to the attraction of the point masses of sun and planets, atmospheric drag, radiation pressure due to solar radiation, earth albedo, and terrestrial infrared radiation, thermic and relativistic effects, and attitude control. But the term dominant in magnitude and computing cost is the gravitational acceleration of the earth which is represented as the gradient of a static potential field with time-varying corrections due to the mass displacements of ocean tides, solid earth tides, and the wobble of the rotation axis. The actual state of free space potential models provides a one-degree-resolution corresponding to a series of about $130\,000$ spherical-harmonic eigenfunctions of the Laplace operator.

The function to be optimised is conventionally modelled as a harmonic potential in spherical coordinates,

$$V(r,\lambda,\varphi) \;=\; \sum_{m=0}^{N}\sum_{n=m}^{N} V_{nm}(r,\lambda,\varphi) \qquad \text{where}$$

$$\begin{aligned}
V_{nm}(\ldots) &= \frac{GM}{a}\Big(\frac{a}{r}\Big)^{n+1}\Big(\bar{C}_{nm}\cos(m\lambda) + \bar{S}_{nm}\sin(m\lambda)\Big)\bar{P}_{nm}(\sin\varphi) \\
&= \frac{GM}{a}\Big(\frac{a}{r}\Big)^{n+1}\mathrm{Re}\Big[(\bar{C}_{nm}-\mathrm{i}\bar{S}_{nm})\mathrm{e}^{\mathrm{i}m\lambda}\Big]\bar{P}_{nm}(\sin\varphi)
\end{aligned} \tag{1}$$

with dimensionless coefficients $\bar{C}_{nm}$ and $\bar{S}_{nm}$. There is $a=a_\oplus$ the major semi-axis of the reference ellipsoid belonging to that model, and $GM=GM_\oplus$ the gravitational constant multiplied with the mass of the earth. The $\bar{P}_{nm}$ denote the fully normalised associated Legendre functions of degree n and order m.

At each integrator-step along a satellite orbit, given as cartesian position, we have to calculate from the potential (1) the gradient, the second-order tensor, both in cartesian coordinates, and the gradient derivatives with respect to the model parameters $\bar{C}_{nm}$ and $\bar{S}_{nm}$, if these parameters belong to p.

2 The Conventional Approach in Spherical Coordinates

Denote the cartesian coordinates of a satellite position by x, y, z. Introduce **spherical coordinates**: the radius $r \geq 0$ and two angles, the latitude $\varphi \in [-\pi/2, +\pi/2]$ and the longitude $\lambda \mod 2\pi$. They are related to cartesian coordinates by

$$x \;=\; r\cos\varphi\cos\lambda\,, \quad y \;=\; r\cos\varphi\sin\lambda\,, \quad z \;=\; r\sin\varphi\,.$$

In the literature one finds instead of φ mostly the **colatitude** or **polar distance** $\vartheta \;=\; \pi/2 - \varphi$. Here we prefer the latitude, because its domain $[-\pi/2, +\pi/2]$ coincides with the range of the inverse trigonometric functions arcsin and most notably arctan. As a, sloppy speaking, fourth cartesian coordinate we make use of the radius of the parallel of latitude

$$\rho \; = \; \sqrt{x^2 + y^2} \; = \; r\cos\varphi\,.$$

Then it holds $x + iy = \rho\exp(i\lambda)$ and $\rho + iz = r\exp(i\varphi)$. Moreover, it proves to be convenient to introduce normalised cartesian coordinates

$$\bar{x} = \frac{x}{r}\,,\quad \bar{y} = \frac{y}{r}\,,\quad \bar{z} = \frac{z}{r} = \sin\varphi\,,\quad \bar{\rho} = \frac{\rho}{r} = \cos\varphi$$

and, for the use in derivatives, the "adjoint" coordinates

$$x^* = -y\,,\quad y^* = +x\,,\quad z^* = 0\,.$$

The first fundamental form of metrics,

$$g_{11} \; = \; 1\,,\quad g_{22} \; = \; r^2\,,\quad g_{33} \; = \; r^2\cos^2\varphi\,,\quad g_{ik} \; = \; 0\quad\text{otherwise,}$$

shows that spherical coordinates are orthogonal coordinates.

The Laplace operator is expressed in spherical coordinates as

$$\triangle f \; = \; \frac{1}{r^2}\left[\frac{\partial}{\partial r}\left(r^2\frac{\partial f}{\partial r}\right) + \frac{1}{\cos\varphi}\frac{\partial}{\partial\varphi}\left(\cos\varphi\frac{\partial f}{\partial\varphi}\right) + \frac{1}{\cos^2\varphi}\frac{\partial^2 f}{\partial\lambda^2}\right]$$

Potential functions such as V_{nm} are solutions of Laplace's equation $\triangle V = 0$. They are harmonic. Expressed in spherical coordinates, Laplace's equation separates into three ordinary differential equations for functions $f_1(r)$, $f_2(\lambda)$, $f_3(\varphi)$ respectively. The base functions that span the general solution (1) of Laplace's equation therefore are products $f_1(r)f_2(\lambda)f_3(\varphi)$. The non-radial part of this product is called **surface (spherical) harmonic**. It is

$$\bar{C}_n^m(\lambda,\varphi) \; = \; \cos(m\lambda)\bar{P}_{nm}(\sin\varphi)\,,\quad \bar{S}_n^m(\lambda,\varphi) \; = \; \sin(m\lambda)\bar{P}_{nm}(\sin\varphi)$$

or in complex notation

$$\bar{C}_n^m(\lambda,\varphi) + i\bar{S}_n^m(\lambda,\varphi) \; = \; e^{im\lambda}\bar{P}_{nm}(\sin\varphi)\,.$$

The complete base functions $r^{-(n+1)}\big(\bar{C}_n^m(\lambda,\varphi) + i\bar{S}_n^m(\lambda,\varphi)\big)$ are called **solid spherical harmonics**. Keep in mind to distinguish the surface harmonic $\bar{C}_n^m$ and the potential coefficient $\bar{C}_{nm}$ by means of the position of the index m.

Despite the separability, the spherical coordinates have disadvantages: the need of trigonometric functions in coordinate transformation and the inherent singularities at the origin (λ and φ undetermined) and at the poles (λ undetermined). To overcome the problems we want to advance in evolutionary steps towards new coordinates.

3 Eliminating Odd Powers of $\bar{\rho} = \cos\varphi$

Outside an arbitrary small neighborhood of the poles the transformation

$$\varphi \;\longmapsto\; \bar{z} = \sin(\varphi)\,, \qquad \frac{\partial}{\partial\varphi} = \cos(\varphi)\frac{\partial}{\partial\bar{z}} = \bar{\rho}\,\frac{\partial}{\partial\bar{z}}$$

is strictly monotonous and continuously differentiable. Thus, we may replace the coordinate φ by $\bar{z} = \sin(\varphi)$. In the following we refer the triplet $(r, \lambda, \bar{z})$ to spherical coordinates too.

The factor $\bar{\rho}$ in the substitute for the latitude derivative causes throughout the applied formulas that the factor $\bar{\rho} = \cos\varphi = \sqrt{\bar{x}^2 + \bar{y}^2} = \sqrt{1-\bar{z}^2}$ appears in even powers only. These formulas are summarised in the following.

The differentiation of the coordinate transformation between cartesian and spherical coordinates yields for each $\xi \in \{x, y, z\}$

$$\frac{\partial r}{\partial\xi} = \bar{\xi}\,, \qquad r\frac{\partial\bar{z}}{\partial\xi} = \delta_{\xi z} - \bar{\xi}\bar{z}\,, \qquad \bar{\rho}^2 r\frac{\partial\lambda}{\partial\xi} = \bar{\xi}^*\,. \tag{2}$$

The second derivatives with respect to an arbitrary $\xi, \eta \in \{x, y, z\}$ read

$$r\frac{\partial^2 r}{\partial\xi\partial\eta} = \delta_{\xi\eta} - \bar{\xi}\bar{\eta}$$

$$r^2\frac{\partial^2\bar{z}}{\partial\xi\partial\eta} = (3\bar{\xi}\bar{\eta} - \delta_{\xi\eta})\,\bar{z} - (\bar{\xi}\delta_{\eta z} + \bar{\eta}\delta_{\xi z}) \tag{3}$$

$$r^2\bar{\rho}^4\frac{\partial^2\lambda}{\partial\xi\partial\eta} = -\left(\bar{\xi}\bar{\eta}^* + \bar{\xi}^*\bar{\eta}\right)(1-\delta_{\xi z})(1-\delta_{\eta z})$$

To transform the potential gradient from spherical to cartesian coordinates, we apply (2) and thereby get

$$r\frac{\partial V}{\partial x} = \bar{x}\left(r\frac{\partial V}{\partial r} - \bar{z}\frac{\partial V}{\partial\bar{z}}\right) - \bar{y}\left(\frac{1}{\bar{\rho}^2}\frac{\partial V}{\partial\lambda}\right)$$

$$r\frac{\partial V}{\partial y} = \bar{y}\left(r\frac{\partial V}{\partial r} - \bar{z}\frac{\partial V}{\partial\bar{z}}\right) + \bar{x}\left(\frac{1}{\bar{\rho}^2}\frac{\partial V}{\partial\lambda}\right) \tag{4}$$

$$r\frac{\partial V}{\partial z} = \bar{z}\left(r\frac{\partial V}{\partial r} - \bar{z}\frac{\partial V}{\partial\bar{z}}\right) + \frac{\partial V}{\partial\bar{z}}$$

In particular, the transformation (4) is invariant under rotations of the basis about the z-axis inserted between the cartesian and the spherical system. But the polar singularity of the spherical coordinates still manifests in the factor $1/\bar{\rho}^2$ of the longitudinal derivative. Since at the poles $\partial V/\partial\lambda$ tends to zero, this singularity is removable.

The transformation of the second order derivatives to cartesian coordinates follows by means of (2) and (3). For arbitrary $\xi, \eta \in \{x, y, z\}$ we get

$$r^2 \frac{\partial^2 V}{\partial\xi\partial\eta} = \bar{\xi}\bar{\eta}\left[\left(r^2\frac{\partial^2 V}{\partial r^2} - 2\bar{z}r\frac{\partial^2 V}{\partial r\partial\bar{z}} + \bar{z}^2\frac{\partial^2 V}{\partial\bar{z}^2}\right) - \left(r\frac{\partial V}{\partial r} - 3\bar{z}\frac{\partial V}{\partial\bar{z}}\right)\right] +$$

$$+ (\bar{\xi}\delta_{\eta z} + \bar{\eta}\delta_{\xi z})\left[\left(r\frac{\partial^2 V}{\partial r\partial\bar{z}} - \bar{z}\frac{\partial^2 V}{\partial\bar{z}^2}\right) - \frac{\partial V}{\partial\bar{z}}\right] + \delta_{\xi z}\delta_{\eta z}\frac{\partial^2 V}{\partial\bar{z}^2} +$$

$$+ (\bar{\xi}\bar{\eta}^* + \bar{\xi}^*\bar{\eta})\left[\frac{1}{\bar{\rho}^2}\left(r\frac{\partial^2 V}{\partial r\partial\lambda} - \bar{z}\frac{\partial^2 V}{\partial\bar{z}\partial\lambda}\right) - (1-\delta_{\xi z})(1-\delta_{\eta z})\frac{1}{\bar{\rho}^4}\frac{\partial V}{\partial\lambda}\right] +$$

$$+ (\delta_{\xi z}\bar{\eta}^* + \delta_{\eta z}\bar{\xi}^*)\frac{1}{\bar{\rho}^2}\frac{\partial^2 V}{\partial\bar{z}\partial\lambda} + \bar{\xi}^*\bar{\eta}^*\frac{1}{\bar{\rho}^4}\frac{\partial^2 V}{\partial\lambda^2} +$$

$$+ \delta_{\xi\eta}\left(r\frac{\partial V}{\partial r} - \bar{z}\frac{\partial V}{\partial\bar{z}}\right) \tag{5}$$

Finally the Laplace operator becomes in these coordinates

$$\triangle f = \frac{1}{r^2}\left[\frac{\partial}{\partial r}\left(r^2\frac{\partial f}{\partial r}\right) + \frac{\partial}{\partial\bar{z}}\left((1-\bar{z}^2)\frac{\partial f}{\partial\bar{z}}\right) + \frac{1}{(1-\bar{z}^2)}\frac{\partial^2 f}{\partial\lambda^2}\right] =$$

$$= \frac{1}{r^2}\left[r^2\frac{\partial^2 f}{\partial r^2} + 2r\frac{\partial f}{\partial r} + (1-\bar{z}^2)\frac{\partial^2 f}{\partial\bar{z}^2} - 2\bar{z}\frac{\partial f}{\partial\bar{z}} + \frac{1}{(1-\bar{z}^2)}\frac{\partial^2 f}{\partial\lambda^2}\right] \tag{6}$$

Making the substitution $f(r,\lambda,\bar{z}) = f_1(r)f_2(\lambda)f_3(\bar{z})$, Laplace's equation separates into three ordinary differential equations

$$f_1 : \quad r^2 f_1''(r) + 2r\, f_1'(r) - n(n+1)f_1(r) = 0$$

$$f_2 : \quad f_2''(\lambda) + m^2 f_2(\lambda) = 0 \tag{7}$$

$$f_3 : \quad (1-\bar{z}^2)f_3''(\bar{z}) - 2\bar{z}\, f_3'(\bar{z}) + \left(n(n+1) - \frac{m^2}{(1-\bar{z}^2)}\right)f_3(\bar{z}) = 0$$

with linking constants or separation constants $n(n+1)$ and m.

4 Digression on Legendre Functions

Remember the orthogonal Legendre Polynomials, defined by the formula of Rodrigues

$$P_n(\bar{z}) = \frac{1}{2^n n!}\frac{\mathrm{d}^n}{\mathrm{d}\bar{z}^n}(\bar{z}^2 - 1)^n \quad \left(\bar{z}\in[-1,+1],\ n=0,1,2,\dots\right).$$

Define to each polynomial P_n the **(associated) Legendre functions** of order m $(m = 0,1,\dots,n)$ by

$$P_{nm}(\bar{z}) = (1-\bar{z}^2)^{\frac{m}{2}}\frac{\mathrm{d}^m}{\mathrm{d}\bar{z}^m}P_n(\bar{z}) = \frac{1}{2^n n!}(1-\bar{z}^2)^{\frac{m}{2}}\frac{\mathrm{d}^{n+m}}{\mathrm{d}\bar{z}^{n+m}}(\bar{z}^2-1)^n. \tag{8}$$

That are polynomials only if $m = 0$. Note that in mathematical literature (e.g. [1, 3]) the Legendre functions are mostly defined with the opposite sign as $P_n^m(\bar{z}) = (-1)^m P_{nm}(\bar{z})$.

The P_{nm} solve the third equation in (7), and, for any fixed order m, the subset $\{P_{nm} \mid n=m, m+1, \ldots\}$ represents a complete orthogonal system in $L^2[-1, +1]$ with

$$\langle P_{km}, P_{nm}\rangle = \frac{1}{2}\int_{-1}^{1} P_{km}(\bar{z})P_{nm}(\bar{z})d\bar{z} = \frac{1}{2n+1}\frac{(n+m)!}{(n-m)!}\delta_{nk}.$$

From the Legendre polynomials they inherit a recurrence relation, thus

$$P_{00} = 1, \qquad P_{mm} = (2m-1)\sqrt{1-\bar{z}^2}\,P_{m-1,m-1}$$

$$P_{m-1,m} = 0, \qquad P_{nm} = \frac{2n-1}{n-m}\bar{z}\,P_{n-1,m} - \frac{n+m-1}{n-m}P_{n-2,m} \tag{9}$$

There are a lot of relations to calculate the derivative, e.g.

$$(1-\bar{z}^2)\,P'_{n,m} = (n+1)\bar{z}P_{n,m} - (n+1-m)P_{n+1,m}$$

$$(1-\bar{z}^2)\,P'_{n,m} = -n\,\bar{z}P_{n,m} + (n+m)\,P_{n-1,m}$$

$$(1-\bar{z}^2)\,P'_{n,m} = m\,P_{n-1,m} - n\,(n+1-m)\bar{\rho}P_{n,m-1}$$

$$(1-\bar{z}^2)\,P'_{n,m} = m\,P_{n+1,m} - (n+1)(n+m)\bar{\rho}P_{n,m-1}$$

For the question of normalisation, look at

$$\max\{P_{\nu\mu}(\bar{z}) \mid |\bar{z}|\leq 1,\, 0\leq\mu\leq\nu\leq m\} = P_{mm}(0) = 1\cdot 3\cdot 5\cdot 7\ldots\cdot(2m-1).$$

The growth with increasing order m produces numeric overflow, in the 8-Byte number model from $m=151$. Since geodesy deals with orders up to $m=360$, normalisation became convention.

It is the surface harmonics that have to be normalised. This means to determine a factor η_{nm} so that

$$\frac{1}{4\pi}\int_{\lambda=-\pi}^{+\pi}\int_{\varphi=-\pi/2}^{+\pi/2} |f_{nm}(\varphi,\lambda)|^2 \cos\varphi\,d\varphi\,d\lambda = 1$$

$$\text{for } f_{nm}(\varphi,\lambda) = \begin{cases} \bar{C}_n^m(\lambda,\varphi) = \eta_{nm}\cos(m\lambda)P_{nm}(\sin\varphi), \\ \bar{S}_n^m(\lambda,\varphi) = \eta_{nm}\sin(m\lambda)P_{nm}(\sin\varphi). \end{cases}$$

It turns out that

$$\eta_{nm} = \sqrt{\frac{2}{1+\delta_{m,0}}(2n+1)\frac{(n-m)!}{(n+m)!}}. \tag{10}$$

Within the surface harmonics the factor is added on the Legendre functions: $\bar{P}_{nm} = \eta_{nm}P_{nm}$. For $m > 0$ the scaled functions adopt their maximum at $\varphi = 0$ with the value

$$\bar{P}_{mm}(0) = \frac{\sqrt{2(2m+1)!}}{2^m\,m!} \longrightarrow \infty \quad \text{as } m \to \infty$$

For $m = 0$

$$\bar{P}_{n0}(1) = \sqrt{2n+1} \longrightarrow \infty \quad \text{as } n \to \infty$$

In both cases the overflow will not be reached within the range of numerically educible integers for n and m.

The recursion of the fully normalised functions is deduced from (9):

$$\bar{P}_{00} = 1, \qquad \bar{P}_{mm} = \nu_{mm} \sqrt{1-\bar{z}^2} \, \bar{P}_{m-1,m-1}$$

$$\bar{P}_{m-1,m} = 0, \qquad \bar{P}_{nm} = \nu_{nm} \left(\bar{z} \, \bar{P}_{n-1,m} - \frac{\bar{P}_{n-2,m}}{\nu_{n-1,m}} \right) \tag{11}$$

with factors

$$\nu_{11} = \frac{\eta_{11}}{\eta_{00}} = \sqrt{3}, \quad \nu_{mm} = (2m-1) \frac{\eta_{m,m}}{\eta_{m-1,m-1}} = \sqrt{\frac{2m+1}{2m}} \quad (m>1)$$

$$\nu_{nm} = \frac{2n-1}{n-m} \frac{\eta_{n,m}}{\eta_{n-1,m}} = \sqrt{\frac{(2n-1)(2n+1)}{(n-m)(n+m)}} \qquad (n>m)$$

An alternative normalisation is the following scaling:

$$\tilde{P}_{nm} = \eta_{nm} P_{nm} \quad \text{with} \quad \eta_{nm} = \sqrt{\frac{(n-m)!}{(n+m)!}} \quad \text{and} \quad \langle \tilde{P}_{km}, \tilde{P}_{nm} \rangle = \frac{\delta_{nk}}{2n+1}.$$

Such scaled Legendre functions have an easy to calculate recursion relation

$$\tilde{P}_{00} = 1, \quad \tilde{P}_{mm} = \sqrt{\frac{2m-1}{2m}} \sqrt{1-\bar{z}^2} \, \tilde{P}_{m-1,m-1}$$

$$\tilde{P}_{m-1,m} = 0, \quad \tilde{P}_{nm} = \frac{1}{\nu_{nm}} \left((2n-1) \bar{z} \, \tilde{P}_{n-1,m} - \nu_{n-1,m} \tilde{P}_{n-2,m} \right) \tag{12}$$

$$\text{with} \quad \nu_{n,m} = \sqrt{(n-m)(n+m)}.$$

5 Transition to Polynomials

As is known, the surface spherical harmonics $C_n^m(\lambda, \varphi)$ and $S_n^m(\lambda, \varphi)$ have an unique representation as polynomials of $\bar{x}, \bar{y}, \bar{z}$ (see Table 1). Polynomials ease the calculation of cartesian derivatives. Regarding the definition (8) of the Legendre functions which are neither polynomials, we recognise the factor $\mathrm{d}^m P_n / \mathrm{d}\bar{z}^m$ being a polynomial of the minimal degree $n-m$. If the remainder $(1-\bar{z})^{m/2} = \cos^m \varphi$ is transfered from $\bar{P}_{nm}(\bar{z})$ to the longitudinal part of V_{nm}, we may obtain polynomials too. Indeed,

$$\left(\bar{C}_{nm}\cos(m\lambda) + \bar{S}_{nm}\sin(m\lambda)\right)\cos^m\varphi \;=$$

$$= \;\mathrm{Re}\left[\left(\bar{C}_{nm} - \mathrm{i}\bar{S}_{nm}\right)\mathrm{e}^{\mathrm{i}m\lambda}\right]\cos^m\varphi \;=$$

$$= \;\mathrm{Re}\left[\left(\bar{C}_{nm} - \mathrm{i}\bar{S}_{nm}\right)\left(\mathrm{e}^{\mathrm{i}\lambda}\cos\varphi\right)^m\right] \;=$$

$$= \;\mathrm{Re}\left[\left(\bar{C}_{nm} - \mathrm{i}\bar{S}_{nm}\right)\left(\bar{x} + \mathrm{i}\bar{y}\right)^m\right] \;=$$

$$= \;\bar{C}_{nm}\,\mathrm{Re}\left[\left(\bar{x} + \mathrm{i}\bar{y}\right)^m\right] + \bar{S}_{nm}\,\mathrm{Im}\left[\left(\bar{x} + \mathrm{i}\bar{y}\right)^m\right].$$

For the new factors of the potential coefficients,

$$\zeta_m(\bar{x}, \bar{y}) \;=\; \mathrm{Re}\left[\left(\bar{x} + \mathrm{i}\bar{y}\right)^m\right] \quad\text{and}\quad \sigma_m(\bar{x}, \bar{y}) \;=\; \mathrm{Im}\left[\left(\bar{x} + \mathrm{i}\bar{y}\right)^m\right],$$

complex multiplication and derivation provide an easy algorithm which will be detailed in (19) and the following.

Concerning now on the polynomials of $\bar{z}$ to be derived from (8):

$$H_{nm}(\bar{z}) \;:=\; \frac{P_{nm}(\bar{z})}{(1-\bar{z}^2)^{m/2}} \;=\; \left(\frac{\mathrm{d}}{\mathrm{d}\bar{z}}\right)^m P_n(\bar{z}). \tag{13}$$

As mth derivative of the polynomial P_n of degree n H_{nm} is a polynomial of degree $n-m$. In particular H_{mm} is a constant:

$$H_{mm}(\bar{z}) \;=\; P_{mm}(0) \;=\; \frac{(2m)!}{2^m\,m!} \;=\; 1{\cdot}3{\cdot}5{\cdot}7{\cdot}\ldots{\cdot}(2m-1) \;\longrightarrow\; \infty \quad\text{as}\quad m \to \infty.$$

Table 1. The first unnormalised surface spherical harmonics $C_n^m(\bar{x}, \bar{y}, \bar{z})$ and $S_n^m(\bar{x}, \bar{y}, \bar{z})$ as functions of cartesian coordinates on $\mathbb{S}^2$.

$C_1^0 \;=\; \bar{z}\,,$	$C_1^1 \;=\; \bar{x}$	
	$S_1^1 \;=\; \bar{y}$	
$C_2^0 \;=\; \dfrac{1}{2}\left(3\bar{z}^2 - 1\right),$	$C_2^1 \;=\; 3\,\bar{x}\bar{z}\,,$	$C_2^2 \;=\; 3\left(\bar{x}^2 - \bar{y}^2\right)$
	$S_2^1 \;=\; 3\,\bar{y}\bar{z}\,,$	$S_2^2 \;=\; 6\,\bar{x}\bar{y}$
$C_3^0 \;=\; \dfrac{1}{2}\,\bar{z}\left(5\bar{z}^2 - 3\right),$	$C_3^1 \;=\; \dfrac{3}{2}\,\bar{x}\left(5\bar{z}^2 - 1\right),$	$C_3^2 \;=\; 15\left(\bar{x}^2 - \bar{y}^2\right)\bar{z}$
	$S_3^1 \;=\; \dfrac{3}{2}\,\bar{y}\left(5\bar{z}^2 - 1\right),$	$S_3^2 \;=\; 30\,\bar{x}\bar{y}\bar{z}$
$C_4^0 \;=\; \dfrac{1}{8}\left(35\bar{z}^4 - 30\bar{z}^2 + 3\right),$	$C_4^1 \;=\; \dfrac{5}{2}\,\bar{x}\bar{z}\left(7\bar{z}^2 - 3\right),$	$C_4^2 \;=\; \dfrac{15}{2}\left(\bar{x}^2 - \bar{y}^2\right)\left(7\bar{z}^2 - 1\right)$
	$S_4^1 \;=\; \dfrac{5}{2}\,\bar{y}\bar{z}\left(7\bar{z}^2 - 3\right),$	$S_4^2 \;=\; 15\,\bar{x}\bar{y}\left(7\bar{z}^2 - 1\right)$

Accordingly, these polynomials inherit the Legendre function's undesirable feature of numeric overflow from the order of $m=151$ (8-Byte arithmetic). They must be normalised.

The recursion of the Legendre functions in (9) is directly carried forward to give

$$H_{00} = 1, \quad H_{mm} = (2m-1) H_{m-1,m-1}$$

$$H_{m-1,m} = 0, \quad H_{nm} = \frac{2n-1}{n-m} \bar{z}\, H_{n-1,m} - \frac{n+m-1}{n-m} H_{n-2,m}$$

For any fixed order m, the functions $\{H_{nm} \,|\, n = m, m+1, \ldots\}$ constitute a complete orthogonal system in $L^2[-1, +1]$ with respect to the scalar product

$$\langle H_{km}, H_{nm} \rangle \;=\; \frac{1}{2} \int_{-1}^{1} (1-t^2)^m H_{km}(t) H_{nm}(t)dt \;=\; \frac{1}{2n+1} \frac{(n+m)!}{(n-m)!} \delta_{nk} \,.$$

From the weight function $w(t) = (1-t^2)^m$ we conclude that the H_{nm} can be represented as a special case of the hypergeometric polynomials or **Jacobi Polynomials**

$$H_{nm} \;=\; \frac{(n+m)!}{2^m\, n!}\, P_{n-m}^{(m,m)} \,.$$

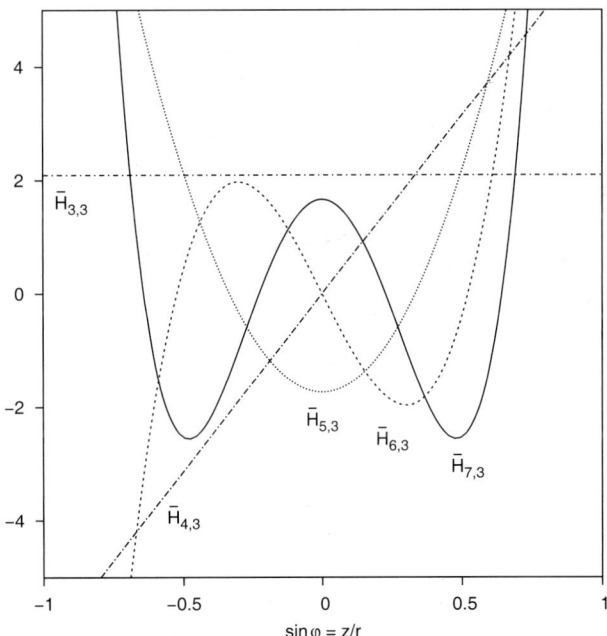

Fig. 1. "Fully" normalised polynomials $\bar{H}_{nm}(\bar{z}) = \bar{P}_n^{(m)}(\bar{z})$ for order $m=3$ (their polynomial degree is $n-m$).

In particular holds $H_{n0} = P_n^{(0,0)} = P_n$.

The most outstanding property of the H_{nm} is their reproduction under derivation. This can be seen from (13):

$$\frac{\mathrm{d}}{\mathrm{d}\bar{z}}H_{nm} = \frac{\mathrm{d}}{\mathrm{d}\bar{z}}\left(\left(\frac{\mathrm{d}}{\mathrm{d}\bar{z}}\right)^m P_n\right) = \left(\frac{\mathrm{d}}{\mathrm{d}\bar{z}}\right)^{m+1} P_n = H_{n,m+1} \qquad (14)$$

Since in (13) the H_{nm} were by differentiation derived from the Legendre polynomials, one gains by $(m-1)$-fold differentiation of equations, which relate the P_n with their derivatives, additional formulas for the H_{nm}, for instance

$$H_{n+1,m} - H_{n-1,m} = (2n+1)\,H_{n,m-1}$$

$$H_{n+1,m} - \bar{z}\,H_{n,m} = (n+m)\,H_{n,m-1}$$

$$(\bar{z}^2-1)\,H_{n,m+1} = (n-m)\bar{z}\,H_{n,m} - (n+m)\,H_{n-1,m}$$

Let's mention three kinds of normalisation which lead to different recursion and derivation formulas.

Table 2. Mathematical properties of the polynomials H_{nm}

Weight function:
$$w(t) = (1-t^2)^m$$

Rodrigues' formula:
$$H_{nm}(t) = \frac{1}{2^n\,n!}\left(\frac{d}{dt}\right)^{n+m}(t^2-1)^n = P_n^{(m)}(t)$$

Standardisation:
$$H_{nm}(1) = \frac{1}{2^m\,m!}\frac{(n+m)!}{(n-m)!} = \frac{(2m)!}{2^m\,m!}\binom{n+m}{n-m}$$

n-Recursion:
$$H_{mm} = \frac{(2m)!}{2^m\,m!}, \qquad H_{nm} = \frac{2n-1}{n-m}\,t\,H_{n-1,m} - \frac{n+m-1}{n-m}\,H_{n-2,m}$$

m-Recursion:
$$(1-t^2)H_{n,m+1} = 2mt\,H_{n,m} - (n-m+1)(n+m)\,H_{n,m-1}$$

Highest coefficient:
$$\hat{k}_{nm} = \frac{1}{(n-m)!}\prod_{\nu=1}^{n}(2\nu-1) = \frac{1}{(n-m)!}\frac{(2n)!}{2^n\,n!}$$

Derivative:
$$\frac{d}{dt}H_{nm}(t) = H_{n,m+1}(t)$$

Differential equation:
$$(1-t^2)f''(t) - 2(m+1)t\,f'(t) + (n-m)(n+m+1)f(t) = 0$$

(a) Full normalisation of the polynomials

Definition:

$$\bar{H}_{nm} = \frac{\bar{P}_{nm}}{(1-t^2)^{m/2}} = \eta_{nm} \, H_{nm}, \qquad \bar{H}_{mm} = \frac{1}{2^m \, m!} \sqrt{\frac{2(2m+1)!}{1+\delta_{m,0}}}$$

Recursion (derived from (11)):

$$\bar{H}_{00} = 1, \qquad \bar{H}_{mm} = \nu_{mm} \, \bar{H}_{m-1,m-1}$$

$$\bar{H}_{m-1,m} = 0, \qquad \bar{H}_{nm} = \nu_{nm} \left(\bar{z} \, \bar{H}_{n-1,m} - \frac{\bar{H}_{n-2,m}}{\nu_{n-1,m}} \right)$$

Derivative (derived from (14)):

$$\frac{\mathrm{d}}{\mathrm{d}\bar{z}} \bar{H}_{nm} = \frac{\eta_{n,m}}{\eta_{n,m+1}} \, \bar{H}_{n,m+1} = \sqrt{\frac{(n-m)(n+m+1)}{1+\delta_{m,0}}} \, \bar{H}_{n,m+1}$$

(b) Partial normalisation of the polynomials

From the normalisation factor η_{nm} of the surface harmonics (10) we use only the last part dependent on $n\pm m$. How to dispose the remainder will be detailed in Sect. 7.

Definition:

$$\tilde{H}_{nm} = \sqrt{\frac{(n-m)!}{(n+m)!}} \, H_{nm} = \sqrt{\frac{1+\delta_{m,0}}{2(2n+1)}} \, \bar{H}_{nm}, \qquad \tilde{H}_{mm} = \frac{\sqrt{(2m)!}}{2^m \, m!}.$$

Recursion (derived from (11)):

$$\tilde{H}_{00} = 1, \qquad \tilde{H}_{mm} = \sqrt{\frac{2m-1}{2m}} \, \tilde{H}_{m-1,m-1},$$

$$\tilde{H}_{m-1,m} = 0, \qquad \tilde{H}_{nm} = \frac{1}{\nu_{nm}} \left((2n-1)\bar{z} \, \tilde{H}_{n-1,m} - \nu_{n-1,m} \, \tilde{H}_{n-2,m} \right)$$

$$\text{with } \nu_{nm} = \sqrt{(n-m)(n+m)}.$$

Derivative (derived from (14)):

$$\frac{\mathrm{d}}{\mathrm{d}\bar{z}} \tilde{H}_{nm} = \sqrt{(n-m)(n+m+1)} \, \tilde{H}_{n,m+1}$$

(c) Normalisation to sectorial unity

For the sake of completeness we mention a very cost-effective normalisation which can be used only for low degree polynomials. This scaling converts all the sectorial polynomials $\hat{H}_{mm}$ into unity. Then the recursions for every order may be run parallelly.

Definition:

$$\hat{H}_{nm} = \frac{2^m m!}{(2m)!} H_{nm} = \frac{P_{nm}}{P_{mm}}, \qquad \hat{H}_{mm} = 1.$$

The root-free n-recursion (9) of the P_{nm} persists for the functions $\hat{H}_{nm}$:

$$\hat{H}_{m-1,m} = 0, \quad \hat{H}_{m,m} = 1, \quad \hat{H}_{n,m} = \frac{2n-1}{n-m}\bar{z}\,\hat{H}_{n-1,m} - \frac{n+m-1}{n-m}\,\hat{H}_{n-2,m}$$

Derivative

$$\frac{\mathrm{d}}{\mathrm{d}\bar{z}}\hat{H}_{nm} = (2m+1)\,\hat{H}_{n,m+1}$$

By (2) we gain the derivatives with respect to cartesian coordinates of $\mathbb{R}^3$

$$r\frac{\partial}{\partial\xi}\hat{H}_{n,m} = (2m+1)\big(\delta_{\xi z} - \bar{\xi}\bar{z}\big)\hat{H}_{n,m+1} \qquad \Big(\xi \in \{x,y,z\}\Big).$$

6 Removing the Polar Singularity

Cartesian coordinates are not afflicted with a polar singularity. The last achieved representation of the surface harmonics,

$$\bar{C}_n^m + \mathrm{i}\bar{C}_n^m = (\bar{x}+\mathrm{i}\bar{y})^m \bar{H}_{nm}(\bar{z}),$$

lets arise the idea to treat them as functions of the three normalised cartesian coordinates on $\mathbb{S}^2$. If we add the coordinate r or rather $1/r$ for the radial part of the solid harmonics, we end up with 4-dimensional coordinates

$$u_0 = \frac{1}{r}, \quad u_1 = \frac{x}{r}, \quad u_2 = \frac{y}{r}, \quad u_3 = \frac{z}{r}. \tag{15}$$

The inverse transformation is

$$x = u_1/u_0, \quad y = u_2/u_0, \quad z = u_3/u_0, \quad 1 = u_1^2 + u_2^2 + u_3^2.$$

In addition, u_1 and u_2 should also be treated as a single complex variable $u_1+\mathrm{i}u_2$. Let us write (x_1, x_2, x_3) for (x, y, z). Then the partial derivatives are written

$$\frac{\partial u_k}{\partial x_i} = u_0\big(\delta_{ik} - u_i u_k\big)$$

$$\frac{\partial^2 u_k}{\partial x_i \partial x_j} = u_0^2\big(3u_i u_j u_k - u_i\delta_{jk} - u_j\delta_{ik} - u_k\delta_{ij}\big)$$

$$\begin{aligned} k &= 0,1,2,3 \\ i,j &= 1,2,3 \end{aligned} \tag{16}$$

The relation to the spherical derivatives is

$$r\frac{\partial f}{\partial r} = -u_0\frac{\partial f}{\partial u_0} = x_1\frac{\partial f}{\partial x_1} + x_2\frac{\partial f}{\partial x_2} + x_3\frac{\partial f}{\partial x_3}$$

$$\frac{\partial f}{\partial \phi} = \sqrt{u_1^2+u_2^2}\left\{\frac{\partial f}{\partial u_3} - \frac{u_3}{u_1^2+u_2^2}\left(u_1\frac{\partial f}{\partial u_1} + u_2\frac{\partial f}{\partial u_2}\right)\right\} =$$

$$= \rho\left\{\frac{\partial f}{\partial x_3} - \frac{x_3}{\rho^2}\left(x_1\frac{\partial f}{\partial x_1} + x_2\frac{\partial f}{\partial x_2}\right)\right\}$$

$$\frac{\partial f}{\partial \lambda} = u_1\frac{\partial f}{\partial u_2} - u_2\frac{\partial f}{\partial u_1} = x_1\frac{\partial f}{\partial x_2} - x_2\frac{\partial f}{\partial x_1}$$

By (16) the cartesian gradient in $\mathbb{R}^3$ and the second derivatives result in

$$\frac{\partial V}{\partial x_i} = \sum_{k=0}^{3}\frac{\partial V}{\partial u_k}\cdot\frac{\partial u_k}{\partial x_i} = u_0\left(\frac{\partial V}{\partial u_i} - u_i\sum_{k=0}^{3}u_k\frac{\partial V}{\partial u_k}\right) \qquad (i=1,2,3) \quad (17)$$

$$\frac{\partial^2 V}{\partial x_i\partial x_j} = \sum_{k=0}^{3}\sum_{l=0}^{3}\left(\frac{\partial u_k}{\partial x_i}\right)\left(\frac{\partial u_l}{\partial x_j}\right)\frac{\partial^2 V}{\partial u_k\partial u_l} + \sum_{k=0}^{3}\left(\frac{\partial^2 u_k}{\partial x_i\partial x_j}\right)\frac{\partial V}{\partial u_k} =$$

$$= u_0^2\left[u_iu_j\left(3\sum_{k=0}^{3}u_k\frac{\partial V}{\partial u_k} + \sum_{k=0}^{3}\sum_{l=0}^{3}u_ku_l\frac{\partial^2 V}{\partial u_k\partial u_l}\right)\right.$$

$$- u_i\left(\frac{\partial V}{\partial u_j} + \sum_{k=0}^{3}u_k\frac{\partial^2 V}{\partial u_k\partial u_j}\right) - u_j\left(\frac{\partial V}{\partial u_i} + \sum_{k=0}^{3}u_k\frac{\partial^2 V}{\partial u_k\partial u_i}\right)$$

$$\left.+ \frac{\partial^2 V}{\partial u_i\partial u_j} - \delta_{ij}\sum_{k=0}^{3}u_k\frac{\partial V}{\partial u_k}\right] \qquad (i,j=1,2,3) \quad (18)$$

As a trial we may use

$$x_1\frac{\partial V}{\partial x_1} + x_2\frac{\partial V}{\partial x_2} + x_3\frac{\partial V}{\partial x_3} = -u_0\frac{\partial V}{\partial u_0}.$$

The harmonic function V_{nm} still separates into a product of three functions $V_{nm}(u_0, u_1, u_2, u_3) = f_1(u_0)f_2(u_1+iu_2)f_3(u_3)$ which satisfy the three ordinary diffential equations

$$f_1 : \quad u_0^2 f_1''(u_0) - n(n+1)f_1(u_0) = 0$$

$$f_2 : \quad (u_1+iu_2)^2 f_2'(u_1+iu_2) - m(m-1)f_2(u_1+iu_2) = 0$$

$$f_3 : \quad (1-u_3^2)f_3''(u_3) - 2(m+1)u_3f_3'(u_3) + (n-m)(n+m+1)f_3(u_3) = 0$$

7 The Forsythe Algorithm

So far we have seen that the potential summand V_{nm} of (1) decomposes in a product of three functions of u_0, u_1+iu_2, and u_3 respectively. It is – subject to the selected algorithm – numerically reasonable not to attach the normalisation factor η_{nm} from (10) to the polynomial H_{nm} only, but to spread it over the three functions:

$$V_{nm}(u) = GM\, a^n u_0^{n+1} \cdot \mathrm{Re}\left[(\bar{C}_{nm} - i\bar{S}_{nm})(u_1 + iu_2)^m\right] \cdot \eta_{nm} H_{nm}(u_3) =$$

$$= \underbrace{\nu_n GM\, a^n u_0^{n+1}}_{R_n(u_0)} \cdot \underbrace{\mu_m\, \mathrm{Re}\left[(\bar{C}_{nm} - i\bar{S}_{nm})(u_1 + iu_2)^m\right]}_{L_m(u_1, u_2)} \cdot \underbrace{\tau_{nm} H_{nm}(u_3)}_{T_{nm}(u_3)}$$

The three factors ν_n, μ_m, τ_{nm} have to fulfil the condition

$$\nu_n \cdot \mu_m \cdot \tau_{nm} = \begin{cases} \eta_{nm} & \text{for fully normalised coefficients } \bar{C}_{nm}, \bar{S}_{nm}, \\ 1 & \text{for unnormalised coefficients } C_{nm}, S_{nm}. \end{cases}$$

In order to recover the three normalisations of the polynomials H_{nm} specified in Sect. 5, we have to set in case of fully normalised coefficients

(a) Full normalisation of polynomials: $T_{nm} = \tau_{nm} H_{nm} = \bar{H}_{nm}$

$$\nu_n = 1, \quad \mu_m = 1, \quad \tau_{nm} = \eta_{nm}$$

(b) partial normalisation of polynomials: $T_{nm} = \tau_{nm} H_{nm} = \tilde{H}_{nm}$

$$\nu_n = \sqrt{2(2n+1)}, \quad \mu_m = \sqrt{\frac{1}{1+\delta_{m,0}}}, \quad \tau_{nm} = \sqrt{\frac{(n-m)!}{(n+m)!}}$$

In case of unnormalised coefficients C_{nm} and S_{nm} we may set

(c) Normalisation to sectorial unity: $T_{nm} = \tau_{nm} H_{nm} = \hat{H}_{nm}$

$$\nu_n = 1, \quad \mu_m = \frac{(2m)!}{2^m m!}, \quad \tau_{nm} = \frac{1}{\mu_m} = \frac{2^m m!}{(2m)!}$$

The first function: The calculation of the radial function is straightforward

$$R_n(u_0) = \nu_n GM a^n u_0^{n+1} \implies u_0 \frac{dR_n}{du_0} = (n+1) R_n(u_0)$$

The second function: The corresponding longitudinal function is converted into $L_m(u_1, u_2) = \bar{C}_{nm}\zeta_m(u_1, u_2) + \bar{S}_{nm}\sigma_m(u_1, u_2)$ with

$$\zeta_m(u_1, u_2) = \mu_m\, \mathrm{Re}\left[(u_1 + iu_2)^m\right], \quad \sigma_m(u_1, u_2) = \mu_m\, \mathrm{Im}\left[(u_1 + iu_2)^m\right].$$

Then, the multiplication $\zeta_m + i\sigma_m = (u_1 + iu_2)^m = (u_1 + iu_2)(u_1 + iu_2)^{m-1}$ results in a recursion for the real and imaginary parts

$$\zeta_0 = \mu_0\,, \qquad \zeta_m = \frac{\mu_m}{\mu_{m-1}}\left(u_1\zeta_{m-1} - u_2\sigma_{m-1}\right),$$

$$\sigma_0 = 0\,, \qquad \sigma_m = \frac{\mu_m}{\mu_{m-1}}\left(u_1\sigma_{m-1} + u_2\zeta_{m-1}\right). \tag{19}$$

The complex derivation of the holomorphic function

$$(u_1 + iu_2) \longmapsto \zeta_m + i\sigma_m = \mu_m(u_1 + iu_2)^m$$

yields all the partial derivatives of ζ_m and σ_m with respect to u_1 and u_2. As a consequence of complex differentiability it will suffice to calculate only the partials with respect to u_1.

$$\frac{\partial \zeta_m}{\partial u_1} = \frac{m\mu_m}{\mu_{m-1}}\zeta_{m-1} = \frac{\partial \sigma_m}{\partial u_2}\,, \qquad\qquad \frac{\partial \sigma_m}{\partial u_1} = \frac{m\mu_m}{\mu_{m-1}}\sigma_{m-1} = -\frac{\partial \zeta_m}{\partial u_2}\,,$$

$$\frac{\partial^2 \zeta_m}{\partial u_1^2} = \frac{m\mu_m}{\mu_{m-1}}\frac{(m-1)\mu_{m-1}}{\mu_{m-2}}\zeta_{m-2} = \frac{m\mu_m}{\mu_{m-1}}\frac{\partial \zeta_{m-1}}{\partial u_1}\,,$$

$$\frac{\partial^2 \zeta_m}{\partial u_1 \partial u_2} = -\frac{\partial^2 \sigma_m}{\partial u_1^2}\,, \qquad \frac{\partial^2 \zeta_m}{\partial u_2^2} = -\frac{\partial^2 \zeta_m}{\partial u_1^2}\,,$$

$$\frac{\partial^2 \sigma_m}{\partial u_1^2} = \frac{m\mu_m}{\mu_{m-1}}\frac{(m-1)\mu_{m-1}}{\mu_{m-2}}\sigma_{m-2} = \frac{m\mu_m}{\mu_{m-1}}\frac{\partial \sigma_{m-1}}{\partial u_1}\,,$$

$$\frac{\partial^2 \sigma_m}{\partial u_1 \partial u_2} = +\frac{\partial^2 \zeta_m}{\partial u_1^2}\,, \qquad \frac{\partial^2 \sigma_m}{\partial u_2^2} = -\frac{\partial^2 \sigma_m}{\partial u_1^2}$$

The third function : $T_{nm}(u_3)$ is evaluated by any of the 3-term-recursions.

$$T_{nm}(u_3) = \tau_{nm}H_{nm}(u_3)\,, \qquad \frac{dT_{nm}}{du_3} = \frac{\tau_{n,m}}{\tau_{n,m+1}}T_{n,m+1}(u_3)$$

First partial derivatives of V_{nm} :

$$u_0\frac{\partial V_{nm}}{\partial u_0} = (n+1)\,R_n\left[\quad C_{nm}\zeta_m \quad + S_{nm}\sigma_m \quad\right]T_{nm} = (n+1)V_{nm}$$

$$\frac{\partial V_{nm}}{\partial u_1} = \frac{m\,\mu_m}{\mu_{m-1}}R_n\left[+C_{nm}\zeta_{m-1} + S_{nm}\sigma_{m-1}\right]T_{nm}$$

$$\frac{\partial V_{nm}}{\partial u_2} = \frac{m\,\mu_m}{\mu_{m-1}}R_n\left[-C_{nm}\sigma_{m-1} + S_{nm}\zeta_{m-1}\right]T_{nm}$$

$$\frac{\partial V_{nm}}{\partial u_3} = \frac{\tau_{n,m}}{\tau_{n,m+1}}R_n\left[\quad C_{nm}\zeta_m \quad + S_{nm}\sigma_m \quad\right]T_{n,m+1}$$

Using the recursion (19) we can supplement the formulas

$$u_1 \frac{\partial V_{nm}}{\partial u_1} + u_2 \frac{\partial V_{nm}}{\partial u_2} = m\, R_n \left[C_{nm}\zeta_m + S_{nm}\sigma_m \right] T_{nm} = m V_{nm}$$

and
$$\sum_{k=0}^{3} u_k \frac{\partial V_{nm}}{\partial u_k} = \frac{\tau_{n,m}}{\tau_{n+1,m+1}} R_n \left[C_{nm}\zeta_m + S_{nm}\sigma_m \right] T_{n+1,m+1}$$

Second partial derivatives of V_{nm} :

$$u_0^2 \frac{\partial^2 V_{nm}}{\partial u_0^2} = (n+1)\, n\, R_n \left[\ \ C_{nm}\zeta_m \ \ + S_{nm}\sigma_m\ \ \right] T_{nm}$$

$$u_0 \frac{\partial^2 V_{nm}}{\partial u_0 \partial u_1} = (n+1) \frac{m\,\mu_m}{\mu_{m-1}} R_n \left[+ C_{nm}\zeta_{m-1} + S_{nm}\sigma_{m-1} \right] T_{nm}$$

$$u_0 \frac{\partial^2 V_{nm}}{\partial u_0 \partial u_2} = (n+1) \frac{m\,\mu_m}{\mu_{m-1}} R_n \left[- C_{nm}\sigma_{m-1} + S_{nm}\zeta_{m-1} \right] T_{nm}$$

$$u_0 \frac{\partial^2 V_{nm}}{\partial u_0 \partial u_3} = (n+1) \frac{\tau_{n,m}}{\tau_{n,m+1}} R_n \left[\ \ C_{nm}\zeta_m \ \ + S_{nm}\sigma_m\ \ \right] T_{n,m+1}$$

$$\frac{\partial^2 V_{nm}}{\partial u_1^2} = \frac{m\,\mu_m}{\mu_{m-1}} \frac{(m-1)\mu_{m-1}}{\mu_{m-2}} R_n \left[+ C_{nm}\zeta_{m-2} + S_{nm}\sigma_{m-2} \right] T_{nm}$$

$$\frac{\partial^2 V_{nm}}{\partial u_1 \partial u_2} = \frac{m\,\mu_m}{\mu_{m-1}} \frac{(m-1)\mu_{m-1}}{\mu_{m-2}} R_n \left[- C_{nm}\sigma_{m-2} + S_{nm}\zeta_{m-2} \right] T_{nm}$$

$$\frac{\partial^2 V_{nm}}{\partial u_1 \partial u_3} = \frac{m\,\mu_m}{\mu_{m-1}} \frac{\tau_{n,m}}{\tau_{n,m+1}} R_n \left[+ C_{nm}\zeta_{m-1} + S_{nm}\sigma_{m-1} \right] T_{n,m+1}$$

$$\frac{\partial^2 V_{nm}}{\partial u_2^2} = - \frac{\partial^2 V_{nm}}{\partial u_1^2}$$

$$\frac{\partial^2 V_{nm}}{\partial u_2 \partial u_3} = \frac{m\,\mu_m}{\mu_{m-1}} \frac{\tau_{n,m}}{\tau_{n,m+1}} R_n \left[- C_{nm}\sigma_{m-1} + S_{nm}\zeta_{m-1} \right] T_{n,m+1}$$

$$\frac{\partial^2 V_{nm}}{\partial u_3^2} = \frac{\tau_{n,m}}{\tau_{n,m+1}} \frac{\tau_{n,m+1}}{\tau_{n,m+2}} R_n \left[\ \ C_{nm}\zeta_m \ \ + S_{nm}\sigma_m\ \ \right] T_{n,m+2}$$

The remaining derivatives are to complete symmetrically. According to the choosen type of normalisation the factors yield

(a) Full normalisation of the polynomials $\left(\nu_n = \mu_m = 1,\ \tau_{nm} = \eta_{nm}\right)$:

$$\frac{m\,\mu_m}{\mu_{m-1}} = m\,, \qquad \frac{\tau_{n,m}}{\tau_{n,m+1}} = \sqrt{\frac{(n-m)(n+m+1)}{1+\delta_{m0}}}$$

(b) Partial normalisation of the polynomials
$$\left(\nu_n = \sqrt{2(2n+1)},\ \mu_m = \sqrt{1/(1+\delta_{m,0})},\ \tau_{nm} = \sqrt{(n-m)!/(n+m)!}\,\right):$$

$$\frac{m\,\mu_m}{\mu_{m-1}} = m\sqrt{1+\delta_{m,1}}\,, \qquad \frac{\tau_{n,m}}{\tau_{n,m+1}} = \sqrt{(n-m)(n+m+1)}$$

(c) Normalisation to sectorial unity $\left(\nu_n=1,\ \mu_m=1/\tau_{nm},\ \tau_{nm}=2^m m!/(2m)!\right)$:

$$\frac{m\,\mu_m}{\mu_{m-1}} = m(2m-1)\,, \qquad \frac{\tau_{n,m}}{\tau_{n,m+1}} = (2m+1)$$

8 Clenshaw Summation

If variational equations with respect to the potential coefficients $\bar{C}_{nm}$, $\bar{S}_{nm}$ are not required, we may apply Clenshaw's algorithm to calculate the sum in (1). The Clenshaw technique holds for functions $T_k(\bar{z})$ satisfying a linear homogeneous recurrence

$$T_1(\bar{z}) = a_1(\bar{z})\,T_0(\bar{z})\,, \qquad T_k(\bar{z}) = a_k(\bar{z})\,T_{k-1}(\bar{z}) + b_k(\bar{z})\,T_{k-2}(\bar{z})\,.$$

with $T_0(\bar{z}) \neq 0$ given. For it the summation algorithm of Clenshaw is written

$$\begin{aligned}
U_N &:= A_N \\
U_{N-1} &:= A_{N-1} + a_N U_N \\
\text{for } k = N-2,\dots,0: \quad U_k &:= A_k + a_{k+1}U_{k+1} + b_{k+2}U_{k+2} \qquad (20)
\end{aligned}$$

$$\text{then}: \ \sum_{k=0}^{N} A_k T_k = U_0 T_0\,.$$

How to profit from the previous results will be shown with the partial sum

$$U_0^{(m)} = \left(\left(\frac{a}{r}\right)^m \bar{H}_{m,m}(\bar{z})\right)^{-1} \sum_{n=m}^{N}\left(\bar{C}_{nm}\zeta_m + \bar{S}_{nm}\sigma_m\right)\left(\frac{a}{r}\right)^n \bar{H}_{n,m}(\bar{z})\,.$$

In this case we have to make the substitutions $k = n-m$ and

$$A_k = \bar{C}_{nm}\zeta_m + \bar{S}_{nm}\sigma_m\,, \qquad\qquad a_k = \nu_{nm}\left(\frac{a}{r}\right)\bar{z}\,,$$

$$T_k(\bar{z}) = \left(\frac{a}{r}\right)^{n-m}\eta_{nm}H_{n,m}(\bar{z})\,, \qquad\qquad b_k = \frac{\nu_{nm}}{\nu_{n-1,m}}\left(\frac{a}{r}\right)^2$$

with ν_{nm} taken from (11). The Clenshaw method can provide the partial derivatives of the sum as well. Let's exemplify that with the variable $u_3 = \bar{z}$. Commonly one differentiates the relation (20) to get (Note that in this case $\partial_{\bar{z}}A_k = \partial_{\bar{z}}b_k = \partial_{\bar{z}\bar{z}}a_k = 0$):

$$\begin{aligned}
U_k &= a_{k+1}\ \ U_{k+1} + b_{k+2}\ \ U_{k+2} + A_k \\
\partial_{\bar{z}}U_k &= a_{k+1}\,\partial_{\bar{z}}U_{k+1} + b_{k+2}\,\partial_{\bar{z}}U_{k+2} + \partial_{\bar{z}}a_{k+1}U_{k+1} \\
\partial_{\bar{z}\bar{z}}U_k &= a_{k+1}\partial_{\bar{z}\bar{z}}U_{k+1} + b_{k+2}\partial_{\bar{z}\bar{z}}U_{k+2} + \partial_{\bar{z}}a_{k+1}\partial_{\bar{z}}U_{k+1}
\end{aligned}$$

Therewith the recursions of derivatives will be annexed to the recursion of the function which amplifies the accumulated rounding error. To get decoupled recursions, write down the n-recursion for the unnormalised polynomials $H_{n,m}$, $H_{n,m+1}$, and $H_{n,m+2}$, and scale each of the equations with the same arbitrary normalisation factor η_{nm} (η_{nm} independent from $\bar{z}$), and set $\nu_{nm} := (a/r)\,\eta_{n,m}/\eta_{n-1,m}$. By (14) we arrive at three independent recurrence relations for

$$
T_{n,m}(\bar{z}) = \left(\frac{a}{r}\right)^{n-m} \eta_{nm} H_{n,m}(\bar{z}), \quad T'_{n,m}(\bar{z}) = \left(\frac{a}{r}\right)^{n-m} \eta_{nm} H_{n,m+1}(\bar{z}),
$$

$$
T''_{n,m}(\bar{z}) = \left(\frac{a}{r}\right)^{n-m} \eta_{nm} H_{n,m+2}(\bar{z}).
$$

They read

$$
T_{nm} = \frac{\nu_{nm}}{n-m} \left((2n-1)\,\bar{z}\,T_{n-1,m} - (n+m-1)\,\nu_{n-1,m}\,T_{n-2,m} \right)
$$

$$
T'_{nm} = \frac{\nu_{nm}}{n-m-1} \left((2n-1)\,\bar{z}\,T'_{n-1,m} - (n+m+0)\,\nu_{n-1,m}\,T'_{n-2,m} \right)
$$

$$
T''_{nm} = \frac{\nu_{nm}}{n-m-2} \left((2n-1)\,\bar{z}\,\tilde{T}''_{n-1,m} - (n+m+1)\,\nu_{n-1,m}\,\tilde{T}''_{n-2,m} \right)
$$

Therefrom we derive decoupled Clenshaw recursions

$$
U_{k-1,m} = A_{k-1,m} + \nu_{nm} \left[\frac{(2n-1)\,\bar{z}}{n-m} U_{k,m} + \frac{n+m+0}{n-m+1} \nu_{n+1,m} U_{k+1,m} \right]
$$

$$
U'_{k-1,m} = A_{k-1,m} + \nu_{nm} \left[\frac{(2n-1)\,\bar{z}}{n-m-1} U'_{k,m} + \frac{n+m+1}{n-m} \nu_{n+1,m} U'_{k+1,m} \right]
$$

$$
U''_{k-1,m} = A_{k-1,m} + \nu_{nm} \left[\frac{(2n-1)\,\bar{z}}{n-m-2} U''_{k,m} + \frac{n+m+2}{n-m-1} \nu_{n+1,m} U''_{k+1,m} \right]
$$

Acknowledgement. This work is partly executed in the frame of SFB 78 "Satellitengeodäsie" located at Technische Universität München and funded by the DFG (Deutsche Forschungsgemeinschaft).

References

[1] Milton Abramowitz and Irene A. Stegun. *Handbook of mathematical functions.* Dover Publications, New York, 10 edition, 1972.

[2] Michael Gerstl. *DOGS-Manual, volume X, Mathematische Grundlagen.* Deutsches Geodätisches Forschungsinstitut, München, 1999.

[3] Robert Sauer and István Szabó. *Mathematische Hilfsmittel des Ingenieurs*, volume I. Springer, Berlin, Heidelberg, New York, 1967.

Integrated Guidance and Control for Entry Vehicles

W. Grimm and W. Rotärmel

Institut für Flugmechanik und Flugregelung, Universität Stuttgart,
Pfaffenwaldring 7a, 70550 Stuttgart, Germany,
`werner.grimm@ifr.uni-stuttgart.de`

Summary. The article presents an integrated approach for the design and control of entry trajectories. The output of the analytical design process is a drag profile together with the nominal sink rate and bank control. All quantities are functions of specific energy. The trajectory design is strongly related to a path control law, which consists of a linear feedback of the drag and sink rate errors. The control design is based on the solution of a globally valid linear model. In simulations, the guidance and control algorithm produces smooth and precise flight paths.

1 Introduction

The entry flight optimization of Apollo by Bulirsch [1] is a milestone in spacecraft mission planning. The result demonstrates the performance of multiple shooting [2] as well as the economic benefit of optimal control. The minimization of the total heat load indicates the necessary size of the thermal protection system. The task of the entry guidance is the on-board correction of the flight path to compensate for disturbances and model uncertainties. Optimal guidance in the sense of neighbouring extremals [3] is naturally related to the multiple shooting method. The paper by Kugelmann and Pesch [4] is a representative example for neighbouring extremals in entry flight.

The original entry guidance of Apollo [5] and the Space Shuttle [6] is based on a drag reference profile as a function of velocity. This is quite natural as entry physically means dissipation of energy by aerodynamic drag. The guidance performs an on-board correction of the drag profile in order to keep the precise range-to-go to the target. The task of the path control is to track the modified drag profile. The classical path controller of the Shuttle is a feedback of sink rate and drag. The Space Shuttle guidance inspired a lot of conceptual work on entry guidance and control. For instance, there are real-time optimization methods for the definition and correction of the drag profile [7], [8], [9], [10].

The present paper describes an integrated approach for trajectory design, guidance, and path control.

1. Path planning consists of an analytical design of a drag-versus-energy profile. The design includes the nominal sink rate and bank control. Both are functions of specific energy, as well.
2. The guidance continuously adapts the drag profile to the required range-to-go. It is based on the observation that a global scale factor on the drag profile has the inverse effect on the predicted range [11].
3. The drag tracking control is a linear feedback of the sink rate and drag errors. The control design is based on a linear, globally valid error dynamics model.

The vehicle model underlying this study is taken from the X-38, which was planned as a rescue vehicle for the International Space Station. Unfortunately, the project was stopped in a rather advanced phase.

2 Dynamical Model

The following equations of motion provide the model for realistic entry simulations.

$$\dot{\delta} = \frac{V \cos\gamma \cos\chi}{R_E + h}, \quad \dot{\lambda} = \frac{V \cos\gamma \sin\chi}{(R_E + h) \cos\delta}, \quad \dot{h} = V \sin\gamma. \quad (1)$$

Equations (1) are kinematic equations describing the position of the spacecraft. δ, λ and h denote declination, longitude, and altitude on a spherical Earth with radius $R_E = 6378$ km. V is the absolute value of the (Earth relative) velocity vector, γ is its inclination with respect to the local horizontal plane ("flight path angle"). χ indicates the flight direction with respect to north ("heading angle").

The following dynamic equations describe the change of the velocity vector.

$$\dot{V} = -\frac{D}{m} - g \sin\gamma, \quad (2)$$

$$\dot{\gamma} = \frac{1}{V} \left(\frac{V^2 \cos\gamma}{R_E + h} + 2\omega_E V \sin\chi \cos\delta + \frac{L}{m} \cos\mu - g \cos\gamma \right), \quad (3)$$

$$\dot{\chi} = \frac{1}{V \cos\gamma} \left[\frac{V^2 (\cos\gamma)^2 \sin\chi \tan\delta}{R_E + h} \right.$$

$$\left. -2\omega_E V (\sin\gamma \cos\chi \cos\delta - \cos\gamma \sin\delta) + \frac{L}{m} \sin\mu \right]. \quad (4)$$

The Earth model consists of the gravity model

$$g = g_0 \left(\frac{R_E}{R_E + h} \right)^2 \quad \text{with} \quad g_0 = 9.80665 \, \text{m/s}^2$$

and the air density

$$\varrho = \varrho_0 \, e^{\sigma(h)} \quad (5)$$

with tabulated data for $\sigma(h)$. $\varrho_0 = 1.225\,\mathrm{kg/m^3}$ denotes air density at sea level. $\omega_E = 7.29 \times 10^{-5}\,\mathrm{rad/s}$ is the turn rate of Earth. The bank angle μ is the only control variable. It is the angle between the lift vector and the vertical plane through the velocity vector. μ is positive if the lift vector is on the righthand side in flight direction and vice versa. Lift L and drag D are modelled as

$$L = q\,S\,c_L, \quad D = q\,S\,c_D \tag{6}$$

with the dynamic pressure $q = \varrho V^2/2$. The model functions and constants are taken from the X-38 project. In (6) $S = 21.31\,\mathrm{m^2}$ denotes the reference wing area of the X-38. The lift and drag coefficients c_L and c_D are tabulated functions of V. It is typical for all entry vehicles that the lift-to-drag ratio

$$E = \frac{L}{D} = \frac{c_L}{c_D}$$

is nearly constant. For the X-38, the value is slightly below 1 all the time. Finally, $m = 9060\,\mathrm{kg}$ is the vehicle's mass.

The most critical vehicle constraint is the heat flux limit:

$$\dot{Q} = C_Q\,\sqrt{\varrho}\,V^{3.15} \leq \dot{Q}_{max} . \tag{7}$$

The assumed data of the X-38 are $C_Q = 10^{-4}$ and $\dot{Q}_{max} = 1.2 \times 10^6\,\mathrm{W/m^2}$. Finally, path planning needs the definition of specific energy:

$$e = \frac{V^2}{2\,g_0} + \frac{R_E}{R_E + h}\,h . \tag{8}$$

Specific energy means total energy per unit weight, its dimension is m. Differentiation of e and substitution of (1), (2) yields

$$\dot{e} = -\frac{V\,D}{m\,g_0} . \tag{9}$$

3 Path Control Design

The task of the path control is to track a predetermined flight path, which is represented as a drag profile in classical entry guidance. In terms of mathematics, the path control law is a mapping from the state space into the control space. In this application, the bank angle μ is controlled such that the vehicle approaches the desired flight path and remains close to it.

3.1 Control Design Model

The classical entry path control [6], [5] is a feedback of the drag and sink rate measurements. In this section a control design model is set up having exactly these outputs as states. The derivation is based on the following simplifications:

1. The air density model (5) reduces to the pure exponential function

$$\varrho(h) = \varrho_0 \, exp(-h/h_s)$$

   with the constant parameter h_s ("scale height"). In reality h_s varies between 7 and 10 km.
2. The drag coefficient c_D is constant.
3. g_0/V^2 may be neglected compared to inverse scale height:

$$\frac{g_0}{V^2} \ll \frac{1}{h_s} \, .$$

4. Specific energy essentially consists of kinetic energy: $e = V^2/(2g_0)$.

Assumptions 1, 2 are standard on entry guidance and control design. Because of assumptions 1–3 the state variables

$$x_1 = \frac{D}{m \, g_0}, \quad x_2 = \frac{\dot h}{h_s}$$

satisfy the differential equations

$$\dot x_1 = x_1 \left[-\frac{2g_0}{V} x_1 - x_2\right], \quad \dot x_2 = \frac{\ddot h}{h_s} \, .$$

x_1 denotes dimensionless normalized drag, x_2 means normalized sink rate. $\ddot h$ takes the role of the control as it is directly related to the bank angle. Instead of time the dissipated energy $\Delta e = e_{init} - e$ will act as independent variable. e_{init} is the initial value of e. According to (9) its rate is

$$\Delta \dot e = V \, x_1 \, .$$

Change of independent variables and application of assumption 4 yields

$$\begin{pmatrix} x_1' \\ x_2' \end{pmatrix} = \begin{pmatrix} -\frac{1}{e} & -\frac{1}{\sqrt{2g_0 e}} \\ 0 & 0 \end{pmatrix} \begin{pmatrix} x_1 \\ x_2 \end{pmatrix} + \begin{pmatrix} 0 \\ 1 \end{pmatrix} \frac{\ddot h}{h_s V x_1} \, .$$

The prime denotes derivatives with respect to Δe. The control transformation $u = \ddot h/(h_s V x_1)$ leads to the ultimate form

$$\begin{pmatrix} x_1' \\ x_2' \end{pmatrix} = \begin{pmatrix} -\frac{1}{e} & -\frac{1}{\sqrt{2g_0 e}} \\ 0 & 0 \end{pmatrix} \begin{pmatrix} x_1 \\ x_2 \end{pmatrix} + \begin{pmatrix} 0 \\ 1 \end{pmatrix} u \, . \tag{10}$$

System (10) is remarkable because of its linearity. However, the coefficients of the system matrix are functions of the independent variable. The purpose of the path control will be to track a nominal trajectory $x_{nom}(e)$. Shaping of $x_{nom}(e)$ will be the task of the path planning module. Both the actual trajectory $x(e)$ and the nominal trajectory $x_{nom}(e)$ satisfy (10) with the controls

$u(e)$ and $u_{nom}(e)$, respectively. As system (10) is linear it is also valid for $\Delta x = x - x_{nom}$ (control error) and the related control $\Delta u = u - u_{nom}$:

$$\begin{pmatrix} \Delta x_1' \\ \Delta x_2' \end{pmatrix} = \begin{pmatrix} -\frac{1}{e} & -\frac{1}{\sqrt{2g_0 e}} \\ 0 & 0 \end{pmatrix} \begin{pmatrix} \Delta x_1 \\ \Delta x_2 \end{pmatrix} + \begin{pmatrix} 0 \\ 1 \end{pmatrix} \Delta u \ . \tag{11}$$

Note that system (11) is a globally valid linear design model for the path controller. This is due to the fact that the original dynamics (10) is linear already; there is no artificial linearization about any operating point.

3.2 Control Law Structure

Tracking of $x_{nom}(e)$ will be accomplished with a state feedback of the form

$$\Delta u = k_1 \frac{\sqrt{2g_0}}{e^{3/2}} \Delta x_1 - k_2 \frac{1}{e} \Delta x_2 \ . \tag{12}$$

Note that $k_{1,2} > 0$ are dimensionless gains. Equation (12) is a command for the artificial control u. However, the actual control of the entry vehicle is the bank angle μ or, more precisely, the vertical lift component, which is proportional to $\cos\mu$. u is directly related to $\cos\mu$ because of

$$\ddot{h} = \dot{V} \sin\gamma + V \cos\gamma \, \dot{\gamma} \ ,$$

and $\cos\mu$ appears in $\dot{\gamma}$ according to (3). Altogether, there is a relation of the form

$$u = A \cos\mu + B \quad \text{with} \quad A = \frac{L\,g}{V\,D\,h_s}$$

and a suitable state dependent function B. The nominal trajectory $x_{nom}(e)$ is flown with the nominal control $\cos\mu_{nom}$, or expressed in terms of u:

$$u_{nom} = A \cos\mu_{nom} + B \ .$$

By definition,

$$\Delta u = u - u_{nom} = A \left(\cos\mu - \cos\mu_{nom} \right) = A \, \Delta \cos\mu \ .$$

Hence, the ultimate control command for $\cos\mu$ is

$$\cos\mu = \cos\mu_{nom} + \frac{\Delta u}{A}. \tag{13}$$

Strictly speaking, the coefficients A, B are state dependent and differ along the actual and nominal flight path. This difference is suppressed here.

3.3 Control Sensitivity

The first step of the control design is to find a reasonable value for k_1, the gain of Δx_1 in the control law (12). The choice of k_1 is based on the consideration that the relative drag error $\Delta x_1/x_1$ shall cause an adequate control action $\Delta \cos \mu$. Inserting control law (12) for Δu in the bank command (13) yields a relation of the form

$$\Delta \cos \mu = F\, k_1\, \frac{\Delta x_1}{x_1} + G\, k_2\, \Delta x_2$$

with

$$F = \frac{4\, g\, h_s\, x_1}{E\, V^2} \tag{14}$$

and some suitable state dependent function G. For $\Delta x_2 = 0$ the factor $F\, k_1$ describes the ratio of the bank command $\Delta \cos \mu$ and the relative drag error $\Delta x_1/x_1$, which induces the command:

$$F\, k_1 = \frac{\Delta \cos \mu}{\Delta x_1/x_1} = C_s\, . \tag{15}$$

The input-output relation (15) of the controller is called control sensitivity. Its dimensionless value C_s is a specification parameter of the controller. For instance, suppose the requirement is that a drag error of 1% shall induce $\Delta \cos \mu = 0.1$. This would correspond to the control sensitivity $C_s = 10$. As soon as the value of C_s is selected, k_1 results from (15):

$$k_1 = C_s/F\, . \tag{16}$$

Note that k_1 given by (16) is a state dependent, time-varying function. Considering the factor F in (14) the V^2-term in the denominator compensates the V^2-dependence of x_1 in the numerator. As all remaining factors are constant or nearly constant, we have

$$k_1 \sim 1/\varrho\, .$$

3.4 Stabilization of the Error Dynamics

Insertion of control law (12) in the error dynamics (11) results in the linear closed-loop model

$$\begin{pmatrix} \Delta x_1' \\ \Delta x_2' \end{pmatrix} = \begin{pmatrix} -\frac{1}{e} & -\frac{1}{\sqrt{2g_0 e}} \\ k_1\, \frac{\sqrt{2g_0}}{e^{3/2}} & -k_2\, \frac{1}{e} \end{pmatrix} \begin{pmatrix} \Delta x_1 \\ \Delta x_2 \end{pmatrix}\, . \tag{17}$$

The selection of the second gain k_2 is based on the requirement that the closed-loop system (17) is stable. That means Δx tends to zero from any initial condition. Fortunately, system (17) can be solved in closed form. We seek a solution of the form

$$\Delta x_1 = e^a, \quad \Delta x_2 = c e^b . \tag{18}$$

Note that e stands for energy, not for Euler's number. Equation (18) represents a solution of system (17) if the following conditions hold:

$$k_2 = \frac{k_1}{\bar{c}} + \bar{c} + \frac{1}{2}, \quad a = 1 + \bar{c}, \quad b = \frac{1}{2} + \bar{c} \quad \text{with} \quad \bar{c} = \frac{c}{\sqrt{2 g_0}} . \tag{19}$$

For given gains $k_{1,2}$ the first equation in (19) represents a quadratic equation for $\bar{c}$:

$$\bar{c}^2 + \left(\frac{1}{2} - k_2 \right) \bar{c} + k_1 = 0 . \tag{20}$$

The two solutions of (20) yield two linearly independent fundamental solutions of the second order linear system (17), having the form (18). Stabilization of the closed loop (17) is equivalent to the following task: For given k_1 find k_2 such that both fundamental solutions tend to zero. As energy decreases solutions of the form (18) vanish if $a, b > 0$, which holds for $\bar{c} > -\frac{1}{2}$. Now, the task is to set k_2 such that $\bar{c}_{1,2} > -\frac{1}{2}$ for both solutions of (20). This is guaranteed by the following strategy: Select $\bar{c}_0 > 0$ and evaluate k_2 according to (19). Then, the two solutions of (20) are

$$\bar{c}_1 = \bar{c}_0 > 0, \quad \bar{c}_2 = \frac{k_1}{\bar{c}_0} > 0 .$$

Note that the consideration above is valid for constant gains $k_{1,2}$, only. It must be checked in simulations that stability is not endangered by time-varying gains as in (16).

4 Path Planning

The path planning module provides the nominal trajectory $x_{nom}(e)$ for $e_f \leq e \leq e_{init}$. e_f denotes the energy value on the transition to the terminal flight phase, which is initiated by a parachute deployment on the X-38. e_{init} denotes initial energy of the guided flight. $x_{nom}(e)$ will be composed of three arcs:

1. $e_f \leq e \leq e_{12}$: $x_{nom}^{(1)}(e)$ is such that $\ddot{h} \approx 0$ throughout.
2. $e_{12} < e \leq e_{23}$: $x_{nom}^{(2)}(e)$ runs close to the heat flux boundary.
3. $e_{23} < e \leq e_{init}$: $x_{nom}^{(3)}(e)$ is such that $\dot{\gamma} \approx 0$ throughout.

The junction points e_{12}, e_{23} are free parameters, which are determined by boundary conditions later on. The design 1–3 is based on profound experience with entry dynamics.

4.1 Sink Rate, Control, and Range

For all three segments the drag histories $x_{nom,1}^{(i)}(e), i = 1, 2, 3$, are defined first. Control law (12), (13) additionally requires the related sink rate $x_{nom,2}^{(i)}(e)$ and the nominal control $\cos \mu_{nom}^{(i)}$ as functions of energy. The sink rate is obtained from the simplified dynamics (10):

$$x_2 = \sqrt{2g_0 e} \, (\frac{dx_1}{de} - \frac{x_1}{e}) \, . \tag{21}$$

Note that the prime in (10) stands for $-d/de$. For the nominal bank angle the differential equation (3) for γ is simplified first:

$$V \dot{\gamma} = \frac{L}{m} \cos \mu - g_0 \, (1 - \frac{e}{e_c}) \, . \tag{22}$$

Equation (22) uses the following simplifications:

1. $\cos \gamma = 1$: This is justified because of the small flight path angle during entry.
2. $\omega_e = 0$: The contribution of Coriolis acceleration is neglected.
3. $R_E + h \approx R_E$: This allows to replace $g \, (R_E + h)$ by the square of circular speed V_c (=7.9 km/s). $e_c = V_c^2/(2g_0)$ is the corresponding energy.

On the other hand, differentiation of $\dot{h} = V \sin \gamma$ gives

$$\ddot{h} = V \cos \gamma \, \dot{\gamma} + \dot{V} \sin \gamma \approx V \dot{\gamma} - (\frac{D}{m} + g_0 \sin \gamma) \sin \gamma \, . \tag{23}$$

Equation (23) also uses simplification 1. Note that all quantities in (23) may be expressed in terms of energy, x_1, and x_2:

$$V = \sqrt{2g_0 e}, \quad \frac{D}{m} = g \, x_1, \quad \sin \gamma = \frac{\dot{h}}{V} = \frac{x_2 \, h_s}{V} \, .$$

$\ddot{h}$ is related to $d\dot{h}/de$ by the chain rule:

$$\ddot{h} = \frac{d\dot{h}}{dt} = \frac{d\dot{h}}{de} \frac{de}{dt} = -\frac{d\dot{h}}{de} V \, x_1 \, .$$

Eliminating $V \dot{\gamma}$ from (22), (23) yields a good estimate for $\cos \mu$:

$$\cos \mu = \frac{1}{E \, x_1} \, [1 - \frac{e}{e_c} - \frac{d\dot{h}}{de} \frac{V \, x_1}{g_0} + (x_1 + \frac{\dot{h}}{V}) \frac{\dot{h}}{V}] \, . \tag{24}$$

(21) and (24) are used to derive the sink rate and the bank angle from a given profile $x_1(e)$. If $x_1(e)$ has a closed-form representation, this applies to $x_2(e)$ and $\cos \mu$, too. The drag history $x_1(e)$ also determines the range along a flight segment $e_1 \le e \le e_2$:

$$s = \int_{t_2}^{t_1} V \, dt = \int_{e_1}^{e_2} \frac{de}{x_1} \, . \tag{25}$$

If the drag history $x_1(e)$ is globally scaled with some constant $d > 0$ the resulting range is multiplied by $1/d$. This observation is essential for the guidance later on.

4.2 Flight in the Low Energy Regime

In the energy interval $e_f \le e \le e_{12}$ the flight path is designed such that the vertical acceleration $\ddot{h}$ is nearly zero. This is just the homogeneous solution of the simplified plant model (10). As the system matrix is the same as for the error dynamics (11) the solution is of the form (18), (19) with $k_{1,2} = 0$. The two fundamental solutions are

$$
\begin{array}{ll}
x_{1,1}(e) = e \, , & x_{2,1}(e) = 0 \, , \\
x_{1,2}(e) = \sqrt{e} \, , & x_{2,2}(e) = -\frac{\sqrt{2g_0}}{2} = \text{const.}
\end{array}
$$

The general homogeneous solution of system (10) is a linear combination of the two fundamental solutions:

$$
\begin{array}{ll}
x_{nom,1}^{(1)}(e) = \alpha_1 \, e + \alpha_2 \sqrt{e} & \text{(normalized drag)} \, , \\
x_{nom,2}^{(1)}(e) = -\alpha_2 \frac{\sqrt{2g_0}}{2} & \text{(normalized sink rate)} \, .
\end{array}
\tag{26}
$$

$x_{nom}^{(1)}(e)$ represents the nominal trajectory in the low energy regime. The sink rate $x_{nom,2}^{(1)}(e)$ also follows from $x_{nom,1}^{(1)}(e)$ using (21). As the sink rate $x_{nom,2}^{(1)}$ is constant the nominal control $\cos\mu_{nom}^{(1)}$ is given by (24) with $d\dot{h}/de = 0$. $\alpha_{1,2}$ are constants of integration, which will be determined by boundary conditions.

4.3 Flight Along the Heat Flux Boundary

In the energy interval $e_{12} < e \le e_{23}$ the flight path is designed such that heat flux is a fraction f of its admissible maximum value: $\dot{Q} = f \, \dot{Q}_{max}$ with $0 < f \le 1$. f is a free parameter, which is adapted to boundary conditions later on. The nominal drag profile $x_{nom,1}^{(2)}(e)$ is derived from the heat flux model (7) as follows. Dynamic pressure appears in the square of $\dot{Q}$:

$$\dot{Q}^2 = C_Q^2 \, \varrho \, V^{6.3} = 2 \, C_Q^2 \, q \, V^{4.3} \, .$$

The term $S \, c_D \, \dot{Q}^2$ contains drag:

$$S \, c_D \, \dot{Q}^2 = 2 \, C_Q^2 \, D \, V^{4.3} \, .$$

The nominal drag profile $x_{nom,1}^{(2)}(e)$ along the heat flux boundary is given by $\dot{Q} = f \, \dot{Q}_{max}$ with $0 < f \le 1$:

$$x^{(2)}_{nom,1}(e) = \frac{S\,c_D\,f^2\,\dot{Q}^2_{max}}{2\,m\,g_0\,C^2_Q\,V^{4.3}} = \frac{S\,c_D\,f^2\,\dot{Q}^2_{max}}{2\,m\,g_0\,C^2_Q\,(2g_0 e)^{2.15}} = \frac{f^2\,P}{e^{2.15}}. \tag{27}$$

P contains all terms except f and energy. These terms are constant except c_D, which depends on Mach number or velocity. However, for entry vehicles the variation of c_D is small. Therefore, it is sufficient to replace P by a suitable mean value on path planning. The normalized sink rate $x^{(2)}_{nom,2}(e)$ and the nominal control $\cos\mu^{(2)}_{nom}$ are derived from the drag profile $x^{(2)}_{nom,1}(e)$ using (21) and (24).

4.4 Flight in the High Energy Regime

In the energy interval $e_{23} < e \le e_{init}$ the flight path is designed such that flight path angle γ and bank angle μ are nearly constant. In this phase, the nominal drag profile $x^{(3)}_{nom,1}(e)$ is given by (22) with $\dot{\gamma} = 0$:

$$\frac{L}{m\,g_0}\cos\mu = 1 - \frac{e}{e_c}.$$

$\cos\mu = \cos\mu_3$ plays the role of a constant parameter in this flight segment, it will be adapted to boundary conditions later on. Lift is expressed by drag and the lift-to-drag ratio E, which is nearly constant. This gives a simple expression for the drag profile:

$$x^{(3)}_{nom,1}(e) = \frac{1}{E\,\cos\mu_3}\left(1 - \frac{e}{e_c}\right). \tag{28}$$

Again, the nominal sink rate $x^{(3)}_{nom,2}(e)$ follows with (21). The nominal control is simply $\cos\mu^{(3)}_{nom} = \cos\mu_3 = \text{const}.$

4.5 Boundary Conditions

The trajectory model $x^{(i)}_{nom}(e), i = 1, 2, 3,$ contains six free parameters:

$$e_{12},\ e_{23},\ \alpha_1,\ \alpha_2,\ \cos\mu_3,\ f\ .$$

For a given value of $f \in (0, 1]$ the first five parameters are determined by the following five boundary conditions:

1. specified value of $x^{(1)}_{nom,1}(e_f)$
2. continuity of $x_{nom,1}(e)$ and $dx_{nom,1}/de$ at e_{12} and e_{23}

Because of the remaining free parameter f the boundary conditions above define a one-parametric set of trajectories. f may be determined by specifying the range s in (25). Note that the drag profiles $x^{(i)}_{nom,1}(e), i = 1, 2, 3,$ are analytic functions of energy with the property that the range integral (25) can be represented in closed form, too. The reader is referred to [12] for details.

   There are interesting conclusions from the continuity conditions at e_{23}:

1. The junction point e_{23} is independent of any boundary values:

$$e_{23} = e_c \frac{2.15}{3.15} \approx 2177\,\text{km} .$$

2. Another result is a relation between f and $\cos \mu_3$:

$$f^2 = \frac{e_{23}^{3.15}}{2.15\,P\,E\,e_c\,\cos \mu_3} .$$

This relation implies a minimum value for f because of $\cos \mu_3 \leq 1$:

$$f_{min}^2 = \frac{e_{23}^{3.15}}{2.15\,P\,E\,e_c} .$$

For the present model we have $f_{min} = 0.88$. I.e. heat flux cannot be reduced below 88% of its maximum value.

Altogether, there is a one-parametric trajectory set, which covers a certain range spectrum. Minimum range is achieved for $f = 1$, maximum range occurs for $f = f_{min}$.

5 Simulation Results

The guidance and control concept is tested with the simulation model (1)–(6) to validate the performance of the tracking control and the precision of the guidance.

5.1 Boundary Conditions and Nominal Trajectory

The "target" is the final destination of the vehicle:

$$\lambda_f = 134.98^o, \quad \delta_f = -28.2425^o . \tag{29}$$

The data belong to some location in Australia, which was foreseen as the landing site of the X-38. The simulation stops as soon as specific energy is less than $e_f = 51\,\text{km}$. The final miss distance to the target indicates the precision of the entry guidance. The initial state is taken from a X-38 mission plan, too:

$$\begin{aligned} \lambda_{init} &= 24.5939^o, & \delta_{init} &= -40.7597^o, & h_{init} &= 121.92\,\text{km} , \\ V_{init} &= 7606.4\,\text{m/s} , & \gamma_{init} &= -1.3705^o, & \chi_{init} &= 126.667^o. \end{aligned} \tag{30}$$

V_{init} and h_{init} determine the initial value $e_{init} = 3070\,\text{km}$ according to (8). First, a nominal flight path $x_{nom}(e)$ is designed according to the procedure in Chap. 4. The heat flux scaling factor f (see Sect. 4.3) is set to $f = 0.9055$. The nominal range

$$s_{nom}(e_{init}) = \int_{e_f}^{e_{init}} \frac{de}{x_{nom,1}(e)}$$

is 9923 km. The great circle distance between the initial position and the target is $s_{init} = 9530\,\text{km}$. The guidance is expected to compensate the difference.

5.2 Guidance and Control Concept

For any instant t, the control command $\cos\mu$ is defined as described below. e_0 denotes the current value $e(t)$ of energy defined in (8). All other variables of the current state are written without any index.

1. Measure the great circle distance s between the current position and the target. Compute the predicted range

$$s_{nom}(e_0) = \int_{e_f}^{e_0} \frac{de}{x_{nom,1}(e)} \ .$$

   Then, the current drag command is

$$x_{ref,1}(e_0) = x_{nom,1}(e_0) \frac{s_{nom}(e_0)}{s} \ . \tag{31}$$

   The nominal sink rate may also be adapted to the scaled drag profile to get some reference value $x_{ref,2}(e_0)$. Details are given in [12].
2. Evaluate the errors in drag and sink rate:

$$\Delta x_1 = x_1 - x_{ref,1}(e_0), \quad \Delta x_2 = x_2 - x_{ref,2}(e_0) \ .$$

3. Evaluate the path control gains $k_{1,2}$ according to (16), (19). The parameters C_s ("control sensitivity") and $\bar{c}$ are set to $C_s = 4$ and $\bar{c} = 5$.
4. Finally, evaluate the control command $\cos\mu$ as given by (12), (13).

The heading angle χ is also affected by the sign of μ as can be seen in (4). If $\sin\mu$ was positive throughout, the vehicle would fly a right turn all the time. To prevent a lateral deviation in target direction, the sign of μ is changed at certain time intervals. This "bank reversal strategy" is a classical part of entry guidance [6] and is not described here in detail.

5.3 Results

Figure 1 depicts the results of the scenario in Sect. 5.1. All quantities are drawn as functions of energy. In the diagrams, the flight starts on the right end at $e_{init} = 3070\,\text{km}$ and terminates on the left end at $e_f = 51\,\text{km}$. The flight duration is 1589 s, the final miss distance to the target is about 2.3 km. In the upper two diagrams the simulated variables (x_1 and $\dot{h}$) are drawn as dashed lines. The nominal values are depicted as solid lines, and the dotted lines indicate the reference values (31). In the x_1-diagram two path constraints are added. The upper constraint (upper dashed line) is the maximum admissible drag along the heat flux limit. The maximum is given by (27) with $f = 1$. The lower constraint for x_1 (lower dashed line) is given by (28) with $\mu_3 = 0$. In Fig. 1 the variation of the lift-to-drag-ratio E along the energy axis is modelled, too. Therefore, the lower constraint is not simply a straight line as expected from (28). The significant difference between $x_{nom,1}$ and $x_{ref,1}$

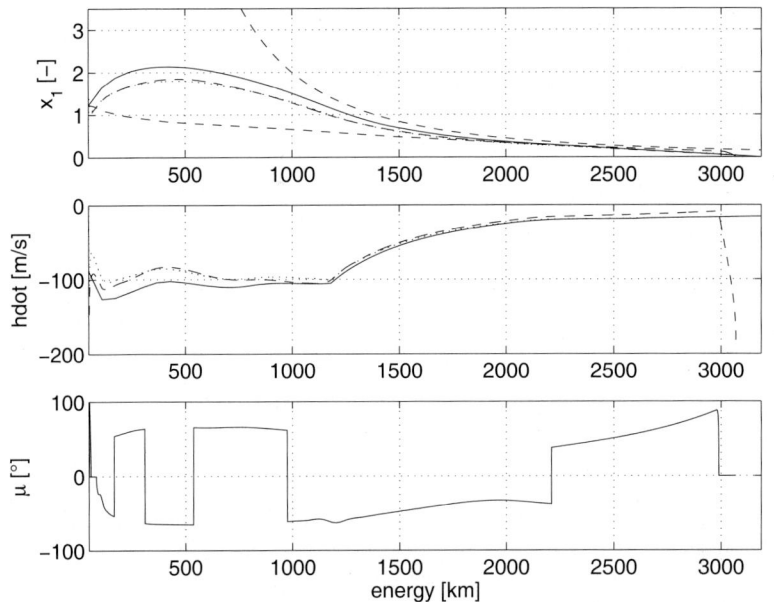

Fig. 1. Simulation result with the X-38 model.

is the consequence of the guidance law (31). The range correction (31) is a simple but effective guidance strategy to maintain the agreement of the actual distance to the target and the predicted range along the remaining flight path [11]. The close agreement of the of the drag command $x_{ref,1}$ and the actual drag demonstrates the tracking performance of the path control. The same holds for the sink rate. In particular, the initial value $\dot{h}(e_{init})$, which corresponds to the initial inclination γ_{init}, is quickly corrected to follow the required command.

The bank angle history shows the bank reversals mentioned in the previous section. Except these commanded jumps $\mu(e)$ is a smooth curve. This is important as μ plays the role of the attitude command for the attitude control in reality.

The flight path itself is depicted by the colorband in Fig. A.29 on page 342. It starts at the entry conditions (30) and terminates at the desired destination (29). The color indicates the heat flux. The maximum value encountered along the simulated flight is $1.057 \times 10^6 \, \text{W/m}^2$, which is about 88% of the limit $\dot{Q}_{max} = 1.2 \times 10^6 \, \text{W/m}^2$. This agrees pretty well with the mission design in Sect. 5.1, where the heat flux was constrained to 90.55% of $\dot{Q}_{max}$. Terminal altitude and speed are given by $h_f = 24.87 \, \text{km}$ and $V_f = 717 \, \text{m/s}$, which is about twice of the speed of sound. This is considered as a suitable initial state for the landing phase with the help of a paraglide.

References

[1] Stoer, J., Bulirsch, R.: Introduction to Numerical Analysis. Springer, Berlin (2002)

[2] Bulirsch, R.: Die Mehrzielmethode zur numerischen Lösung von nichtlinearen Randwertproblemen und Aufgaben der optimalen Steuerung. Report of the Carl-Cranz Gesellschaft, DLR, Oberpfaffenhofen, Germany, 1971

[3] Bryson, A. E., JR., and Ho, Y. C.: Applied Optimal Control. Hemisphere, New York (1975)

[4] Kugelmann, B., and Pesch, H. J.: New General Guidance Method in Constrained Optimal Control, Part 1: Numerical Method, Part 2: Application to Space Shuttle Guidance. Journal of Optimization Theory and Applications, **67**, 421–446 (1990)

[5] Graves, C.A., and Harpold, J.C.: Apollo Experience Report – Mission Planning for Apollo Entry. NASA TN D-6275, March 1972

[6] Harpold, J.C., Graves, C.A.: Shuttle Entry Guidance. The Journal of the Astronautical Sciences, **27**, 239–268 (1979)

[7] Roenneke, A.J.: Adaptive On-Board Guidance for Entry Vehicles. AIAA Paper 2001-4048, Aug. 2001

[8] Lu, P., and Hanson, J.M.: Entry Guidance for the X-33 Vehicle. Journal of Spacecraft and Rockets, **35**, 342–349 (1998)

[9] Lu, P.: Entry Guidance and Trajectory Control for Reusable Launch Vehicle. Journal of Guidance, Control, and Dynamics, **20**, 143–149 (1997)

[10] Grimm, W., van der Meulen, J.G., and Roenneke, A.J.: An Optimal Update Scheme for Drag Reference Profiles in an Entry Guidance. Journal of Guidance, Control, and Dynamics, **26**, 695–701 (2003)

[11] Hanson, J.M., Coughlin, D.J., Dukeman, G.A., Mulqueen, J.A., and McCarter, J.W.: Ascent, Transition, Entry, and Abort Guidance Algorithm Design for the X-33 Vehicle. AIAA Paper 98-4409, Aug. 1998

[12] Rotärmel, W.: Entry Guidance and Path Control of the X-38. Thesis, Universität Stuttgart, Stuttgart (2005)

A Note on Nonsmooth Optimal Control Problems

Hans Joachim Oberle

Department of Mathematics, University of Hamburg, Germany,
oberle@math.uni-hamburg.de

Summary. The paper is concerned with general optimal control problems (OCP) which are characterized by a nonsmooth ordinary state differential equation. More precisely, we assume that the right-hand side of the state equation is piecewise smooth and that the switching points, which separate these pieces, are determined as roots of a state- and control dependent (smooth) switching function. For this kind of optimal control problems necessary conditions are developed. Special attention is payed to the situation that the switching function vanishes identically along a nontrivial subarc. Such subarcs, which are called singular state subarcs, are investigated with respect to the necessary conditions and to the junction conditions. In extension to earlier results, cf. [5], in this paper the case of a zero-order switching function is considered.

1 Nonsmooth Optimal Control Problems, Regular Case

We consider a general OCP with a piecewise defined state differential equation. The problem has the following form.

Problem (P1). Determine a piecewise continuous control function $u : [a, b] \to \mathbb{R}$, such that the functional

$$I = g(x(b)) \tag{1}$$

is minimized subject to the following constraints (state equations, boundary conditions, and control constraints)

$$x'(t) = f(x(t), u(t)), \quad t \in [a, b] \quad \text{a.e.,} \tag{2a}$$
$$r(x(a), x(b)) = 0, \tag{2b}$$
$$u(t) \in U = [u_{\min}, u_{\max}] \subset \mathbb{R}. \tag{2c}$$

The right-hand side of the state equation (2a) may be of the special form

$$f(x, u) = \begin{cases} f_1(x, u), & \text{if} \quad S(x, u) \leq 0, \\ f_2(x, u), & \text{if} \quad S(x, u) > 0, \end{cases} \tag{3}$$

where the functions $S : \mathbb{R}^{n+1} \to \mathbb{R}$, $f_k : \mathbb{R}^n \times \mathbb{R} \to \mathbb{R}^n$ $(k = 1, 2)$, and $r : \mathbb{R}^n \times \mathbb{R}^n \to \mathbb{R}^\ell$, $\ell \in \{0, \ldots, 2n\}$, are assumed to be sufficiently smooth. S is called the *switching function* of Problem (P1).

Our aim is to derive necessary conditions for Problem (P1). To this end, let (x^0, u^0) denote a solution of the problem with a piecewise continuous optimal control function u^0.

Further, we assume that the problem is *regular* with respect to the minimum principle, that is: For each $\lambda, x \in \mathbb{R}^n$ both *Hamiltonians*

$$H_j(x, u, \lambda) := \lambda^{\mathrm{T}} f_j(x, u), \quad j = 1, 2, \tag{4}$$

possess a unique minimum u_j^0 with respect to the control $u \in U$.

Finally, for this section, we assume that the following regularity assumption holds.

Regularity Condition (R). There exists a finite grid $a =: t_0 < t_1 < \ldots < t_s < t_{s+1} := b$ such that the optimal switching function $S[t] := S(x^0(t), u^0(t))$ is either positive or negative in each open subinterval $]t_{j-1}, t_j[$, $j = 1, \ldots, s+1$.

Note, that the one-sided-limits $u(t_j^\pm)$ exist due to the assumption of the piecewise continuity of the optimal control. Now, we can summarize the necessary conditions for Problem (P1). Here, on each subinterval $[t_j, t_{j+1}]$, we denote $H(x, u, \lambda) := H_k(x, u, \lambda)$ where $k \in \{1, 2\}$ is chosen according to the sign of S in the corresponding subinterval.

Theorem 1. With the assumptions above the following necessary conditions hold. There exist an adjoint variable $\lambda : [a, b] \to \mathbb{R}^n$, which is a piecewise C^1–function, and Lagrange multipliers $\nu_0 \in \{0, 1\}$, $\nu \in \mathbb{R}^\ell$, such that (x^0, u^0) satisfies

$$\lambda'(t) = -H_x(x^0(t), u^0(t), \lambda(t)), \quad t \in [a, b] \text{ a.e.} \quad \text{(adjoint equations),} \tag{5a}$$

$$u^0(t) = \operatorname{argmin}\{H(x^0(t), u, \lambda(t)) : u \in U\} \quad \text{(minimum principle),} \tag{5b}$$

$$\lambda(a) = -\frac{\partial}{\partial x^0(a)}[\nu^T r(x^0(a), x^0(b))] \quad \text{(natural boundary conditions),} \tag{5c}$$

$$\lambda(b) = \frac{\partial}{\partial x^0(b)}[\nu_0 g(x^0(b)) + \nu^T r(x^0(a), x^0(b))], \tag{5d}$$

$$\lambda(t_j^+) = \lambda(t_j^-), \quad j = 1, \ldots, s, \quad \text{(continuity condition),} \tag{5e}$$

$$H[t_j^+] = H[t_j^-], \quad j = 1, \ldots, s, \quad \text{(continuity condition).} \tag{5f}$$

Proof. Without loss of generality, we assume, that there is just *one* point $t_1 \in]a, b[$, where the switching function $S[\cdot]$ changes sign. Moreover, we assume that the following *switching structure* holds

$$S[t] \quad \begin{cases} < 0, & \text{if} \quad a \le t < t_1 \\ > 0, & \text{if} \quad t_1 < t \le b. \end{cases} \tag{6}$$

We compare the optimal solution (x^0, u^0) only with those admissible solutions (x, u) of the problem which have the same switching structure (6). Each candidate of this type can by associated with its separated parts $(\tau \in [0, 1])$

$$x_1(\tau) := x(a + \tau(t_1 - a)), \; x_2(\tau) := x(t_1 + \tau(b - t_1)),$$
$$u_1(\tau) := u(a + \tau(t_1 - a)), \; u_2(\tau) := u(t_1 + \tau(b - t_1)). \tag{7}$$

Now, $(x_1, x_2, t_1, u_1, u_2)$ performs an admissible and $(x_1^0, x_2^0, t_1^0, u_1^0, u_2^0)$ an optimal solution of the following auxillary optimal control problem.

Problem (P1'). Determine a piecewise continuous control function $u = (u_1, u_2) : [0, 1] \to \mathbb{R}^2$, such that the functional

$$I = g(x_2(1)) \tag{8}$$

is minimized subject to the constraints

$$x_1'(\tau) = (t_1 - a)f_1(x_1(\tau), u_1(\tau)), \quad \tau \in [0, 1], \quad \text{a.e.,} \tag{9a}$$
$$x_2'(\tau) = (b - t_1)f_2(x_2(\tau), u_2(\tau)), \tag{9b}$$
$$t_1'(\tau) = 0, \tag{9c}$$
$$r(x_1(0), x_2(1)) = 0, \tag{9d}$$
$$x_2(0) - x_1(1) = 0, \tag{9e}$$
$$u_1(\tau), u_2(\tau) \in U \subset \mathbb{R}. \tag{9f}$$

Problem (P1') is a classical optimal control problem with a smooth right-hand side, and $(x_1^0, x_2^0, t_1^0, u_1^0, u_2^0)$ is a solution of this problem. Therefore, we can apply the well–known necessary conditions of optimal control theory, cf. [2], [3], and [4], i.e there exist continuous and piecewise continuously differentiable adjoint variables $\lambda_j : [0, 1] \to \mathbb{R}^n, j = 1, 2$, and Lagrange-multipliers $\nu_0 \in \{0, 1\}, \nu \in \mathbb{R}^\ell$, and $\nu_1 \in \mathbb{R}^n$, such that with the Hamiltionian

$$\widetilde{H} := (t_1 - a) \lambda_1^{\mathrm{T}} f_1(x_1, u_1) + (b - t_1) \lambda_2^{\mathrm{T}} f_2(x_2, u_2), \tag{10}$$

and the augmented performance index

$$\Phi := \nu_0 g(x_2(1)) + \nu^{\mathrm{T}} r(x_1(0), x_2(1)) + \nu_1^{\mathrm{T}} (x_2(0) - x_1(1)), \tag{11}$$

the following conditions hold

$$\lambda_1' = -\widetilde{H}_{x_1} = -(t_1 - a)\frac{\partial}{\partial x_1}\left(\lambda_1^{\mathrm{T}} f_1(x_1, u_1)\right), \tag{12a}$$

$$\lambda_2' = -\widetilde{H}_{x_2} = -(b - t_1)\frac{\partial}{\partial x_2}\left(\lambda_2^{\mathrm{T}} f_2(x_2, u_2)\right) \tag{12b}$$

$$\lambda_3' = -\widetilde{H}_{t_1} = -\lambda_1^{\mathrm{T}} f_1(x_1, u_1) + \lambda_2^{\mathrm{T}} f_2(x_2, u_2), \tag{12c}$$

$$u_k(\tau) = \operatorname{argmin}\{\lambda_k(\tau)^{\mathrm{T}} f_k(x_k(\tau), u) : u \in U\}, \quad k = 1, 2 \tag{12d}$$

$$\lambda_1(0) = -\Phi_{x_1(0)} = -\frac{\partial}{\partial x_1(0)}(\nu^{\mathrm{T}} r), \lambda_1(1) = \Phi_{x_1(1)} = -\nu_1, \tag{12e}$$

$$\lambda_2(0) = -\Phi_{x_2(0)} = -\nu_1, \lambda_2(1) = \Phi_{x_2(1)} = \frac{\partial}{\partial x_2(1)}(\nu_0 \, g + \nu^{\mathrm{T}} \, r), \tag{12f}$$

$$\lambda_3(0) = \lambda_3(1) = 0. \tag{12g}$$

Now, due to the autonomy of the state equations and due to the regularity assumptions above, both parts $\lambda_1^{\mathrm{T}} f_1$ and $\lambda_2^{\mathrm{T}} f_2$ of the Hamiltonian are constant on $[0, 1]$. Thus, λ_3 is a linear function which vanishes due to the boundary conditions (12g). Together with the relation (12c) one obtains the continuity of the Hamiltonian (5f).

If one recombines the adjoints

$$\lambda(t) := \begin{cases} \lambda_1\left(\dfrac{t - a}{t_1 - a}\right), & t \in [a, t_1[, \\ \lambda_2\left(\dfrac{t - t_1}{b - t_1}\right), & t \in [t_1, b], \end{cases} \tag{13}$$

one obtains the adjoint equation (5a) from Eq. (12a-b), the minimum principle (5b) from Eq. (12d), and the natural boundary conditions and the continuity conditions (5c-e) from Eq. (12e-f). □

It should be remarked that the results of Theorem 1 easily can be extended to nonautonomous optimal control problems with nonsmooth state equations. This holds too, if the performance index contains an additional integral term $I = g(x(t_b)) + \int_{t_a}^{t_b} f_0(t, x(t), u(t))dt$. Both extensions can be treated by standard transformation techniques which transform the problems into the form of Problem (P1). The result is, that for the extended problems, one simply has to redefine the Hamiltonian by

$$H(t, x, u, \lambda, \nu_0) := \nu_0 \, f_0(t, x, u) + \lambda^{\mathrm{T}} f(t, x, u). \tag{14}$$

Example 1. The following example is taken from the well-known book of Clarke. It describes the control of an electronic circuit which encludes a diode and a condensor. The control u is the initializing voltage, the state variable x denotes the voltage at the condensor. The resuling optimal control problem is given as follows.

Minimize the functional

$$I(u) = \frac{1}{2} \int_0^2 u(t)^2 dt \tag{15}$$

with respect to the state equation

$$x'(t) = \begin{cases} a\,(u - x), & \text{if} \quad S = x - u \leq 0, \\ b\,(u - x), & \text{if} \quad S = x - u > 0, \end{cases} \tag{16}$$

and the boundary conditions

$$x(0) = 4, \qquad x(2) = 3. \tag{17}$$

First, we consider the smooth case, i.e. we choose $a = b = 2$. The solution easily can be found applying the classical optimal control theory. The Hamiltonian is given by

$$H = u^2/2 + a\lambda(u - x),$$

which yields the adjoint equation $\lambda' = a\lambda$, the optimal control $u = -a\lambda$. Thus, we obtain the linear *two-point boundary value problem*

$$\begin{aligned} x' &= -a^2\,\lambda - a\,x, & x(0) &= 4, \\ \lambda' &= a\lambda, & x(2) &= 3. \end{aligned} \tag{18}$$

The (unique) solution for the parameter $a = 2$ is given in Fig. 1.

For the nonsmooth case, $a \neq b$, we assume that there is just one point $t_1 \in \,]0, 2[$ where the switching function changes sign. Further, due to the results for the smooth problem shown in Fig. 1, we assume the solution structure

$$S[t] \begin{cases} > 0, & \text{if} \quad 0 \leq t < t_1, \\ < 0, & \text{if} \quad t_1 < t \leq 2. \end{cases} \tag{19}$$

According to Theorem 1 we obtain the following necessary conditions for the solution (x^0, u^0):

(i) $t \in [0, t_1]$: $H = H_2 = \dfrac{1}{2}u^2 + b\lambda(u - x),$

$\lambda' = b\lambda, \quad u = -b\lambda.$

(ii) $t \in [t_1, 2]$: $H = H_1 = \dfrac{1}{2}u^2 + a\lambda(u - x),$

$\lambda' = a\lambda, \quad u = -a\lambda.$

The continuity condition (5f) yields

$$H[t_1^+] - H[t_1^-] = (b - a)\lambda(t_1)\left[\frac{a + b}{2}\lambda(t_1) + x(t_1)\right] = 0.$$

So, we obtain the following *three-point boundary value problem*

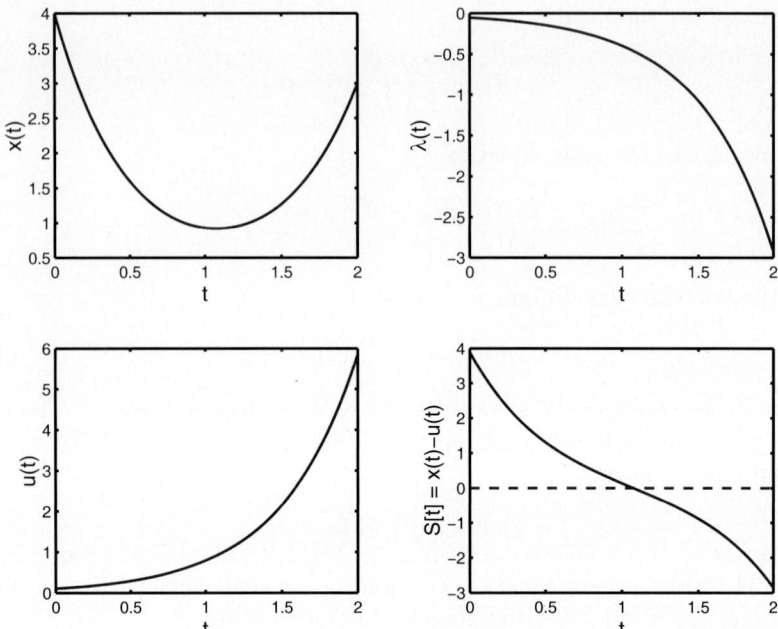

Fig. 1. Example 1: Smooth Case.

$$x' = \begin{cases} -b(b\lambda + x) : t \in [0, t_1], \\ -a(a\lambda + x) : t \in [t_1, 2], \end{cases}$$

$$\lambda' = \begin{cases} b\lambda : t \in [0, t_1], \\ a\lambda : t \in [t_1, 2], \end{cases} \tag{20}$$

$$x(0) = 4, \quad x(2) = 3, \quad \frac{a+b}{2}\lambda(t_1) + x(t_1) = 0.$$

In Fig. 2 the numerical solution of this boundary value problem is shown for the parameters $a = 4$ and $b = 2$. The solution is obtained via the multiple shooting code BNDSCO, cf. [6, 7]. One observes that the preassumed sign distribution of the switching function is satisfied. Further, the optimal control and the optimal switching function is discontinous at the switching point t_1.

For the parameters $a = 2$ and $b = 4$ the solution of the boundary value problem (20) is shown in Fig. 3. Here, the preassumed sign distribution of the switching function is not satisfied. So, the estimated switching structure for these parameters is not correct and we have to consider the singular case, i.e. the switching function vanishes identically along a nontrivial subarc.

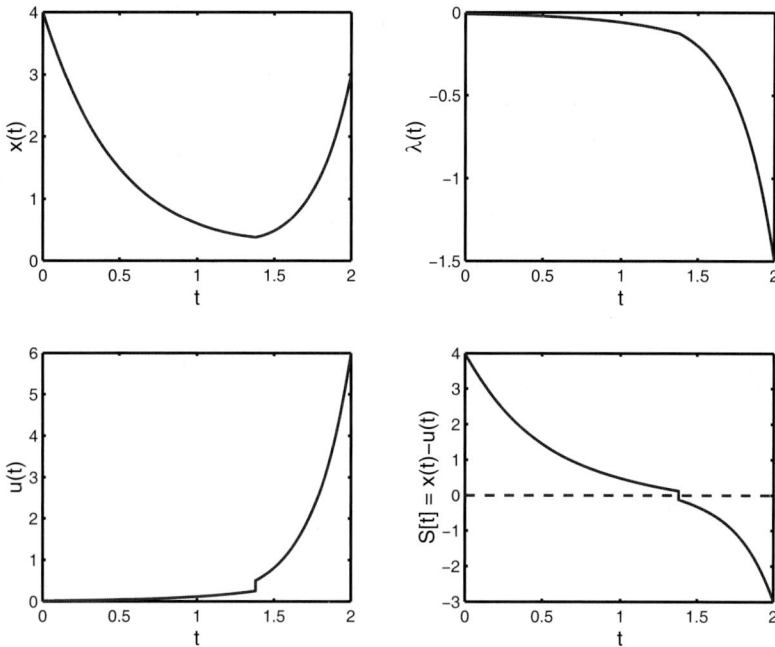

Fig. 2. Example 1: Nonsmooth and Regular Case, $a = 4$, $b = 2$.

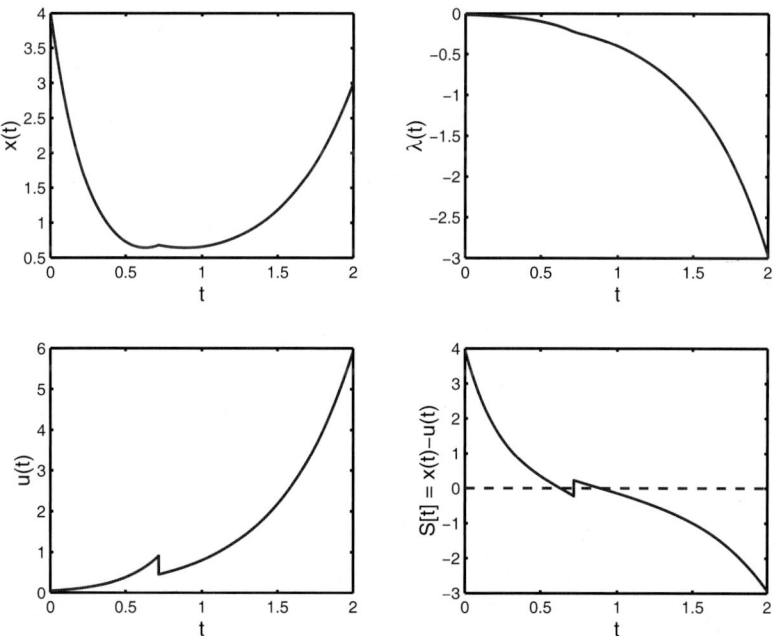

Fig. 3. Example 1: Nonsmooth and Regular Case, $a = 2$, $b = 4$.

2 Nonsmooth Optimal Control Problems, Singular Case

In this section we continue the investigation of the general optimal control problem (P1). However, we drop the regularity condition (R). More precisely, we assume that a solution (x^0, u^0) of the optimal control problem contains a finite number of nontrivial subarcs, where the switching function vanishes identically. These subarcs are called *singular state subarcs*, cf. the analogous situation of singular control subarcs, cf. [1]. In order to have a well-defined problem, we now have to consider the dynamics on the singular manifold $S(x, u) = 0$. Therefore, we generalize the problem formulation (P1) a bit, and allow the system to possess an independent dynamic on the singular subarcs.

Problem (P2). Determine a piecewise continuous control function $u : [a, b] \to \mathbb{R}$, such that the functional

$$I = g(x(b)) \tag{21}$$

is minimized subject to the following constraints

$$x'(t) = f(x(t), u(t)), \quad t \in [a, b] \quad \text{a.e.,} \tag{22a}$$

$$r(x(a), x(b)) = 0, \tag{22b}$$

$$u(t) \in U = [u_{\min}, u_{\max}] \subset \mathbb{R}, \tag{22c}$$

where the right-hand side f is of the special form

$$f(x, u) = \begin{cases} f_1(x, u), & \text{if} \quad S(x, u) < 0, \\ f_2(x, u), & \text{if} \quad S(x, u) = 0, \\ f_3(x, u), & \text{if} \quad S(x, u) > 0, \end{cases} \tag{23}$$

with smooth functions $f_k : \mathbb{R}^n \times \mathbb{R} \to \mathbb{R}^n$, $k = 1, 2, 3$. All other assumptions with respect to Problem (P1) may be satisfied also for (P2).

Again, our aim is to derive necessary conditions for (P2). To this end, we assume that there exists a finite grid $a < t_1 < \ldots < t_s < b$ such that the t_j are either isolated points where the switching function $S[t] := S(x^0(t), u^0(t))$ changes sign or entry or exit points of a singular state subarc.

We assume, that the switching function is of order zero with respect to the control u, i.e.

$$S_u(x^0(t), u^0(t)) \neq 0 \tag{24}$$

holds along each singular state subarc. By the implicit function theorem, the equation $S(x, u) = 0$ can be solved (locally unique) for u. Thus, we assume that there exists a continuously differentiable function $u = V(x)$ which solves the equation above. With this, we define

$$\widehat{f}_2(x) := f_2(x, V(x)). \tag{25}$$

For the regular subarcs we introduce the Hamiltonian

$$H(x, u, \lambda) := H_j(x, u, \lambda) := \lambda^{\mathrm{T}} f_j(x, u), \tag{26}$$

where $j \in \{1, 3\}$ is chosen in the corresponding regular subinterval $[t_k, t_{k+1}]$ according to the sign of S. For the singular subarcs we set

$$H(x, u, \lambda) := H_2(x, u, \lambda) := \lambda^{\mathrm{T}} \widehat{f_2}(x). \tag{27}$$

In the following, we summarize the necessary conditions for Problem (P2).

Theorem 2. *With the assumptions above the following necessary conditions hold. There exist an adjoint variable* $\lambda : [a, b] \to \mathbb{R}^n$, *which is a continuous and piecewise* C^1-*function, and Lagrange multipliers* $\nu_0 \in \{0, 1\}$, $\nu \in \mathbb{R}^\ell$, *such that* (x^0, u^0) *satisfies the conditions*

$$\lambda'(t) = -H_x(x^0(t), u^0(t), \lambda(t)), \quad t \in [a, b], \text{ a.e.} \tag{28a}$$

$$u^0(t) = \begin{cases} \operatorname{argmin}\{H(x^0(t), u, \lambda(t)) : u \in U\} & \text{on regular arcs,} \\ V(x^0(t)) & \text{on singular subarcs,} \end{cases} \tag{28b}$$

$$\lambda(a) = -\frac{\partial}{\partial x^0(a)} [\nu^T r(x^0(a), x^0(b))], \tag{28c}$$

$$\lambda(b) = \frac{\partial}{\partial x^0(b)} [\nu_0 g(x^0(b)) + \nu^T r(x^0(a), x^0(b))], \tag{28d}$$

$$\lambda(t_j^+) = \lambda(t_j^-), \quad j = 1, \ldots, s, \tag{28e}$$

$$H[t_j^+] = H[t_j^-], \quad j = 1, \ldots, s. \tag{28f}$$

Note, that on a singular subarc there holds no minimum principle for the control which is completely determined by the switching equation $S(x, u) = 0$.

Proof of Theorem 2. For simplicity, we assume, that the switching function $S[\cdot]$ along the optimal trajectory has just *one* singular subarc $[t_1, t_2] \subset]a, b[$, and that the following *switching structure* holds

$$S[t] \quad \begin{cases} < 0, & \text{if} \quad a \leq t < t_1, \\ = 0, & \text{if} \quad t_1 \leq t \leq t_2, \\ > 0, & \text{if} \quad t_2 < t \leq b. \end{cases} \tag{29}$$

Again, we compare the optimal solution (x^0, u^0) with those admissible solutions (x, u) of the problem which have the same switching structure. Each candidate is associated with its separated parts ($\tau \in [0, 1]$, $t_0 := a$, $t_3 := b$)

$$\begin{aligned} x_j(\tau) &:= x(t_{j-1} + \tau(t_j - t_{j-1})), \quad j = 1, 2, 3, \\ u_j(\tau) &:= u(t_{j-1} + \tau(t_j - t_{j-1})), \quad j = 1, 3. \end{aligned} \tag{30}$$

Now, $(x_1, x_2, x_3, t_1, t_2, u_1, u_3)$ performs an abmissible solution and $(x_1^0, x_2^0, x_3^0, t_1^0, t_2^0, u_1^0, u_3^0)$ an optimal solution of the following auxillary optimal control problem.

Problem (P2'). Determine a piecewise continuous control function $u = (u_1, u_3) : [0,1] \to \mathbb{R}^2$, such that the functional

$$I = g(x_3(1)) \tag{31}$$

is minimized subject to the constraints ($t_0 := a, t_3 := b, \tau \in [0,1]$)

$$x_j'(\tau) = \begin{cases} (t_j - t_{j-1}) f_j(x_j(\tau), u_j(\tau)), & \text{a.e.,} \quad j = 1, 3, \\ (t_2 - t_1) \widehat{f}_2(x_2(\tau)), & \text{a.e.,} \quad j = 2, \end{cases} \tag{32a}$$

$$t_k'(\tau) = 0, \quad k = 1, 2, \tag{32b}$$

$$r(x_1(0), x_3(1)) = 0, \tag{32c}$$

$$x_2(0) - x_1(1) = x_3(0) - x_2(1) = 0, \tag{32d}$$

$$u_1(\tau), u_2(\tau), u_3(\tau) \in U \subset \mathbb{R}. \tag{32e}$$

Problem (P2') again is a classical optimal control problem with a smooth right-hand side. We can apply the classical necessary conditions of optimal control theory, cf. Hestenes, [4]. If S satisfies the constraint qualification (36), there exist continuous and continuously differentiable adjoint variables λ_j, $j = 1, 2, 3$, and Lagrange-multipliers $\nu_0 \in \{0, 1\}$, $\nu \in \mathbb{R}^\ell$, and $\nu_1, \nu_2 \in \mathbb{R}^n$, such that with the Hamiltonian

$$\widetilde{H} := (t_1 - a) \lambda_1^{\mathrm{T}} f_1(x_1, u_1) + (t_2 - t_1) \lambda_2^{\mathrm{T}} \widehat{f}_2(x_2) + (b - t_2) \lambda_3^{\mathrm{T}} f_3(x_3, u_3) \tag{33}$$

and the augmented performance index

$$\Phi := \nu_0 \, g(x_3(1)) + \nu^{\mathrm{T}} r(x_1(0), x_3(1)) + \nu_1^{\mathrm{T}} (x_2(0) - x_1(1)) + \nu_2^{\mathrm{T}} (x_3(0) - x_2(1)), \tag{34}$$

the following conditions hold

$$\lambda_1' = -\widetilde{H}_{x_1} = -(t_1 - a) \left(\lambda_1^{\mathrm{T}} f_1 \right)_{x_1}, \tag{35a}$$

$$\lambda_2' = -\widetilde{H}_{x_2} = -(t_2 - t_1) \left(\lambda_2^{\mathrm{T}} \widehat{f}_2 \right)_{x_2} \tag{35b}$$

$$\lambda_3' = -\widetilde{H}_{x_3} = -(b - t_2)(\lambda_3^{\mathrm{T}} f_3)_{x_3} \tag{35c}$$

$$\lambda_4' = -\widetilde{H}_{t_1} = -\lambda_1^{\mathrm{T}} f_1 + \lambda_2^{\mathrm{T}} \widehat{f}_2, \tag{35d}$$

$$\lambda_5' = -\widetilde{H}_{t_2} = -\lambda_2^{\mathrm{T}} \widehat{f}_2 + \lambda_3^{\mathrm{T}} f_3, \tag{35e}$$

$$u_j(\tau) = \mathrm{argmin}\{\lambda_j(\tau)^{\mathrm{T}} f_j(x_j(\tau), u) : u \in U\}, \quad j = 1, 3, \tag{35f}$$

$$\lambda_1(0) = -\Phi_{x_1(0)} = -(\nu^{\mathrm{T}} r)_{x_1(0)}, \quad \lambda_1(1) = \Phi_{x_1(1)} = -\nu_1, \tag{35g}$$

$$\lambda_2(0) = -\Phi_{x_2(0)} = -\nu_1, \quad \lambda_2(1) = \Phi_{x_2(1)} = -\nu_2, \tag{35h}$$

$$\lambda_3(0) = -\Phi_{x_3(0)} = -\nu_2, \quad \lambda_3(1) = \Phi_{x_3(1)} = (\ell_0 \, g + \nu^{\mathrm{T}} r)_{x_3(1)}, \tag{35i}$$

$$\lambda_4(0) = \lambda_4(1) = \lambda_5(0) = \lambda_5(1) = 0. \tag{35j}$$

Due to the autonomy of the optimal control problem, all three parts $\lambda_1^{\mathrm{T}} f_1$, $\lambda_2^{\mathrm{T}} \widehat{f}_2$, and $\lambda_3^{\mathrm{T}} f_3$ of the Hamiltonian are constant. Therefore, the adjoints λ_4

and λ_5 vanish and we obtain the global continuity of the augmented Hamiltonian (33). If one recombines the adjoints

$$\lambda(t) := \begin{cases} \lambda_1\left(\dfrac{t-a}{t_1-a}\right), & t \in [a, t_1[, \\[2mm] \lambda_2\left(\dfrac{t-t_1}{t_2-t_1}\right), & t \in [t_1, t_2], \\[2mm] \lambda_3\left(\dfrac{t-t_2}{b-t_2}\right), & t \in]t_2, b], \end{cases} \tag{36}$$

the state and control variables accordingly, one obtains all the necessary conditions of the Theorem. □

Again, we mention that the results of Theorem 2 easily can be extended to nonautonomous nonsmooth optimal control problems and to optimal control problems with performance index of Bolza type, as well.

Example 2. Again, we consider the example of Clarke, [3], but now we try to find solutions which contain a singular state subarc. The optimal control problem is given as follows.
Minimize the functional

$$I(u) = \frac{1}{2} \int_0^2 u(t)^2 dt \tag{37}$$

with respect to the state equation

$$x'(t) = \begin{cases} a\,(u-x), & \text{if} \quad S = x - u \leq 0, \\ b\,(u-x), & \text{if} \quad S = x - u > 0, \end{cases} \tag{38}$$

and the boundary conditions

$$x(0) = 4, \qquad x(2) = 3. \tag{39}$$

If we assume that there is exactly on singular state subarc, or more precisely

$$S[t] \begin{cases} > 0, & \text{if} \quad 0 \leq t < t_1, \\ = 0, & \text{if} \quad t_1 \leq t \leq t_2, \\ < 0, & \text{if} \quad t_2 < t \leq 2, \end{cases} \tag{40}$$

we obtain the following necessary conditions due to Theorem 2.

(i) $t \in [0, t_1]$: $H = H_3 = \dfrac{1}{2}u^2 + b\lambda(u-x),$
$\lambda' = b\lambda, \quad u = -b\lambda.$

(ii) $t \in [t_1, t_2]$: $H = H_2 = \dfrac{1}{2}x^2,$
$\lambda' = -x, \quad u = V(x) = x.$

(iii) $t \in [t_2, 2]$: $H = H_1 = \dfrac{1}{2}u^2 + a\lambda(u-x),$
$\lambda' = a\lambda, \quad u = -a\lambda.$

The continuity of the Hamiltonian yields with

$$H[t_1^-] = \frac{1}{2}b^2\lambda(t_1)^2 + b\lambda(t_1)(-b\,\lambda(t_1) - x(t_1))$$
$$= -\frac{1}{2}b\lambda(t_1)(b\lambda(t_1) + 2x(t_1))$$
$$H[t_1^+] = \frac{1}{2}x(t_1)^2.$$

the interior boundary condition $x(t_1) + b\lambda(t_1) = 0$. The analogous condition holds at the second switching point t_2. Altogether we obtain the following *multipoint boundary value problem.*

$$x' = \begin{cases} -b(b\,\lambda + x), & t \in [0, t_1], \\ 0, & t \in [t_1, t_2], \\ -a(a\,\lambda + x), & t \in [t_2, 2], \end{cases}$$

$$\lambda' = \begin{cases} b\lambda, & t \in [0, t_1], \\ -x, & t \in [t_1, t_2], \\ a\lambda, & t \in [t_2, 2], \end{cases} \tag{41}$$

$$x(0) = 4, \qquad x(2) = 3,$$
$$x(t_1) + b\lambda(t_1) = 0, \qquad x(t_2) + a\lambda(t_2) = 0.$$

For the parameters $a = 2$, and $b = 4$ the numerical solution is shown in Fig. 4. One observes the singular subarc with the switching points $t_1 \doteq 0.632117$, $t_2 \doteq 0.882117$.

3 Conclusions

In this paper optimal control problems with nonsmooth state differential equations are considered. Two solution typs are distinguished. In the first part of the paper regular solutions have been considered. The regularity is characterized by the assumption that the switching function changes sign only at isolated points. In the second part so called singular state subarcs are admitted. These are nontrivial subarcs, where the switching function vanishes identically. For both situations necessary conditions are derived from the classical (smooth) optimal control theory.

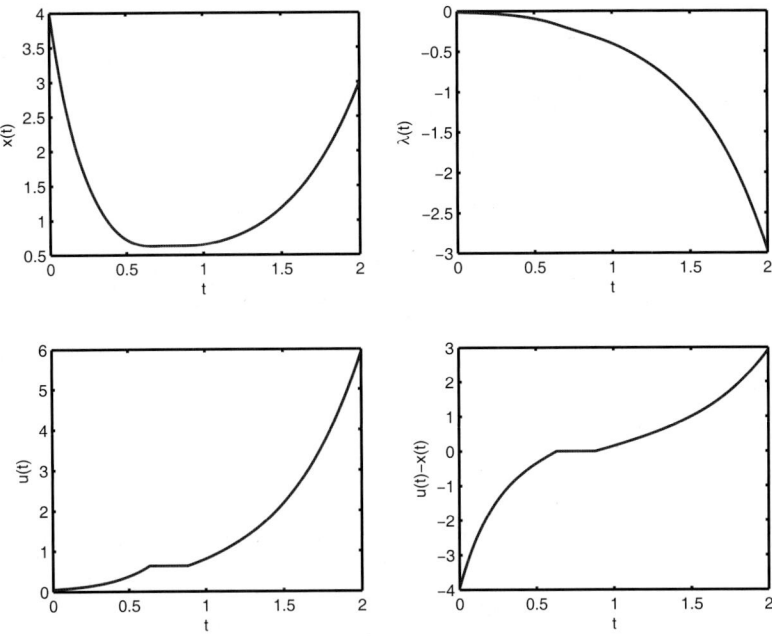

Fig. 4. Example 2: Nonsmooth and Singular Case, $a = 2$, $b = 4$.

References

[1] Bell, D.J., Jacobson, D.H.: Singular Optimal Control Problems. Academic Press, New York, (1975)

[2] Bryson, A.E., Ho, Y.C.: Applied Optimal Control. Ginn and Company. Waltham, Massachusetts (1969)

[3] Clarke, F.H.: Optimization and Nonsmooth Analysis. Wiley, New York (1983)

[4] Hestenes, M.R.: Calculus of Variations and Optimal Control Theory. John Wiley a. Sons, Inc., New York (1966)

[5] Oberle, H.J., Rosendahl,R.: Numerical computation of a singular-state subarc in an economic optimal control problem. Optimal Control Application and Methods, **27**, 211–235 (2006)

[6] Oberle, H.J., Grimm,W.: BNDSCO – A Program for the numerical solution of optimal control problems. Report No. 515, Institut for Flight Systems Dynamics, Oberpfaffenhofen, German Aerospace Research Establishment DLR (1989)

[7] Stoer, J., Bulirsch R.: Introduction to Numerical Analysis. Texts in Applied Mathematics, Springer, New York, Vol. 12 (1996)

A

Color Figures

Fig. A.1. Prof. Dr. Dr.h.c.mult. Roland Bulirsch

Fig. A.2. Academic Genealogy of Prof. Dr. R. Bulirsch (cf. p. 3)

Fig. A.3. Applied Mathematics book for undergraduate students (1986) which inspired this anthology 20 years later: Frontside, see also the flipside text in Breitner "Epilogue" (p. 227).

A.1 Color Figures to Denk, Feldmann (pp. 11–26)

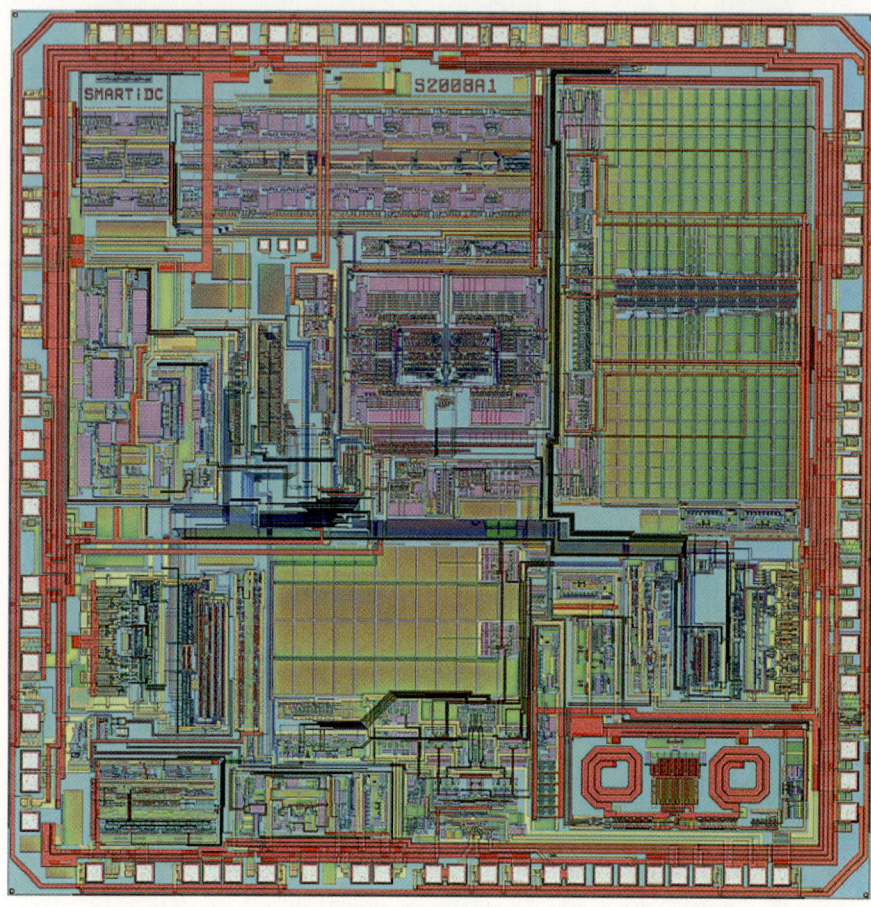

Fig. A.4. GSM transceiver for mobile phones. On the bottom right the antenna structures can be seen, the regular structures (bottom middle and top right) are memory cells, while the irregular structures mark the digital signal processing part of the chip. More details are discussed on page 18.

A.2 Color Figures to Pulch (pp. 27–42)

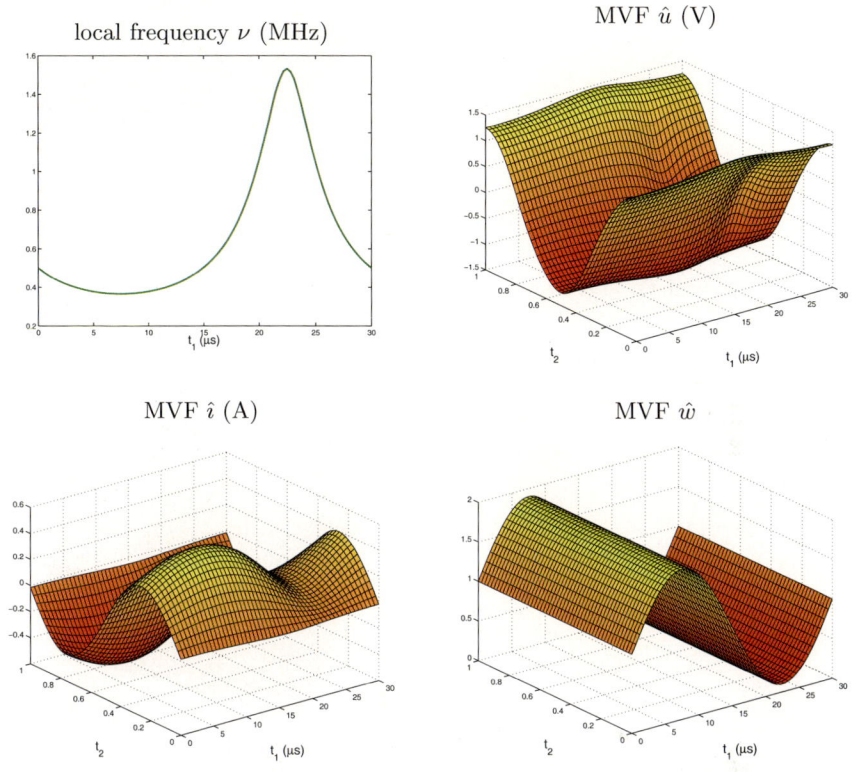

Fig. A.5. MPDAE solution for rate $T_1 = 30$ μs (cf. p. 40).

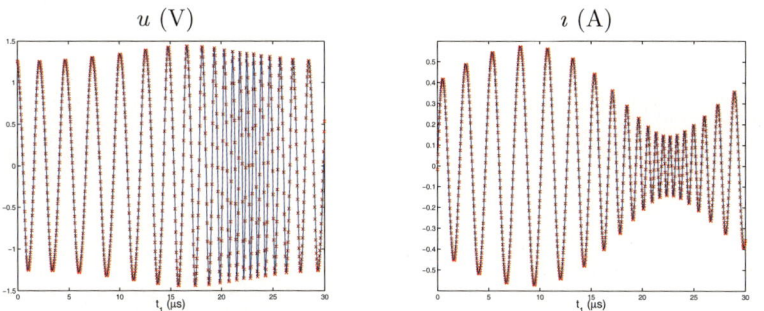

Fig. A.6. DAE solution for rate $T_1 = 30$ μs reconstructed by MPDAE solution (red) and computed by transient integration (blue) (cf. p. 41).

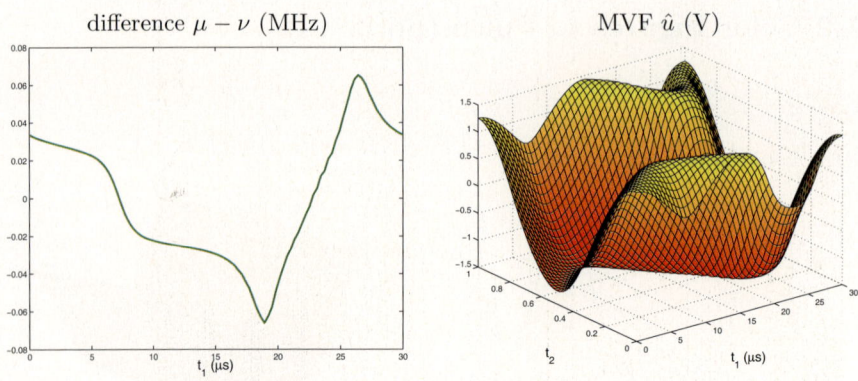

Fig. A.7. Part of MPDAE solution for rate $T_1 = 30\ \mu$s with modified phase condition (cf. p. 41).

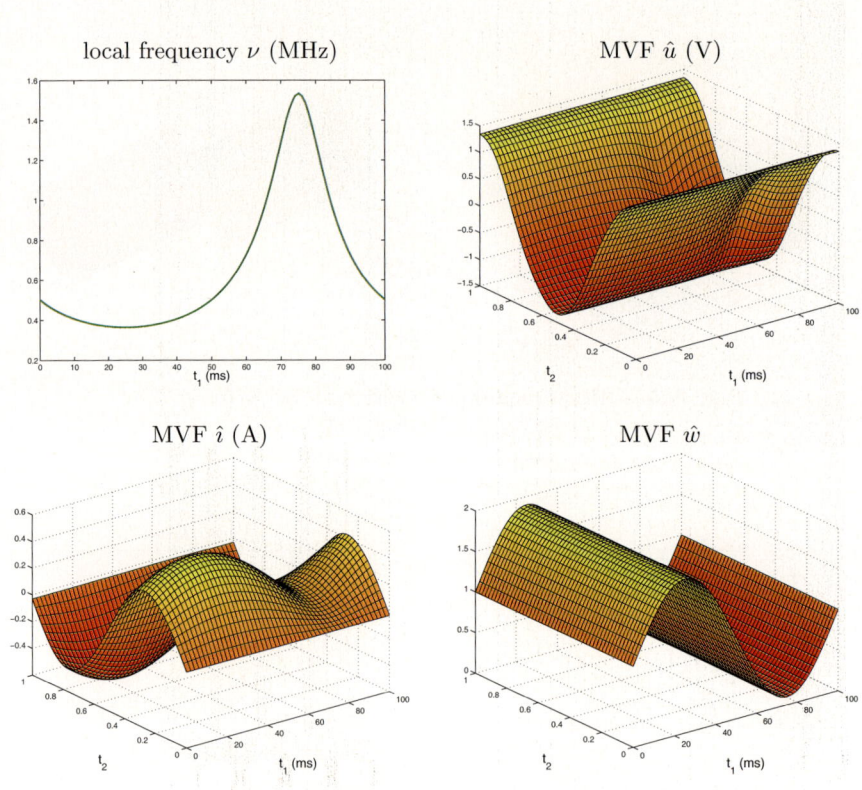

Fig. A.8. MPDAE solution for rate $T_1 = 100$ ms (cf. p. 41).

A.3 Color Figures to Lachner (pp. 71–90)

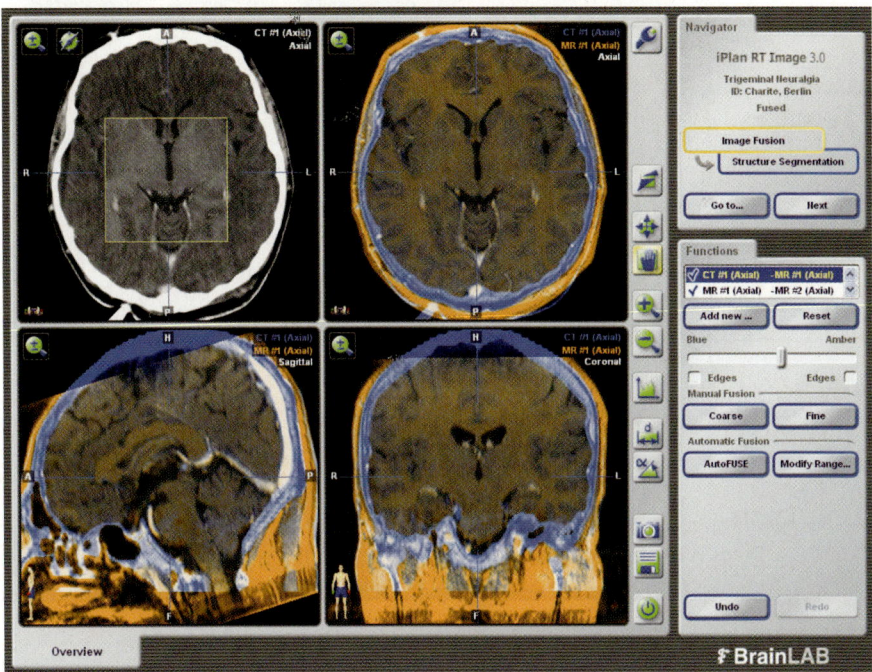

Fig. A.9. Graphical user interface of BrainLAB planning software iPlan RT. Registration of CT and MRI. For CT, the patient wears a headframe which is rigidly fixed to skull, and whose screws cause local skin distortions and swellings. Hence these regions should be excluded from image registration. This can be accomplished by the "Modify Range…" dialog. The automatic registration result is obtained by a click on "AutoFUSE". Axial, coronal and sagittal cuts through the image volumes are superimposed by means of an amber-blue color-representation. The forth ("spyglass") view shows a part of the floating image (MRI) superimposed to the reference image (CT). Since the user must approve the registration result for quality assurance, there are multiple tools for changing and manipulating the combined display of the images (cf. page 6).

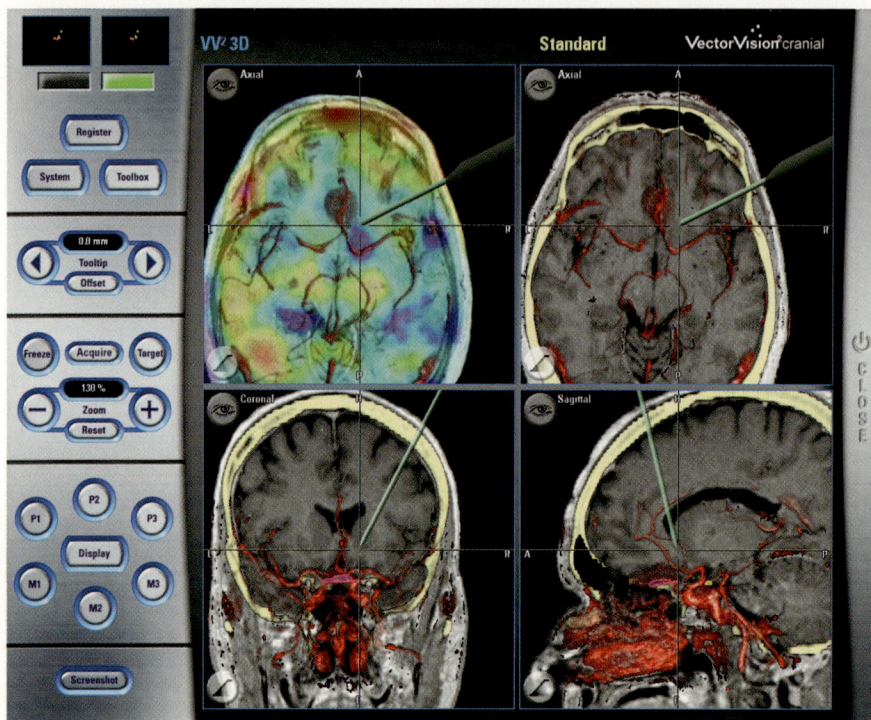

Fig. A.10. Graphical user interface of BrainLAB image guided surgery software VectorVision® cranial. Real-time display of surgical instruments (here a pointing device) during surgery is performed in an image volume created by the combination of several image modalities. Three of the views show orthogonal cuts through the integrated image volume consisting of the skull bones (yellow; from CT) and the major vessels (red; from angio-MRI) superimposed to a high-resolution anatomical MRI. The forth view shows functional SPECT data using a color/intensity palette. Correct simultaneous display of the different image modalities is only possible by means of image registration (cf. page 88).

A.4 Color Figures to Stolte, Rentrop (pp. 99–110)

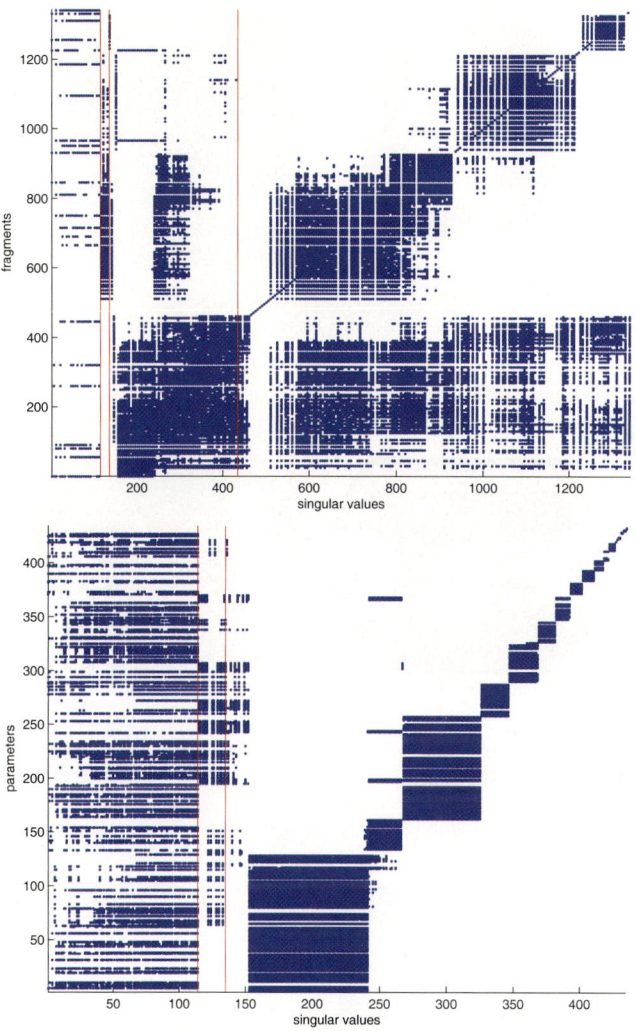

Fig. A.11. In this figure, the entries (with values bigger than 10^{-5}) of the matrix U (in extracts) and the matrix V are plotted. The matrices are from the SVD of the Jacobian of the model function for the data set Enolase and for the initial values initial3. The positions 114, 135 and 435 are highlighted as they correspond to the stages within the singular values. With the interrelations $\{u_{i,j} = $ part from i-th fragment of j-th singular value$\}$ and $\{v_{i,j} = $ part from i-th parameter of j-th singular value$\}$ obviously, these figure allow the identification of the essential input data (fragments) and parameters (probabilities) (cf. p. 109).

A.5 Color Figures to Wimmer et al. (pp. 111–124)

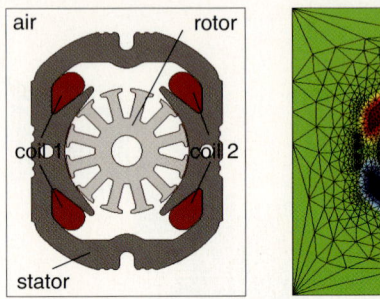

Fig. A.12. Left: geometry of a motor; right: contour plot of the scalar magnetic potential A_z (2804 dof) (cf. p. 122).

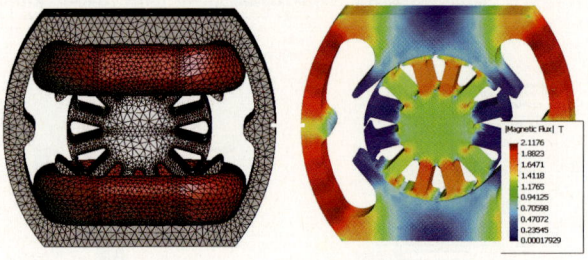

Fig. A.13. Left: discretization of an electric motor model (379313 dof); right: plot of the absolute value of the magnetic flux density distribution. (cf. p. 123).

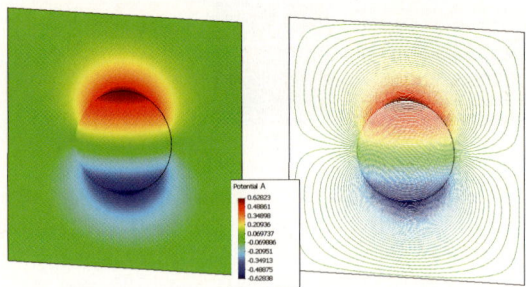

Fig. A.14. Left: contour plot of the magnetic scalar potential ψ_m of a magnetized sphere (66128 dof); right: contour lines of the magnetic scalar potential ψ_m (cf. p. 113).

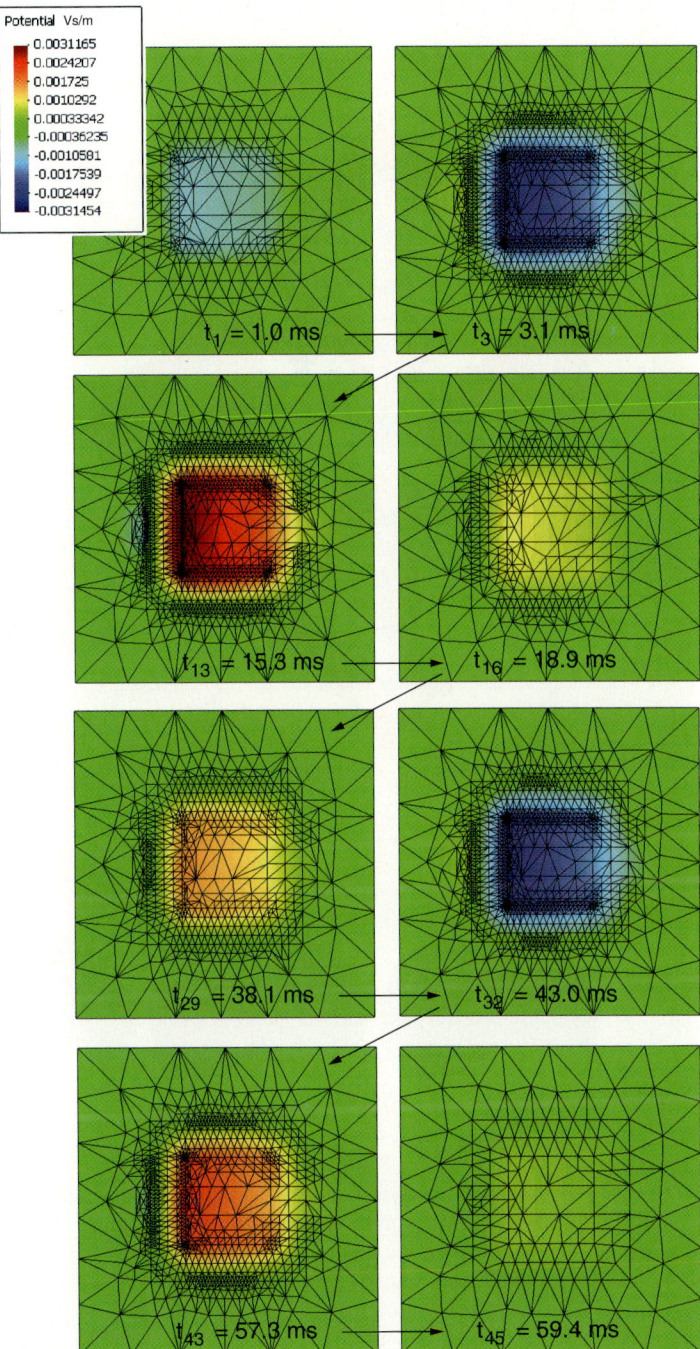

Fig. A.15. Magnetic scalar potential A_z at different time steps.

A.6 Color Figures to Haber et al. (pp. 127–143)

Fig. A.16. Remaining Bernstein-Bézier coefficients are determined due to C^1-conditions in the order: red, blue, green, and yellow. See p. 135 for more details.

12,380 coeff.	6,010 coeff.	2,880 coeff.
compr. ratio 2.7 : 1	compr. ratio 5.6 : 1	compr. ratio 11.7 : 1

Fig. A.17. Scattered data compression for an arbitrarily shaped domain: Distribution of $N = 12,359$ scattered data points and approximating spline surfaces with a varying number of degrees of freedom (left to right). For a rectangular domain, the compression ratio is about three times higher than the ratio N : #coefficients, cf. p. 138. For an arbitrarily shaped domain, we need to store one additional integer number for every triangle $T \in \mathcal{T}$ (see p. 131) to uniquely determine the position of the Bernstein-Bézier coefficients. Therefore, the compression ratios given above are slightly lower than for the rectangular domain case.

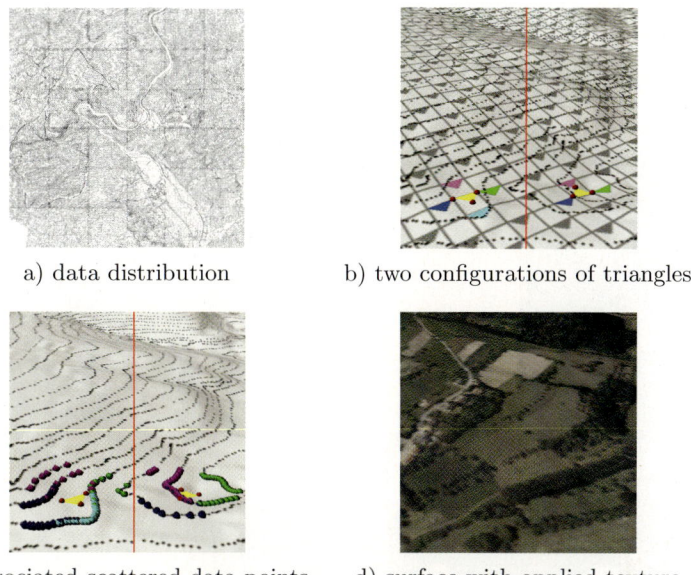

a) data distribution   b) two configurations of triangles

c) associated scattered data points   d) surface with applied texture

Fig. A.18. Stages of the approximation process. a) Distribution of 736,577 scattered data points. b) Perspective view onto the surface with projected spline grid and data distribution. Two different sets of triangles $T \in \mathcal{T}$ (blue, magenta, green, cyan) are shown on the left and on the right. c) Corresponding scattered data points (color coded) that have been used in the local least squares fitting for each triangle. (see p. 135 for details). d) Final result with applied texture.

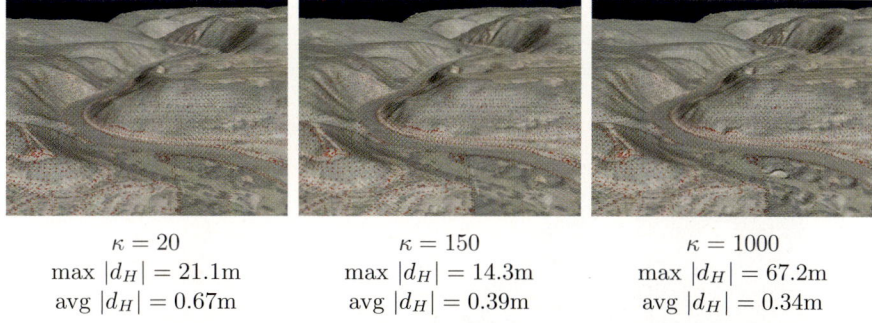

$\kappa = 20$	$\kappa = 150$	$\kappa = 1000$
max $\|d_H\| = 21.1$m	max $\|d_H\| = 14.3$m	max $\|d_H\| = 67.2$m
avg $\|d_H\| = 0.67$m	avg $\|d_H\| = 0.39$m	avg $\|d_H\| = 0.34$m

Fig. A.19. Influence of the condition number κ on the approximation quality of the surface. Left: κ is too low, the average local approximation order of the spline decreases; Middle: κ is chosen appropriately; Right: κ is chosen far too high, thereby reducing the average approximation error but exhibiting high errors at individual data points in regions of extremely sparse data (see bottom right corner of image). See p. 133 and 138 for details on κ and the Hausdorff distance d_H, respectively.

A.7 Color Figures to Herzog et al. (pp. 161–174)

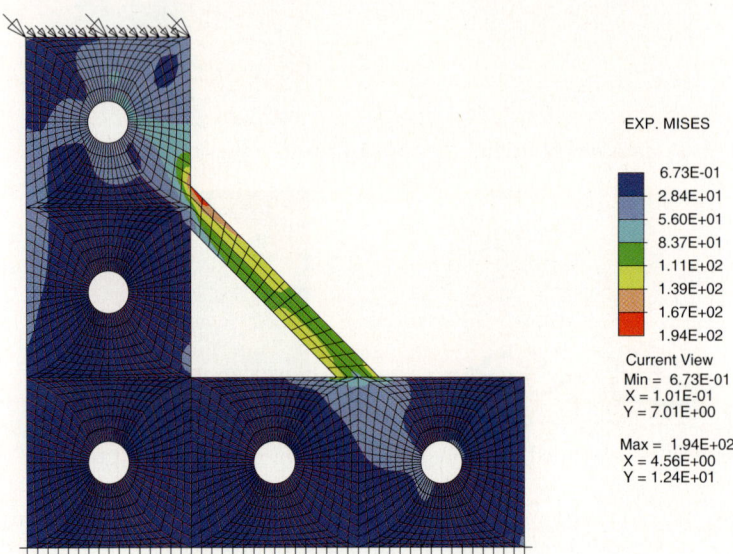

Fig. A.20. Expectation Value of von Mises stress (cf. page 171)

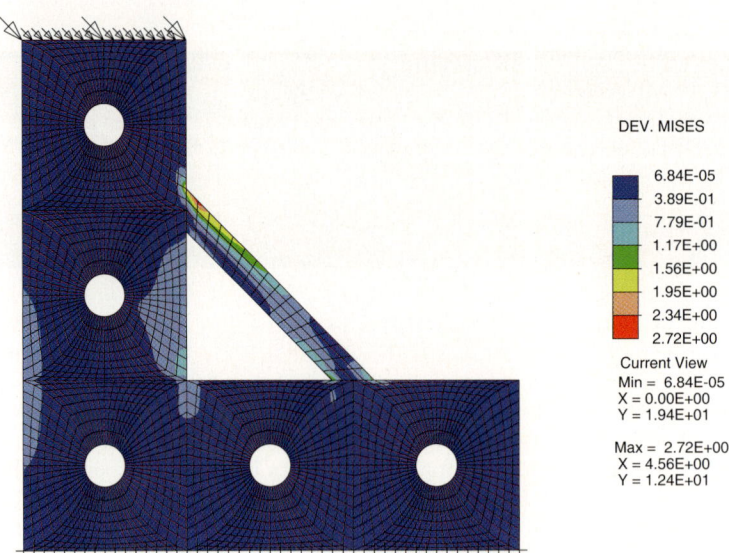

Fig. A.21. Deviation of von Mises stress (cf. page 171)

A.8 Colored Figures to Stelzer, Stryk (pp. 175–192)

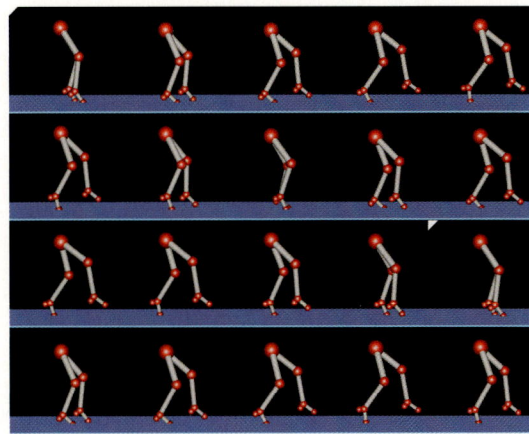

Fig. A.22. Bipedal walking with passively compliant three-segmented legs: Resulting gait sequence for parameter set optimized for low hip torques and bounded minimum walking speed (page 186).

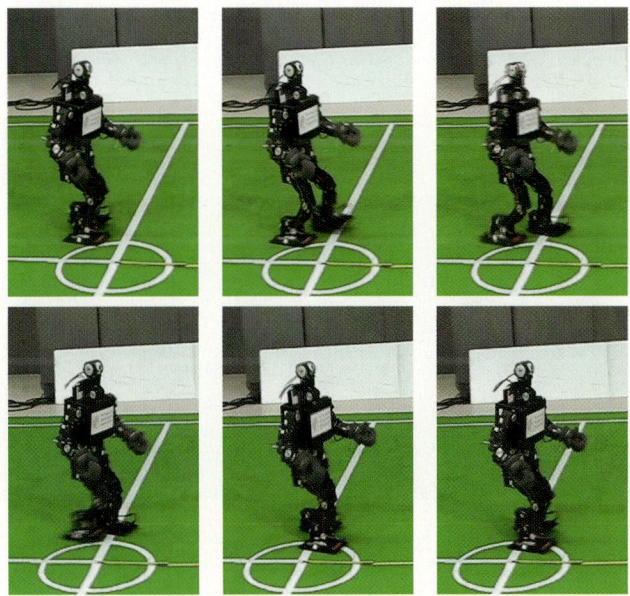

Fig. A.23. Optimized walking of the autonomous humanoid robot Bruno (page 189).

A.9 Color Figures to Teichelmann, Simeon (pp. 193–208)

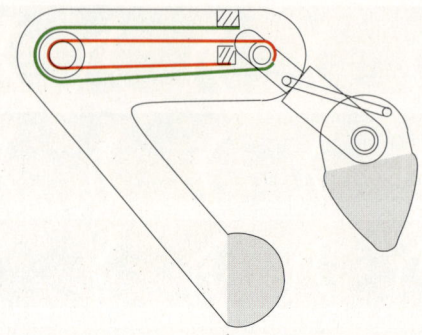

Fig. A.24. Sketch of a manipulator operated by a spring (red) and a SMA wire (green) (cf. p. 203).

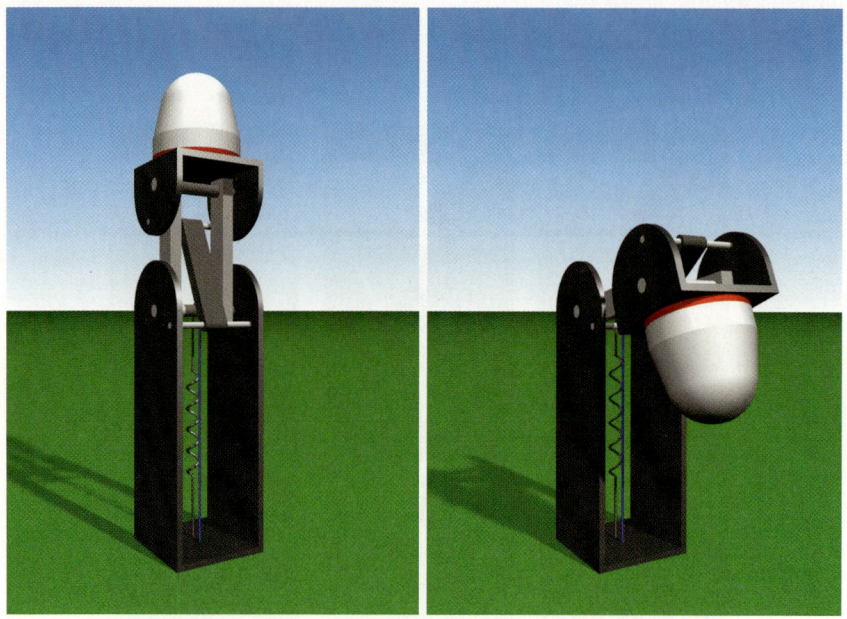

Fig. A.25. Visualisation of a roboter finger. The finger is closed by rising the temperature of the (blue) SMA wire. The spring takes the role of an antagonist that opens the finger when the SMA wire is cooled down (cf. p. 204).

A.10 Color Figures to Breitner (pp. 211–227)

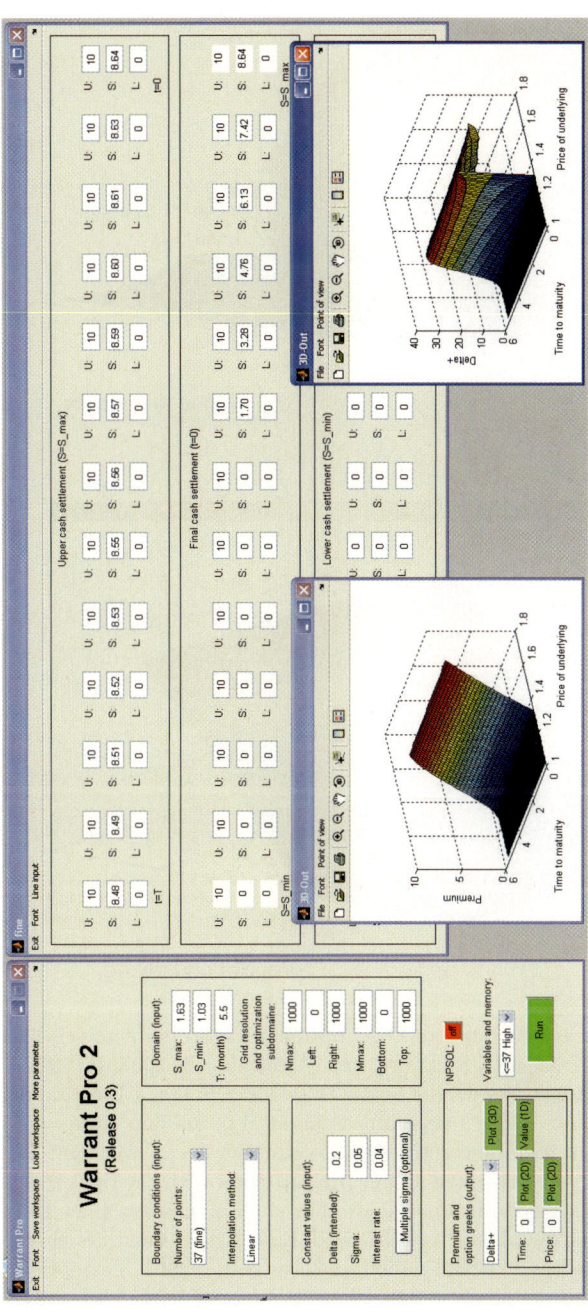

Fig. A.26. The MATLAB GUI (graphical user interface) of WARRANT-PRO-2 (Release 0.3) is intuitive and user friendly. A stand-alone application for all WINDOWS, LINUX and UNIX computers is available. The GUI has several windows: The main control window, the boundary conditions' window, the multiple σ window and the 1-, 2- and 3-dimensional graphics windows for a derivatives price and its "greeks", see also p. 219.

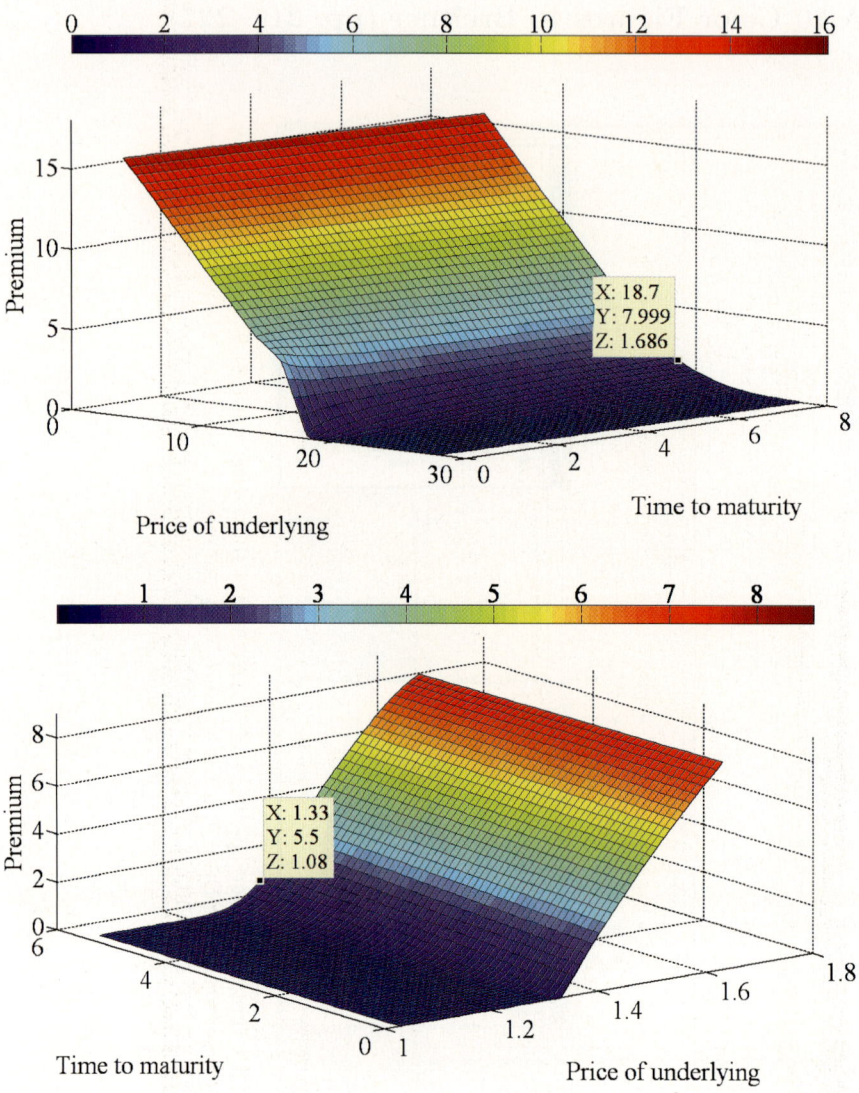

Fig. A.27. Top: The Porsche AG buys an OTC option with the right to sell 62.4 million USD, i. e. revenues from an US export, until December 3, 2007, for 46.917 million Euro. Thus the exchange rate will be better than 1.33 USD for 1 Euro, comp. all cash settlements, too. The option's premium is about 1.08 million Euro = 2.3 % about 5.5 month earlier in mid June 2007, see also Breitner Sec. 5. Bottom: The **Hannoversche Leben buys optimized OTC put options** with the right to sell 7.166 million TUI shares for 18.70 Euro per share. For share prices less than 18.70 Euro, i. e. the relevant scenarios, $\Delta \approx -1$ holds. Mid June 2007 8 month is chosen for the time to maturity to cover a critical period until the end of 2007. The option's premium is about 1.686 Euro per share = 9.0 % in mid June 2007, see also p. 221.

A.11 Colored Figures to Callies, Sonner (pp. 261–276)

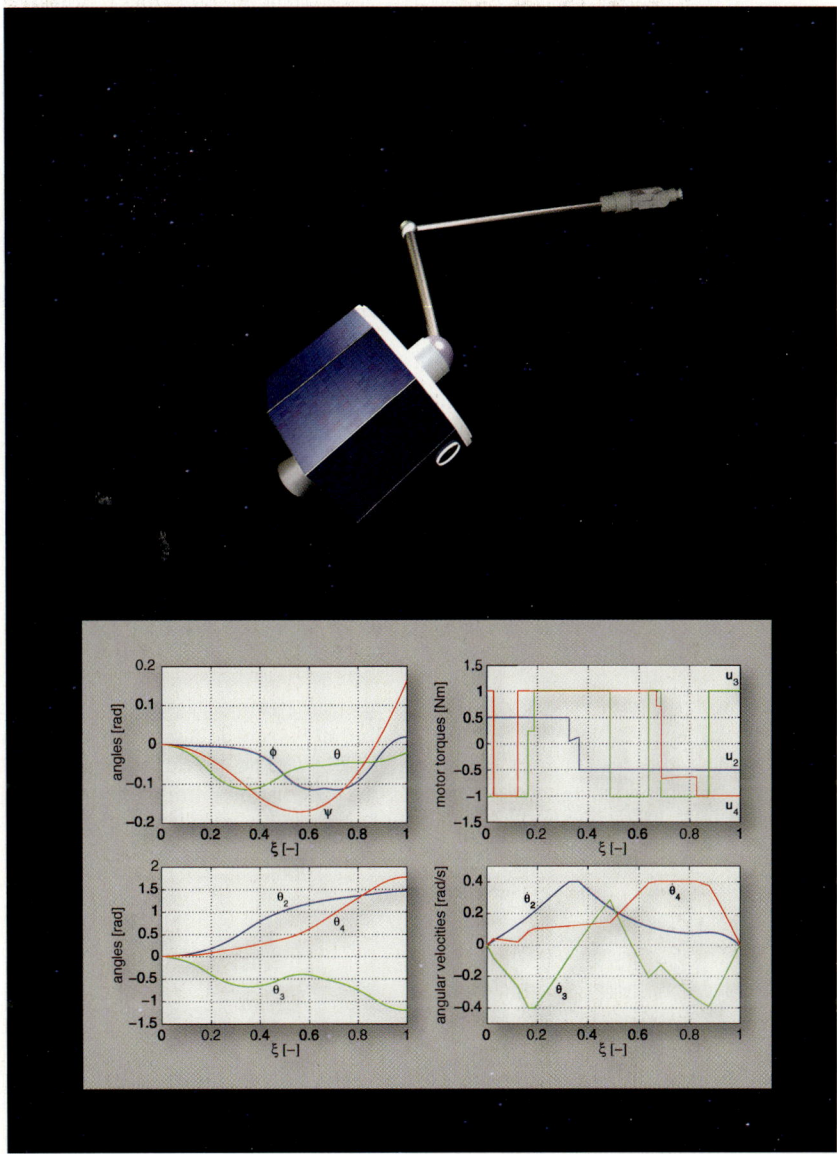

Fig. A.28. Free-floating spin-stabilized space robotic system: 6-DOF example system consisting of a small spin-stabilized satellite (body 1) and a 3-DOF manipulator (bodies 2,3,4) with revolute joints, mounted on a despun platform on top of the satellite; satellite orientation (ψ, ϑ, ϕ), optimal controls, joint angles $\theta_2, \theta_3, \theta_4$ and joint velocities $\dot{\theta}_2, \dot{\theta}_3, \dot{\theta}_4$ vs. normalized time ξ are shown.

A.12 Color Figures to Grimm, Rotärmel (pp. 295–308)

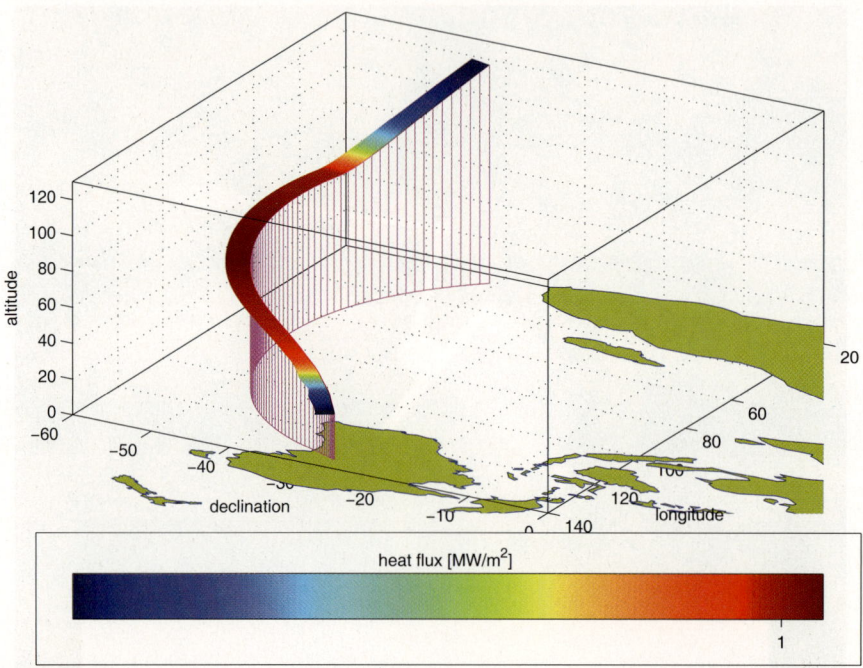

Fig. A.29. Entry flight of the X-38: The guided entry starts at the south of Africa at an altitude of about 120 km. The specified landing site is on the Australian continent. The colorband indicates the heat flux along the flight. The maximum is reached at about half of the covered range and amounts to 88% of the admissible limit. More details are discussed on page 307.